Introduction to Impact Dynamics

Introduction to Impact Dynamics

T.X. Yu
The Hong Kong University of Science and Technology
Kowloon, Hong Kong

XinMing Qiu
Tsinghua University
Beijing, China

This edition first published 2018 by John Wiley & Sons Singapore Pte. Ltd under exclusive licence granted by Tsinghua University Press for all media and languages (excluding simplified and traditional Chinese) throughout the world (excluding Mainland China), and with non-exclusive license for electronic versions in Mainland China.

Registered Offices
John Wiley & Sons, Inc., 111 River Street, Hoboken, NJ 07030, USA
John Wiley & Sons Singapore Pte. Ltd, 1 Fusionopolis Walk, #07-01 Solaris South Tower, Singapore 138628

Editorial Office
1 Fusionopolis Walk, #07-01 Solaris South Tower, Singapore 138628

For details of our global editorial offices, customer services, and more information about Wiley products visit us at www.wiley.com.

Wiley also publishes its books in a variety of electronic formats and by print-on-demand. Some content that appears in standard print versions of this book may not be available in other formats.

Library of Congress Cataloging-in-Publication Data

Names: Yu, T. X. (Tongxi), 1941– author. | Qiu, XinMing, author.
Title: Introduction to impact dynamics / by T.X. Yu, Prof. XinMing Qiu.
Description: Hoboken, NJ ; Singapore : John Wiley & Sons, 2018. | Includes bibliographical references and index. |
Identifiers: LCCN 2017033421 (print) | LCCN 2017044755 (ebook) | ISBN 9781118929858 (pdf) | ISBN 9781118929865 (epub) | ISBN 9781118929841 (cloth)
Subjects: LCSH: Materials–Dynamic testing.
Classification: LCC TA418.34 (ebook) | LCC TA418.34 .Y8 2018 (print) | DDC 620.1/125–dc23
LC record available at https://lccn.loc.gov/2017033421

Cover design by Wiley
Cover image: © silavsale/Shutterstock

Set in 10/12pt Warnock by SPi Global, Pondicherry, India
Printed in Singapore by C.O.S. Printers Pte Ltd

10 9 8 7 6 5 4 3 2 1

Contents

Preface

Various impact events occur every day and everywhere in the physical world, in engineering and in people's daily lives. Our universe and planet were formed as a result of a series of impact and explosion events. With the rapid development of land vehicles, ships, and aircraft, traffic accidents have become a serious concern of modern society. Landing of spacecraft, safety in nuclear plants and offshore structures, as well as protection of human bodies during accidents and sports, all require better knowledge regarding the dynamic behavior of structures and materials.

Obviously, impact dynamics is a big subject, which looks at the dynamic behavior of all kinds of materials (e.g. metals, concrete, polymers, and composites), and the structures under study range from small objects (e.g., a mobile phone being dropped on the ground) to complex systems (e.g., a jumbo jet or the World Trade Center before 9/11). The impact velocity may vary from a few meters per second (as seen in ball games) to several kilometers per second (as seen in military applications). Driven by the needs of science and engineering, the dynamic response, impact protection, crashworthiness, and energy absorption capacity of materials and structures have attracted more and more attention from researchers and engineers. Numerous research papers and monographs have appeared in the literature, and it is not possible for anyone to condense the huge amount knowledge out there into a single book.

This book is mainly meant as a textbook for graduate students (and probably also for senior undergraduates), aiming to provide fundamental knowledge of impact dynamics. Instead of covering all aspects of impact dynamics, the contents are organized so as to consider only its three major aspects: (i) wave propagation in solids; (ii) materials' behavior under high-speed loading; and (iii) the dynamic response of structures to impact. The emphasis here is on theoretical models and analytical methods, which will help readers to understand the fundamental issues raised by various practical situations. Numerical methods and software are not the main topic of this textbook. Readers who are interested in numerical modeling related to impact dynamics will have to consult other sources for the relevant knowledge.

The audience for this textbook may also include those engineers working in the automotive, aerospace, mechanical, nuclear, marine, offshore, and defense sectors. This textbook will provide them with fundamental guidance on the relevant concepts, models, and methodology, so as to help them face the challenges of selecting materials and designing/analyzing structures under intensive dynamic loading.

The contents of this textbook have been used in graduate courses at a number of universities. The first author (T.X. Yu) taught Impact Dynamics as a credit graduate course at Peking University, UMIST (now University of Manchester), and the Hong Kong University of Science and Technology. In recent years, he has also used part of the contents to deliver a short course for graduate students in many universities, including Tsinghua University, Zhejiang University, Wuhan University, Xi'an Jiaotong University, Taiyuan University of Technology, Hunan University, and Dalian University of Technology. The second author (X.M. Qiu) has also taught Impact Dynamics as a credit graduate course at Tsinghua University over recent years.

Using the content developed for these graduate courses, we authored a textbook in Chinese, entitled *Impact Dynamics* and published by Tsinghua University Press in 2011. Although the current English version is mainly based on this Chinese version, we have made many changes. For instance, some contents in Chapters 3 and 4 have been rewritten, and Chapter 10 containing case studies is entirely new for this English version.

As a textbook, we have adopted much content from relevant monographs, such as Meyer (1994) (for Chapters 3 and 4) and Stronge and Yu (1993) (for Chapter 6), including a number of figures. This is because that content clearly elaborated the respective concepts and methods with carefully selected examples and illustrations, which are particularly suitable for a graduate course. Those monographs have been cited accordingly in the relevant places, and we would like to express our sincere gratitude to the original authors.

We would also like to thank Ms. Lixia Tong of Tsinghua University Press, who gave us a great deal of help in preparing this book.

T.X. Yu and XinMing Qiu
July 2017

Introduction

With the rapid development of all kinds of transport vehicles, the lives lost and high cost of traffic accidents are of serious concern to modern society. The public is becoming increasingly aware of the safe design of components and systems with the objective of minimizing human suffering as well as the financial burdens on society. At the same time, many other issues in modern engineering, e.g., nuclear plants, offshore structures, and safety gear for humans, also require us to understand the dynamic behavior of structures and materials.

Driven by the needs of engineering, the dynamic response, impact protection, crashworthiness, and energy absorption capacity of various materials and structures have attracted more and more attention from researchers and engineers. As a branch of applied mechanics, impact dynamics aims to reveal the fundamental mechanisms of large dynamic deformation and failure of structures and materials under impact and explosive loading, so as to establish analytical models and effective tools to deal with various complex issues raised from applications.

In the classical theories of elasticity and plasticity, usually only static problems are of concern, in which the external load is assumed to be applied to the material or structure slowly, and the corresponding deformation of the material or structure is also slow. The acceleration of material is very small and thus the inertia force is negligible compared with the applied external load; hence the whole deformation process can be analyzed under an equilibrium state.

However, it is known that the material behavior and structural response under dynamic loading are quite different from those under quasi-static loading. In engineering applications, the external load may be intensive and change rapidly with time, termed intense dynamic loading; consequently, the deformation of material or of a structure has to be quick enough under intense dynamic loading. Some examples are given in the following.

The collision of vehicles. Cars, trains, ships, aircraft, and other vehicles may collide with each other or with surrounding objects during accidents. These accidents will lead to the failure or deformation of the structures as well as personnel casualties, resulting in serious economic losses. As the number of cars has rapidly increased in many countries, car accidents have become the number one cause of death in the world. Collisions between ships and collisions between ships and rocks/bridges all cause huge economic loss as well as environmental pollution. Along with the development of high-speed rail transportation, the safety of occupants is also of greater public concern. It is very dangerous if a bird impinges on the cockpit or engine of an airplane, as the relative velocity between bird and airplane could be high even though the speed of the bird is not great. More and more space debris has been produced as a result of human activities, and the relative velocity of space debris to spacecraft can be as high as 10 km/s, so there would be great damage in the case of a collision.

Damage effects of explosive. Buildings, bridges, pipelines, vehicles, ships, aircrafts, and protective structures could be subjected to intensive explosion loading, due to industrial accident, military action, or terrorist attack. Typically, these structures would be suddenly loaded by a shock wave propagating in the air.

The effects of natural disasters. Natural disasters, such as earthquakes, tsunamis, typhoons, floods and so on will produce intensive dynamic loads to structures, e.g., dams, bridges, and high-rise buildings. These intense dynamic loads are likely to cause damage to the structures.

The strong dynamic loads caused by the local rupture of the storage structures. In nuclear power plants or chemical plants, if there is local damage to a pipeline, the jet of high-pressure liquid that would escape from the broken section exerts a lateral reaction force (the blowdown force) on the broken pipe, causing rapid acceleration and large deformation, termed "pipe whip". After local damage, the consequences from a pressure vessel or a dam could be disastrous.

Load of high-speed forming. During a dynamic metal forming process, such as explosive forming and electromagnetic forming, the work-piece is subjected to intensive dynamic loading and deforms rapidly. Similar situations take place in the process of forging or high-speed stamping.

Impact or collision in daily life and sports. For example, falling objects, falling on the ice, collision between moving people, a football or golf ball hitting the head or body with high speed.

All kinds of the above-mentioned problems encountered in engineering or daily life require the understanding and study of the behavior of the solid materials and structures subjected to intensive dynamic loads. First of all, why is the dynamic behavior of materials and structures usually different from the quasi-static behavior? This is the result of three major attributes in mechanics, as briefly illustrated here.

Stress wave propagation in material and structure. When a dynamic load is applied to the surface of a solid, the stress and generalized deformation will propagate in the form of stress wave. If the disturbance is weak, it is an elastic wave; but if the stress level of wave is higher than the yield strength of the material, it will be plastic wave.

Suppose a solid medium has a characteristic scale of L, and the wave speed of its material is c. It is subjected to an external dynamic loading that has a characteristic time t_c, e.g., the time period for the external load to reach its maximum value or time duration of the impulse. If $t_c << L/c$, the stress and deformation distribution in this solid are not uniform; hence, the effect of stress wave propagation must be considered. For example, the characteristic scale of the crust is very large, so the effects of earthquake or underground explosion are mainly presented in the form of stress waves.

In a piling machine or a split Hopkinson pressure bar device (SHPB for short – an important experimental technique in studying the dynamic properties of materials; refer to Chapter 3 for more details), the perturbation is along the longitudinal (large scale) direction of a long bar rather than in the radial (small scale) direction. Hence wave reflection, transmission, and dispersion are important factors that need to be analyzed carefully.

By contrast, some other structural components that are widely used in engineering, such as beams, plates, and shells, are usually subjected to lateral loads along their thickness direction, which is the smallest scale direction of the structure. The elastic wave speed in metals is usually in the order of several kilometers per second (e.g., 5.1 km/s for steel). Therefore, in several micro-seconds all the particles in the thickness direction of the structure will be affected by the external disturbance, and then the entire section of structure will be accelerated and then move together. This global motion of the whole cross-sections of the structure is classified as the elastic-plastic dynamic response of the structure; this is discussed in detail in Part 3 of this book. This subsequent global structural response may last several milliseconds or even several seconds, depending on the type of structure and loading, before the structure reaches its maximum deformation.

Because the effective time of stress wave propagation is usually several orders of magnitude smaller than that of the long-term structural response, the total response of the structure can be divided into two decoupled separate stages. That is, in the analysis of wave propagation, the structure is assumed to remain in its original configuration, which is regarded as the reference frame for geometric relations and equations of motion, while in the analysis of structural response, the early time wave propagation is disregarded and only its global deformation is considered.

Rate-dependency of a material's properties. The material in a solid or structure will deform rapidly under intensive dynamic loading. Depending on the microscopic deformation mechanism, the resistance of material to rapid deformation is generally higher than that to slow deformation, as revealed by numerous experiments on materials. For example, the mechanism of plastic deformation of metals is mainly attributed to the movement of dislocations. The resistance to the dislocation motion will be much higher when the dislocation passes through the metal lattice at a high speed than at a low speed, and this will lead to the higher yield stress and the high flow stress of metals during high-speed deformation.

An important task in the study of dynamic properties of materials is to summarize the effect of strain rate on the stress–strain relationship, based on the experimental data, so as to establish the strain rate-dependent constitutive relation of materials. As the strain history and instantaneous strain rate of the material elements inside a structure vary with position and time, the dynamic constitutive relation has to be simplified to a large extent when it is applied to dynamics analysis of structures.

Inertia effect in structural response. In the analysis of dynamic response of a structure, usually both elastic deformation and plastic deformation exist, and the boundary between elastic-plastic regions changes with time. Therefore, different constitutive relations should be employed in different regions. Further, the complicated moving boundary has to be dealt with. In order to reduce the complexity, the constitutive relation of material also needs to be simplified in the theoretical modeling of structural dynamics. The most successful idealization commonly adopted in theoretical modeling is to assume that the structure is made of a rigid-perfectly plastic material, which neglects all the effects of elasticity, strain hardening and strain rate.

The basis of the hypothesis is that the structure usually experiences considerable large plastic deformation under intensive dynamic loading, and thus most of the work done by the external load will be dissipated by plastic deformation, while only a small amount of external work will be stored in the form of elastic deformation energy. Therefore, to ignore the elastic deformation and the corresponding energy will only result in minor influence over the final deformation and failure mode of the structure. As will be seen in the book, this idealization largely simplifies the analysis.

Due to the similarity of material idealization, the dynamic analysis of rigid-perfectly plastic structures is closely related to the limit analysis of structures, especially in its concept and methodology. For example, the widely accepted concepts in limit analysis in the kinematically admissible velocity field can be successfully extended to construct dynamic deformation mechanisms containing stationary or traveling plastic hinges.

At the same time, it should be noted that the main difference between dynamic analysis and limit analysis lies with the intervention of the inertia effect in dynamic analysis. The limit analysis based on the theory of plasticity reveals that a structure made of rigid-perfectly plastic material under external load must have a limit state, i.e., if the external load reaches a certain limit value, the structure will become a mechanism and lose the load-carrying capacity. On the other hand, from the dynamic analysis of the same structure, it is found that if the dynamic load exceeds the static limit load, i.e., the collapse load, the structure will be accelerated. According to the D'Alembert principle, it is the inertia force of the structure that is in equilibrium with the external load and that resists deformation. The greater the external load, the greater the acceleration, so the greater the inertia force. Thus, the structure can bear a much higher external load than the static limit load in a short time, which is a notable feature of the structural dynamic response and is different from the static limit analysis.

Generally speaking, dynamic loading and dynamic response always become significant when accidents occur. Nowadays, alongside the development of computing capability and software, different kinds of numerical tools and methods have been developed very rapidly, so they have wider and wider applications. Therefore, some researchers think that it is appropriate to employ numerical simulations in handling problems of impact dynamics. However, even for numerical simulations, a proper understanding of the basic principles, concepts, and theoretical models used in dynamic analyses is crucial if simulation methods and models are to give reliable results. Furthermore, the numerical simulations will result in a huge amount data, so our understanding of impact dynamics will greatly help us to digest the data and discover the underlying physical significance and engineering implications.

This textbook aims to demonstrate the fundamental features of the dynamic behavior of materials and structures, to clearly illustrate the widely applicable theoretical models and analytical methods, and to highlight the most important factors that affect dynamic behavior. The textbook is highly relevant to education programs at both graduate and senior undergraduate levels. For those engineers who are working in the automotive, aerospace, mechanical, nuclear, marine, offshore, and defense sectors, the book will also provide fundamental guidance on relevant concepts, models, and methodology, to help them face the challenges of understanding the dynamic behavior of materials and of analyzing and designing structures under various types of intensive dynamic loading.

Part 1

Stress Waves in Solids

1

Elastic Waves

In a deformable solid medium, the disturbance to mechanical equilibrium is represented by the change in particle velocity and the corresponding changes in stress and strain states. When some parts of a solid are first disturbed, finite time durations are required for this disequilibrium to be felt by other parts of the body, due to the deformable properties of the body. This kind of propagation as a result of the disturbance in stress and strain through a solid body is termed a *stress wave*.

1.1 Elastic Wave in a Uniform Circular Bar

1.1.1 The Propagation of a Compressive Elastic Wave

Consider a uniform circular bar made of isotropic material, as shown in Figure 1.1. Let x denote the *longitudinal coordinate* measured from an origin O, which is fixed in the space; and let $u(x)$ denote the *displacement* undergone by a plane AB in the bar, which is initially at a distance x from O. Then $u + \frac{\partial u}{\partial x}\delta x$ is the *displacement* of plane $A'B'$ which is parallel to AB but is initially at a distance $x + \delta x$ from O.

A force applied rapidly at time $t = 0$, over the end plane at $x = 0$, will cause a *disturbance* to propagate elastically along the bar, so that a compressive normal stress, $-\sigma_0$, will pass through plane AB at time t.

It should be noted that the slender bar assumption is adopted here, i.e. the pulse length is at least six times the typical cross-sectional dimension of the bar. In this case, the strain and inertia in the *transverse* direction can be neglected. The gravitational force and damping of the material are also ignored in the following analysis.

The equilibrium of a representative element of the bar is illustrated in Figure 1.2. Here A_0 is the initial cross-section area of the bar, ρ_0 is the initial density of the material, and $-\sigma_0$ is the stress transmitted, with the negative sign reflecting the fact that the stress is compressive, as shown in the figure.

From Newton's second law, the equation of motion of the representative element is $-\frac{\partial \sigma_0}{\partial x}\delta x\, A_0 = \rho_0 A_0 \delta x \frac{\partial^2 u}{\partial t^2}$, which could be simplified as

$$\frac{\partial \sigma_0}{\partial x} = -\rho_0 \frac{\partial^2 u}{\partial t^2} \tag{1.1}$$

Introduction to Impact Dynamics, First Edition. T.X. Yu and XinMing Qiu.

Published 2018 by John Wiley & Sons Singapore Pte. Ltd.

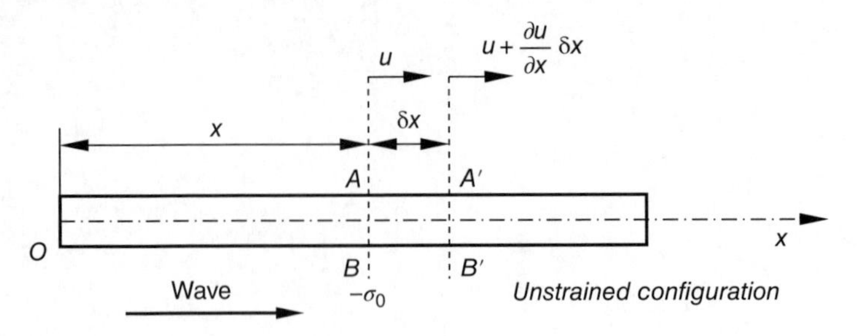

Figure 1.1 The propagation of a compressive elastic wave in a uniform circular bar.

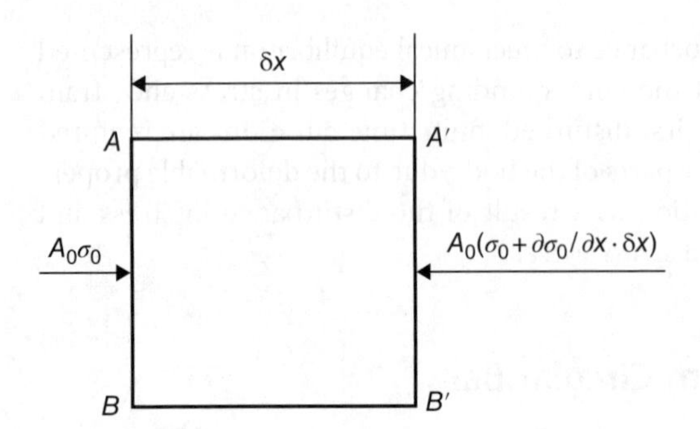

Figure 1.2 The equilibrium of a representative element of the bar.

To analyze the deformation of the representative element of length δx, it is clear that the strain of the element is

$$\varepsilon = \frac{\partial u}{\partial x} \tag{1.2}$$

Assuming the solid material has Young's modulus E, then, according to Hooke's law, in the linear elastic stage we have

$$-\sigma_0 = E\frac{\partial u}{\partial x} \tag{1.3}$$

The stress variation over the element is obtained from the partial differential of Eq. (1.3) with respect to x:

$$\frac{\partial \sigma_0}{\partial x} = -E\frac{\partial^2 u}{\partial x^2} \tag{1.4}$$

Substituting Eq. (1.1) into Eq. (1.4) leads to

$$\rho_0\frac{\partial^2 u}{\partial t^2} = E\frac{\partial^2 u}{\partial x^2} \tag{1.5}$$

With the notation of $c_L = \sqrt{E/\rho_0}$, Eq. (1.5) is rewritten as

$$\frac{\partial^2 u}{\partial t^2} = c_L^2\frac{\partial^2 u}{\partial x^2} \tag{1.6}$$

Obviously, Eq. (1.6) is a typical one-dimensional (1D) wave equation of the following form:

$$\frac{\partial^2 u}{\partial t^2} = c^2 \frac{\partial^2 u}{\partial x^2} \tag{1.7}$$

Considering the general solution of Eq. (1.7), $u(x, t)$, in the following form:

$$u(x,t) = f_1(x - ct) + f_2(x + ct) \tag{1.8}$$

and substituting Eq. (1.8) into the wave equation (Eq. 1.7), we find

$$\frac{\partial u}{\partial t} = -cf_1'(x-ct) + cf_2'(x+ct), \quad \frac{\partial^2 u}{\partial t^2} = c^2 f_1''(x-ct) + c^2 f_2''(x+ct)$$
$$\frac{\partial u}{\partial x} = f_1'(x-ct) + f_2'(x+ct), \quad \frac{\partial^2 u}{\partial x^2} = f_1''(x-ct) + f_2''(x+ct)$$

Thus it can be verified that Eq. (1.8) satisfies the wave equation, so it gives a general solution to Eq. (1.7). In order to understand the mechanical meaning of Eq. (1.8), only one term is studied here, $u(x,t) = f_1(x - ct)$, i.e., $f_2 = 0$ (see Figure 1.3).

At $t = t_1$, the particle at position $x = x_1$ has a displacement $u = s$, and at $t = t_2$, the particle at position $x = x_2$ also has a displacement $u = s$. Thus from Eq. (1.8), the displacement should satisfy $s = f_1(x_1 - ct_1) = f_1(x_2 - ct_2)$, which results in $x_1 - ct_1 = x_2 - ct_2$, leading to the following speed of wave propagation:

$$c = \frac{x_2 - x_1}{t_2 - t_1} \tag{1.9}$$

This confirms that for the wave propagation governed by Eq. (1.6), $c_L = \sqrt{E/\rho_0}$ precisely represents the speed of the longitudinal waves (compressive or tensile).

There are two terms on the right-hand side of the general solution of wave propagation, Eq. (1.8). The term $f_1(x - ct)$ denotes the wave traveling in the $+x$ direction, i.e., a forward-traveling wave, and the term $f_2(x + ct)$ denotes the wave traveling in the $-x$ direction, i.e., a backward-traveling wave. Both the traveling waves in Eq. (1.8), $f_1(x - ct)$ and $f_2(x + ct)$, have the following characteristics: the waves are traveling at a constant speed with no change in their shape or magnitude, i.e. the 1D longitudinal waves are *non-dispersive*.

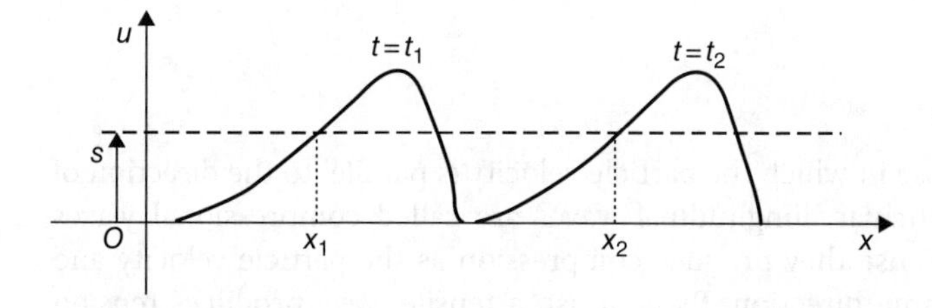

Figure 1.3 Particle displacements produced by a forward wave.

Table 1.1 Typical longitudinal wave speed in solid materials

	Steel	Aluminum	Glass	Polystyrene
E (GPa)	205	75	95	
ρ_0 (g/cm^3)	7.8	2.7	2.5	
c_L (m/s)	5100	5300	6200	2300

E, Young's modulus; ρ_0, density; c_L, wave speed.

The speed of longitudinal (compressive or tensile) waves in four typical materials are given in Table 1.1. It should be emphasized that the wave speed depends on both Young's modulus and density of the material. Therefore, although the density of aluminum is only about one-third of that of steel, the wave speeds are similar.

For a 1D longitudinal wave, there are two ways in which you can distinguish between a compressive and a tensile wave:

- *Look at the sign of the stress.* A compressive wave produces a negative normal stress and a tensile wave produces a positive stress.
- *Look at the directions of the particle velocity and the wave propagation.* For a compressive wave, the particle velocity is in the same direction as the wave propagation, whereas for a tensile wave, the particle velocity is in the opposite direction to the wave propagation.

1.2 Types of Elastic Wave

Different types of elastic wave can propagate in solids. These waves are classified according to the relationship between the motion of the particles and the direction of propagation of the waves and also according to the boundary conditions. The most common types of elastic wave in solids are:

- Longitudinal (irrotational) waves
- Transverse (shear) waves
- Surface (Rayleigh) waves
- Interfacial (Stoneley) waves
- Bending (flexural) waves (in beams and plates).

1.2.1 Longitudinal Waves

Longitudinal waves are those in which the particle velocity is parallel to the direction of travel of the wave. In particular, longitudinal waves are called compressional waves or compression waves, because they produce compression as the particle velocity and wave velocity are in the same direction. By contrast, a tensile wave produces tension as the particles and waves travel in opposite directions.

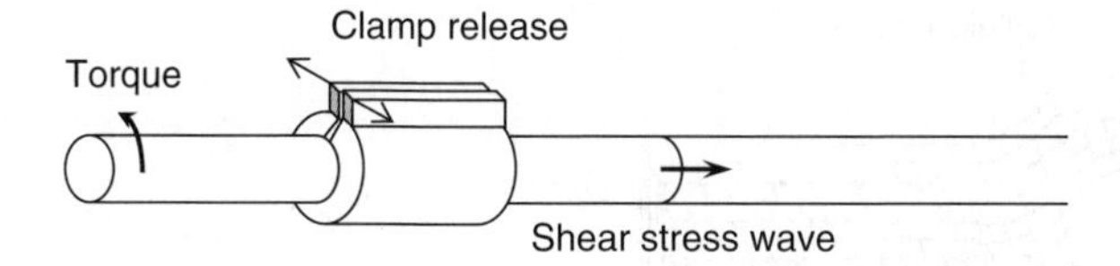

Figure 1.4 The transverse wave in a circular bar produced by suddenly releasing the clamp.

Longitudinal waves are also known as *irrotational waves.* In seismology, they are known as P-waves. This is because a longitudinal wave travels at the fastest speeds and so arrives at seismic stations first (**p**rimary waves); the rock moves forward and backward in the same direction as the wave is traveling (**p**ush–pull wave, **p**arallel to propagation), and the wave is able to **p**ass through solids and liquids. In infinite and semi-infinite media, longitudinal waves are also known as *dilatational waves,* due to the changes in the volume of the media.

1.2.2 Transverse Waves

Transverse waves are another common type of wave, in which the particle velocity is perpendicular to the wave's propagation. An example of a transverse wave is shown in Figure 1.4. In this arrangement a circular bar is clamped at a given position and a torque is applied to the left end of the bar. Thus there is shear stress on the left side of the clamp and zero stress on the right-hand side. When the clamp is suddenly released, the stress disturbance will propagate, i.e., a wave will travel from the left side to the right side of the bar. The particle velocity is within the cross-sectional plane of the bar, while the wave propagation direction is along the bar; hence, as the particle velocity is perpendicular to the wave velocity, this torsional wave is a transverse wave. The normal strains are all zero, with no resulting change in density, while the shear strains are non-zero, producing a change in shape. Thus, transverse waves are called *shear waves,* and are also known as *distortional* or *equivolumal waves.*

For an elastic material with shear modulus G and density ρ_0, the speed of the transverse wave is derived as $c_S = \sqrt{G/\rho_0}$, which is slower than that of the longitudinal wave in the same solid, as $\frac{c_S}{c_L} = \sqrt{\frac{G}{E}} = \frac{1}{\sqrt{2(1+\nu)}} < 1$.

In seismology, transverse waves are known as **S**-waves, because they do not travel as quickly as P-waves (**s**low wave) and will arrive at a seismic station second (**s**econdary wave); the rocks move from **s**ide to side (**s**hear wave) and the waves only travel through **s**olids.

1.2.3 Surface Wave (Rayleigh Wave)

Surface waves are analogous to gravitational waves on the surface of water. As shown in Figure 1.5, the material particles move up and down as well as back and forth, tracing elliptical paths. The surface wave is restricted to the region adjacent to the surface. The particle velocity decreases very rapidly (exponentially) as one moves away from

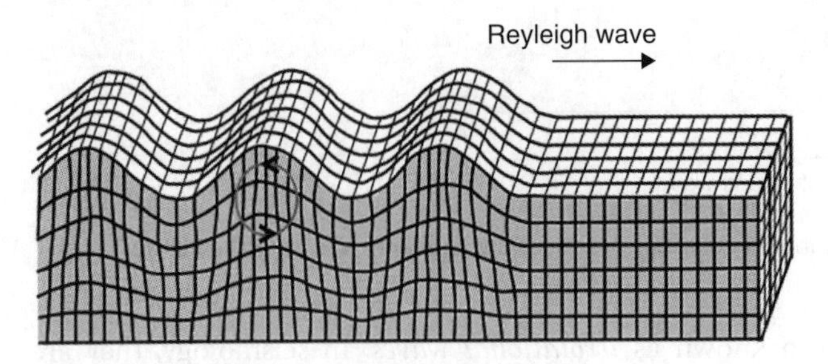

Figure 1.5 Diagram of a Rayleigh wave.

the surface. Surface waves in solids (called *Rayleigh waves*) are a particular case of interfacial waves where one of the materials has negligible density and elastic wave speed.

If a hammer hits the surface of a semi-infinite solid, several waves are propagated after the blow – among these the speed of the longitudinal wave (P-wave) is greater than that of the transverse wave (S-wave), and the surface wave (Rayleigh wave) only affects the solid for a finite distance under the surface.

1.2.4 Interfacial Waves

When two semi-infinite media with different material properties are in contact, special waves form at their interface in the case of a disturbance. The surface wave in a solid (Rayleigh wave) could be regarded as a special case of the interfacial wave – that is, the density and elastic wave in one of the contacting media, such as air, could be omitted.

1.2.5 Waves in Layered Media (Love Waves)

The earth is composed of layers with different properties, and so special wave patterns emerge. This was first studied by Love. As a result of Love waves, the horizontal component of displacement produced by earthquakes can be significantly larger than the vertical component, which is a behavior that is not consistent with Rayleigh waves.

1.2.6 Bending (Flexural) Waves

These waves involve propagation of flexure in 1D (beams and arches) or 2D (plates and shells) configurations. Made from a material of density ρ_0 and elastic modulus E, a straight beam with cross-sectional area A_0 and principal moment of inertia I is shown in Figure 1.6. A coordinate system with the x-axis along the beam length and the z-axis in the direction of deflection is adopted. When a bending moment M and shear force Q are applied, a transverse deflection w is produced in the beam.

By considering the equilibrium of a small element of length δx as shown in Figure 1.6 (b), the equation of motion of this element gives

$$-(\rho_0 A_0 \delta x)\frac{\partial^2 w}{\partial t^2} = \frac{\partial Q}{\partial x}\delta x \tag{1.10}$$

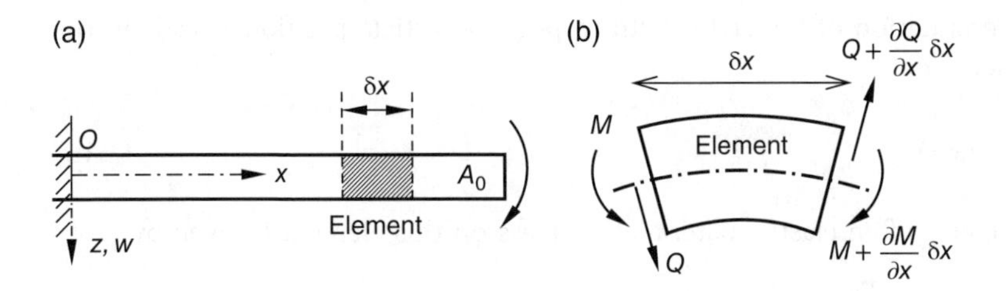

Figure 1.6 A straight beam under bending. (a) Beam under bending; (b) free body diagram of an element.

From the relation given by elastic mechanics,

$$EI\frac{\partial^3 w}{\partial x^3} = Q \tag{1.11}$$

The combination of Eqs (1.10) and (1.11) leads to the wave equation of a bending wave:

$$\rho_0 A_0 \frac{\partial^2 w}{\partial t^2} = -EI\frac{\partial^4 w}{\partial x^4} \tag{1.12}$$

Or in an equivalent form:

$$\frac{\partial^2 w}{\partial t^2} = -c_L^2 k^2 \frac{\partial^4 w}{\partial x^4} \tag{1.13}$$

where $c_L = \sqrt{E/\rho_0}$ is the speed of the longitudinal wave, and k denotes the radius of gyration of the cross-section about the neutral axis, i.e., $I = A_0 k^2$ holds.

Obviously, solutions in the form of $w(x, t) = f_1(x - ct)$ or $w(x, t) = f_1(x + ct)$, which are the general solutions of the regular wave equation, do not satisfy the bending wave equation (Eq. 1.13). This implies that flexural disturbance of arbitrary form always propagates with dispersion.

1.3 Reflection and Interaction of Waves

1.3.1 Mechanical Impedance

Let us focus on the longitudinal waves again. As shown in Sections 1.1 and 1.2, for a forward longitudinal wave, i.e., the wave moving in the positive x-direction, from the general solution of the wave equation, Eq. (1.8), the displacement of particle is

$$u(x,t) = f(x - ct) \tag{1.14}$$

Differentiating Eq. (1.14) with respect to time t leads to the particle velocity:

$$v_0 = \frac{\partial u}{\partial t} = -cf'(x,t) \tag{1.15}$$

A partial differentiation of Eq. (1.14) with respect to particle position x leads to the strain of this element:

$$\varepsilon = \frac{\partial u}{\partial x} = f'(x,t) \tag{1.16}$$

From the property of an elastic material, the stress on this element is given by

$$\sigma = -E\varepsilon = -Ef' = \frac{Ev_0}{c} \tag{1.17}$$

Using the expression of wave speed, $c = \sqrt{E/\rho_0}$, Eq.(1.17) can be rewritten as

$$\sigma = \frac{Ev_0}{c} = \rho_0 c v_0 = v_0\sqrt{E\rho_0} \tag{1.18}$$

where the quantity $\rho_0 c$ is termed the *mechanical impedance*, or *sonic/sound impedance*, of the material. This expression can also be applied to the propagation of the tensile wave:

$$v_0 = \frac{\sigma}{\sqrt{E\rho_0}} = \frac{c}{E}\sigma = \frac{\sigma}{\rho_0 c} \tag{1.19}$$

In Eq. (1.19), the particle velocity is related to the current stress. For example, for steel, if the stress is 100 MPa, then the particle velocity is

$$v_0 = \frac{c}{E}\sigma = \frac{5100\text{m/s} \times 100\text{MPa}}{205\text{GPa}} \approx 2.49\,\text{m/s} \tag{1.20}$$

The mechanical impedance of steel is

$$\rho_0 c = 7800\,\text{kg/m}^3 \times 5100\,\text{m/s} \approx 4 \times 10^7\text{Ns/m}^3 \tag{1.21}$$

Wave speed and *mechanical impedance* are two very important concepts for stress waves. The wave speed indicates the velocity of the disturbance propagating in a deformable solid, while the mechanical impedance represents the degree of resistance of the deformable solid to the disturbance.

1.3.2 Waves When they Encounter a Boundary

We will briefly describe the interaction of waves when they encounter a boundary. Figure 1.7 shows the longitudinal waves that are reflected and refracted at the boundary as well as the two transverse waves that are generated at the interface. These effects, reflection and refraction, occur when the wave encounters a medium with different mechanical impedance, which is defined as the product of the medium density and its elastic wave speed. These refraction and reflection angles are given by a simple relationship of the form:

$$\frac{\sin\theta_1}{c_\text{L}} = \frac{\sin\theta_2}{c_\text{S}} = \frac{\sin\theta_3}{c_\text{L}} = \frac{\sin\theta_4}{c'_\text{L}} = \frac{\sin\theta_5}{c'_\text{S}} \tag{1.22}$$

The interactions of a wave with an interface are very simple when the incidence is normal ($\theta_1 = 0$). In this case, a longitudinal wave refracts/transmits and reflects longitudinal

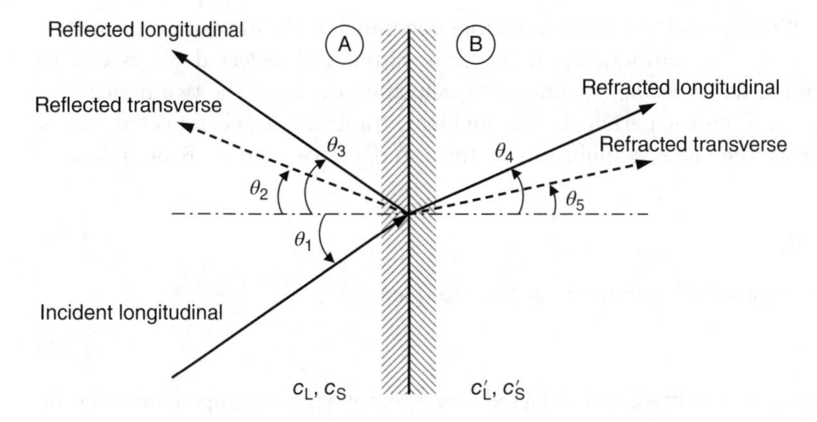

Figure 1.7 Reflection and refraction when a longitudinal wave encounters an interface.

waves, and a shear wave refracts/transmits shear waves. In this way, it becomes a 1D wave reflection and transmission problem.

1.3.3 Reflection and Transmission of 1D Longitudinal Waves

Figure 1.8(a) shows the fronts of a wave propagating along a cylinder in a medium in which the wave speed is c_A. The particle velocity is v and the stress is σ. Figure 1.8(b) illustrates the stresses at the interface related to incident, transmitted, and reflected

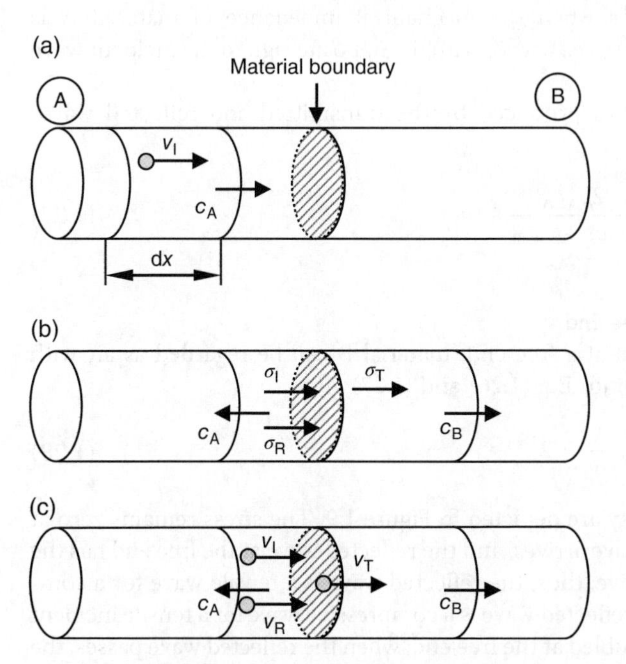

Figure 1.8 The reflection and transmission of a longitudinal wave in a one-dimensional cylinder.

waves. Figure 1.8(c) depicts the particle velocities generated by the incident, transmitted, and reflected waves. The amplitudes of the transmitted and reflected waves can be calculated from the densities, ρ_A, ρ_B, and wave velocities, c_A, c_B, of the two media.

Let subscripts I, T and R pertain to the incident, transmitted and reflected waves, respectively. From the force equilibrium at the interface, i.e., the A–B boundary of the bar,

$$\sigma_I + \sigma_R = \sigma_T \tag{1.23}$$

Then, from the material's continuity at the interface,

$$v_I + v_R = v_T \tag{1.24}$$

For the incident, transmitted, and reflected waves, employing the impedance relation between particle velocity and stress, i.e., Eq.(1.19),

$$v_I = \frac{\sigma_I}{\rho_A c_A}, \quad v_R = -\frac{\sigma_R}{\rho_A c_A}, \quad v_T = \frac{\sigma_T}{\rho_B c_B} \tag{1.25}$$

The stresses produced by the transmitted and reflected waves are obtained by combining Eqs (1.23)– (1.25):

$$\frac{\sigma_T}{\sigma_I} = \frac{2\rho_B c_B}{\rho_B c_B + \rho_A c_A}, \quad \frac{\sigma_R}{\sigma_I} = \frac{\rho_B c_B - \rho_A c_A}{\rho_B c_B + \rho_A c_A} \tag{1.26}$$

From Eq. (1.26), the amplitudes of transmitted and reflected waves are all determined by the mechanical impedances of materials. When the mechanical impedance of material B is larger, then $\rho_B c_B > \rho_A c_A$, $\sigma_R/\sigma_I > 0$, i.e., a wave with the same sign as the incident wave is reflected; whereas when the mechanical impedance of material A is larger, then $\rho_B c_B < \rho_A c_A$, $\sigma_R/\sigma_I < 0$, i.e., a wave with the opposite sign to the incident wave is reflected.

Similarly, the particle velocities produced by the transmitted and reflected waves can be calculated as

$$\frac{v_R}{v_I} = \frac{\rho_A c_A - \rho_B c_B}{\rho_A c_A + \rho_B c_B}, \quad \frac{v_T}{v_I} = \frac{2\rho_A c_A}{\rho_A c_A + \rho_B c_B} \tag{1.27}$$

Special Case 1: Reflection at a Free End

For stress waves being reflected at a free end, material B can be regarded as air with $\rho_B c_B = 0$. Substituting $\rho_B c_B = 0$ into Eq. (1.26) and (1.27) gives

$$\frac{\sigma_T}{\sigma_I} = 0, \quad \frac{\sigma_R}{\sigma_I} = -1, \quad \frac{v_R}{v_I} = 1, \quad \frac{v_T}{v_I} = 2 \tag{1.28}$$

The stress and particle velocity are depicted in Figure 1.9. The stress remains zero at the free end when an incident wave arrives, and the reflected wave at the free end has the opposite sign to the incident wave; thus, the reflected wave is a tensile wave for a compressive incident wave, and the reflected wave is a compressive wave for a tensile incident wave. The particle velocity is doubled at the free end; when the reflected wave passes, the particle velocity is the same as that when the incident wave passes.

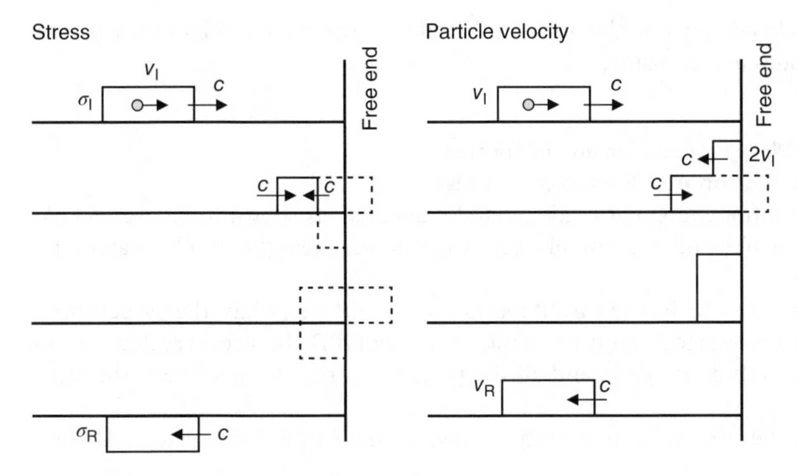

Figure 1.9 One-dimensional wave reflection at a free end.

Special Case 2: Reflection at a Fixed End

For stress wave being reflected at a fixed (clamped) end, material B could be regarded as a rigid body, $E_B = \infty$, and this leads to $\rho_B c_B = \infty$. Substituting $\rho_B c_B = \infty$ into Eq.(1.26) and (1.27) results in

$$\frac{\sigma_T}{\sigma_I} = 2, \quad \frac{\sigma_R}{\sigma_I} = 1, \quad \frac{v_R}{v_I} = -1, \quad \frac{v_T}{v_I} = 0 \tag{1.29}$$

The stress and particle velocity are depicted in Figure 1.10. The stress is doubled at a fixed end when an incident wave arrives, and the reflected wave at the fixed end has the same sign as the incident wave; thus, the reflected wave is also a compressive wave for a compressive incident wave, and the reflected wave is a tensile wave for a tensile incident wave. The particle velocity remains at zero at the fixed end; when the reflected wave

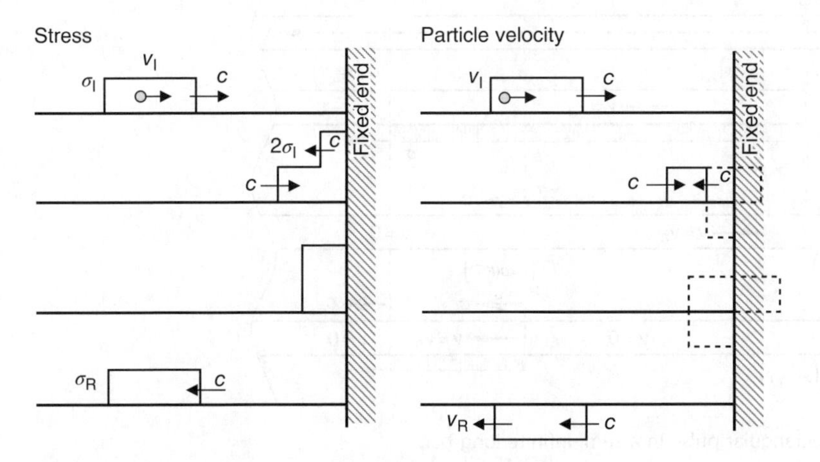

Figure 1.10 One-dimensional wave reflection at a fixed end.

passes, the particle velocity is in the opposite direction to when the incident wave passes, but the magnitudes are the same.

Examples of 1-D Wave Propagation and Interaction

Example 1 Formation of a Rectangular Pulse

Consider a semi-infinite long bar made of elastic material, as shown in Figure 1.11. At time $t = 0$, give the free end A an initial velocity v_0 towards the *right*, which creates a 1D compressive wave.

Then at an instant $t > 0$, the wave front reaches B with $AB = ct$, whilst the particle originally located at A moves to A' with $AA' = v_0 t$. In segment $A'B$, the stress created is compressive with magnitude $\sigma = \rho c v_0$, and all the particles in this segment have the same velocity v_0.

At $t = T$, give the free end A a velocity v_0 towards the *left*, which creates a 1D tensile wave.

At an instant $t > T$, the tensile wave front reaches D with $A'D = c(t - T)$, while the free end moves to A''. These two elastic waves superpose on each other, resulting in a rectangular stress pulse of length cT and magnitude $\sigma = \rho c v_0$. This pulse (in segment DB) is a compressive wave and moves towards the right at speed c. Apart from the particles within segment DB, the particles in the bar possess no stress and no velocity, so they are in their undisturbed state.

Example 2 The Interaction of a Compressive Pulse and a Tensile Pulse

Consider a slender bar A_1A_2 with a center point B, as shown in Figure 1.12. In the bar segment A_1B, a compressive pulse of stress σ is propagating from the left towards the right, while the particles within the pulse are traveling at velocity $+v_0$ (to the right).

At the same time, in the bar segment A_2B that is symmetrical to segment A_1B, a tensile wave with the same stress magnitude σ is propagating from the right towards the left, while the particles within the pulse are traveling at velocity $+v_0$ (to the right).

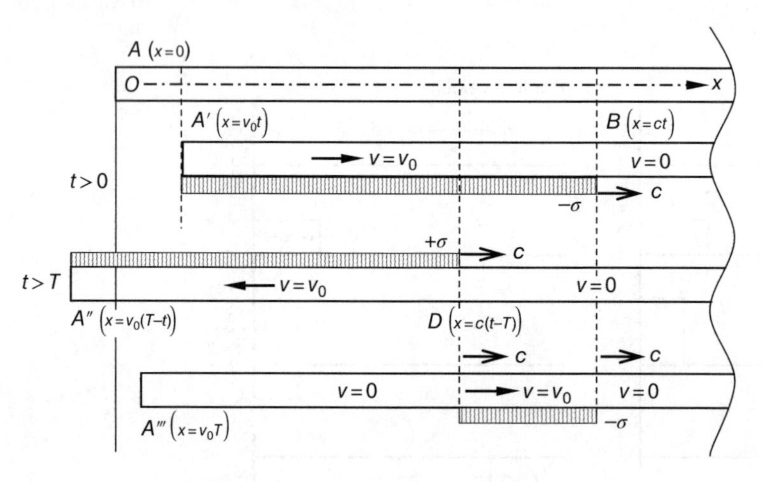

Figure 1.11 A rectangular pulse in a semi-infinite long bar.

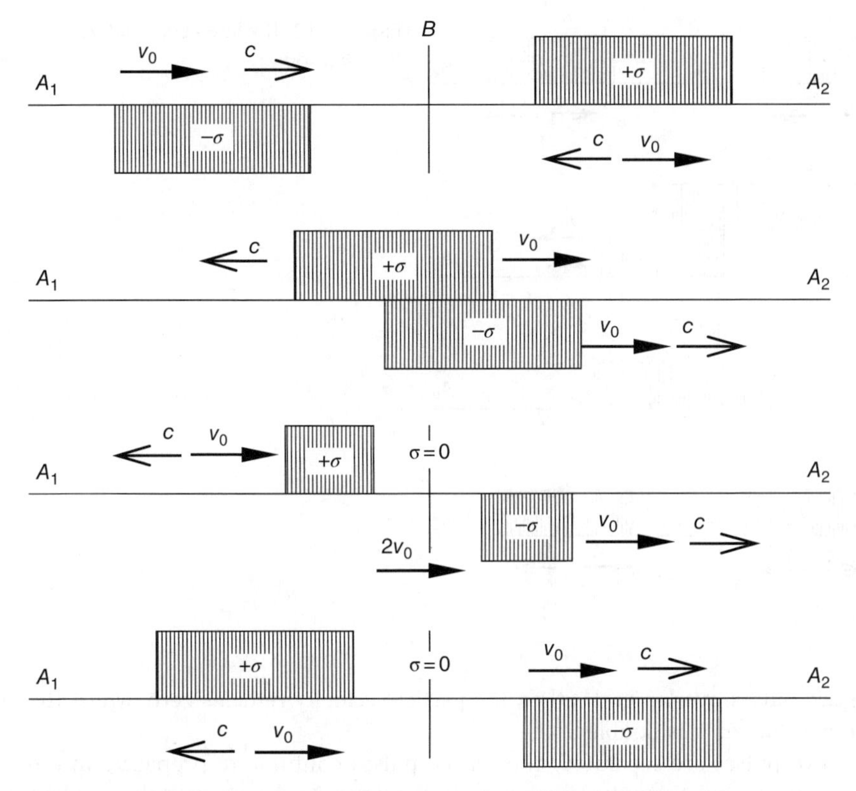

Figure 1.12 The interaction of a compressive pulse and a tensile pulse.

When the two pulses meet each other at cross-section B, the stress at B is reduced to zero, while the particle velocity is doubled, i.e., $2v_0$. After that, in the region where the two pulses overlap each other, the stress remains zero, while the particle velocity becomes $2v_0$, i.e., it is doubled.

After the two pulses are separated, a compressive pulse propagates in segment A_2B, while a tensile wave propagates in segment A_1B. Considering the fact that at any instant the stress at B remains at zero, imagine the bar is cut at B and becomes the free end of both bars A_1B and A_2B. Then the above propagation picture represents the wave reflection at the free end B, i.e., when a 1D wave is reflected at a free end, the stress changes its sign while the particle velocity is doubled.

Example 3 The Interaction of Two Tensile Pulses

Consider a slender bar A_1A_2 with center point D, as shown in Figure 1.13. In segments A_1D and A_2D, which are symmetrical to each other, two identical tensile pulses of stress σ are propagating head to head, while the particles within the pulses are traveling at velocity $+v_0$ (to the right) and $-v_0$ (to the left), respectively.

When the two pulses meet each other at cross-section D, the particle velocity at D is reduced to zero, while the stress becomes 2σ, i.e., it is doubled. After that, in the region

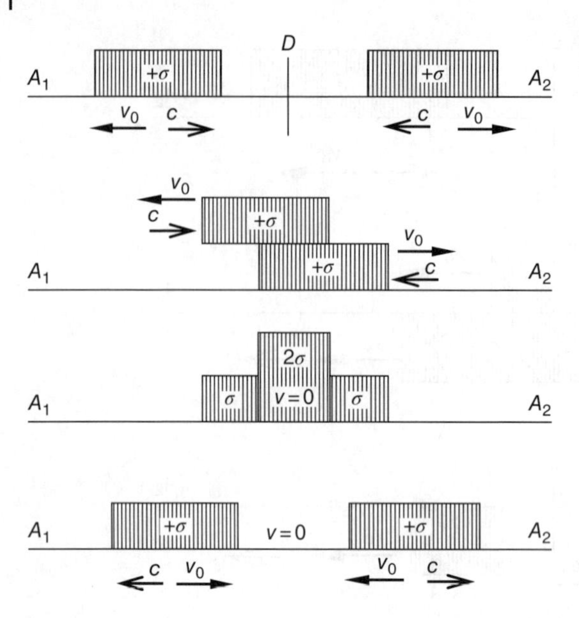

Figure 1.13 The interaction of two tensile pulses.

where the two pulses overlap each other, the particle velocity remains zero, whilst the stress becomes 2σ, i.e., it is doubled.

After the two pulses are separated, two tensile pulses continue to propagate in the respective segments, but the pulses propagate in the opposite direction to the incident pulses. Considering the fact that at any instant the particle velocity at D remains zero, imagine the bar is cut at D and becomes the fixed ends of both bars A_1D and A_2D. Then the above propagation picture represents the wave reflection at fixed end D; when a 1D wave is reflected at a fixed end, the particle velocity will remain zero there, while the stress is doubled. In the rest of the bars, both the wave propagation and the particle velocity change their signs.

Example 4 The Normal Collinear Collision of Two Identical Bars

Consider two bars of identical material and size traveling in opposite directions at the same speed, v_0, as shown in Figure 1.14. A collinear collision of these two bars occurs at $t = 0$. Immediately after the collision, compressive waves of stress $\sigma = \rho c v_0$ will propagate along the two bars. The velocity of the particles inside the region passed by the wave fronts becomes zero.

In Figure 1.14(b), the stress wave propagations and reflections are plotted in the plane of position and time $(x - t)$, which is termed the *space–time diagram*, or the *Lagrange diagram*. As the speed of an elastic wave is constant, in the space–time diagram each wave is represented by a straight line of slope $1/c$. Thus, after the normal collinear collision, the two compression waves generated are represented by two straight lines starting from the origin O, with a slope of $1/c$, moving towards the upper-left and upper-right directions, respectively.

At time $t = L/c$, with L denoting the length of each bar, everywhere in the two bars has zero particle velocity, i.e., the two bars are at rest, but they still have compressive stress

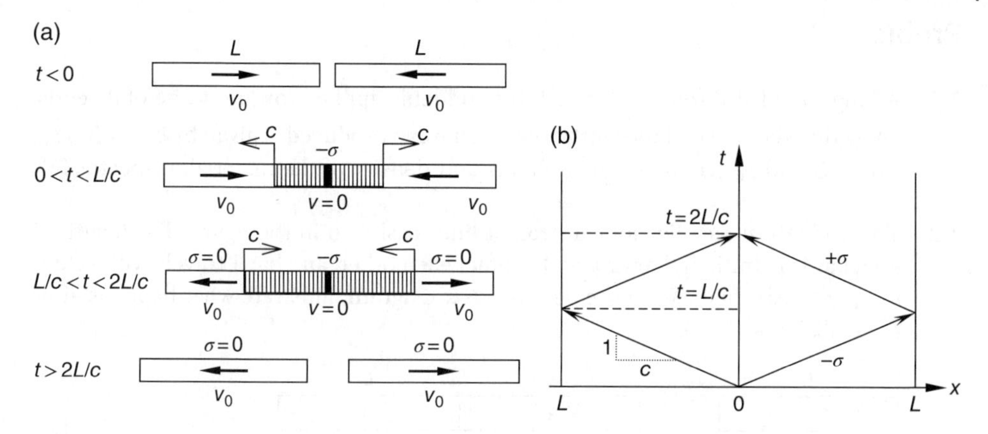

Figure 1.14 The normal collinear collision of two identical bars. (a) Stress distribution; (b) space–time diagram.

$\sigma = \rho c v_0$. These compressive pulses are reflected at the free ends and become tensile ones. When these two reflected tensile pulses travel inwards, in the regions behind the wavefronts the stresses reduce to zero, while the particles gain an outward velocity v_0. In the space–time diagram (Figure 1.14b), the two reflected tensile waves are represented by two straight lines starting from the boundary with a slope of $1/c$.

At time $t = 2L/c$, two reflected tensile wavefronts meet at the interface, $x = 0$, so that the particles at the interface also gain an outward velocity v_0; consequently, the two bars are separated. From the analysis of this example, it is evident that the space–time diagram (Lagrange diagram) is a powerful tool with which to display wave propagations and interactions.

For the bars of cross-sectional area A, the initial kinetic energy of one bar is $K_0 = AL\rho v_0^2/2$. At time $t = L/c$, the initial kinetic energy is converted entirely to the elastic strain energy of the bar, $W^e = AL\sigma^2/2E = AL\rho v_0^2/2$. After time $t = 2L/c$, however, the strain energy is converted entirely back to kinetic energy. There is no energy loss in the whole process, so the *coefficient of restitution* (COR) is $e = 1$ in this perfectly elastic collision case, which is unlikely to happen in real collisions.

In summary, first, the governing equation of elastic waves was derived in this chapter, and then various types of elastic wave were classified. Finally, the wave reflection and interaction were elaborated with several examples.

Questions 1

1 Please list the basic assumptions of 1D elastic wave theory, and point out its limitations.

2 For engineering applications, in which cases should the effects of elastic wave propagation be considered?

Problems 1

1.1 A long cylindrical rod is subjected to a suddenly applied torque at one of its ends. Show that the speed of the elastic torsional wave produced is given by $c_T = \sqrt{G/\rho_0}$, where G and ρ_0 are the shear modulus and density of the material, respectively.

1.2 Three identical bars lie along a straight line, as shown in the figure. The length of each bar is L. Initially, bars 2 and 3 contact each other, and bar 1 travels with velocity v_0 towards them. With the help of a t–x diagram, illustrate what happens after the collision at $t = 0$.

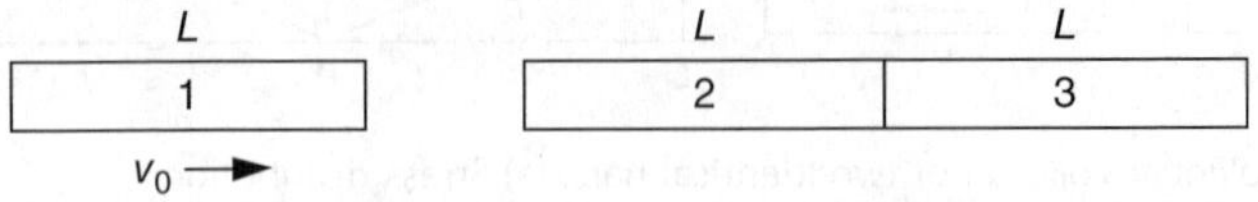

2

Elastic-Plastic Waves

2.1 One-Dimensional Elastic-Plastic Stress Wave in Bars

As shown in Figure 2.1, a semi-infinite long bar with uniform cross-section and density ρ_0 is suddenly loaded at its free end ($x = 0$) causing stress σ or initial velocity v.

If the stress produced is lower than the material's yield stress, $\sigma < Y$, the stress wave is an elastic wave which will propagate at the longitudinal wave speed $c_0 = \sqrt{E/\rho_0}$.

As indicated in Figure 2.2, if the solid is a non-linear elastic or an elastic-plastic material, the slope of the stress–strain curve decreases when $\sigma > Y$, and so does the speed of the stress wave. It should be noted that the difference between a non-linear elastic material and an elastic-plastic material is apparent in the unloading process. For a non-linear elastic material, the unloading follows the loading curve in the opposite direction, while for an elastic-plastic material, the slope of the unloading curve is the same as that during elastic loading. Here we will focus on the one-dimensional (1D) longitudinal wave propagating in an elastic-plastic material.

It is assumed that all the plane cross-sections remain plane during the wave propagation process, and so all the variables are functions of position x and time t only. For a continuous function of displacement $u(x, t)$, the strain and the particle velocity are given by the partial differentiation of $u(x, t)$ as

$$\varepsilon = \frac{\partial u}{\partial x}, \; v = \frac{\partial u}{\partial t}$$

From the continuity condition of the displacement function, exchanging the variable orders of the partial differentiation leads to

$$\frac{\partial v}{\partial x} = \frac{\partial \varepsilon}{\partial t} \tag{2.1}$$

On the other hand, the equation of motion results in

$$\rho_0 \frac{\partial v}{\partial t} = \frac{\partial \sigma}{\partial x} \tag{2.2}$$

Introduction to Impact Dynamics, First Edition. T.X. Yu and XinMing Qiu.

Published 2018 by John Wiley & Sons Singapore Pte. Ltd.

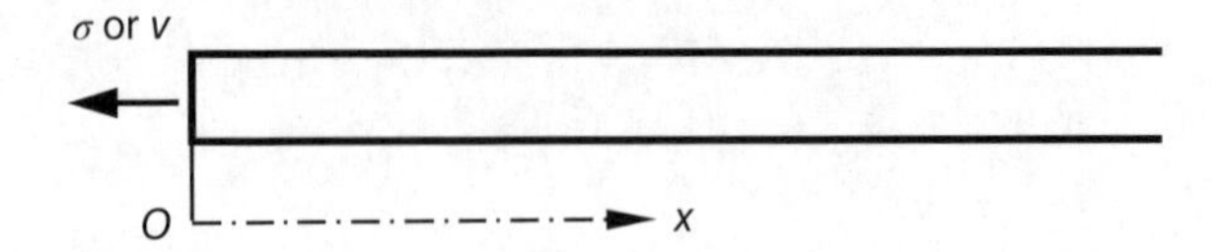

Figure 2.1 Stress wave in a semi-infinite long bar.

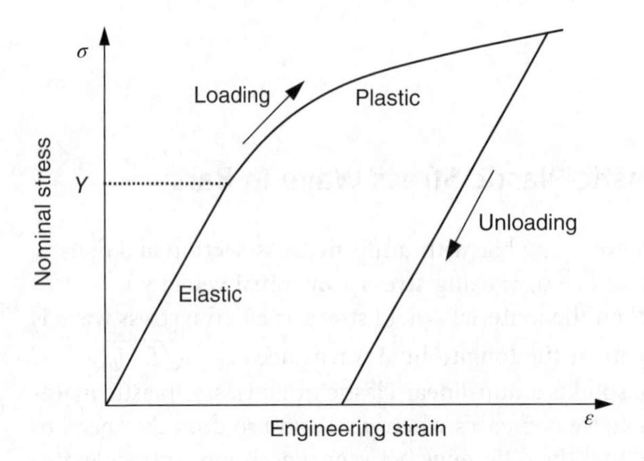

Figure 2.2 The stress–strain curve of an elastic plastic material.

If the solid material does not exhibit a strain rate effect and a temperature effect, i.e., stress σ is independent of strain rate $\dot{\varepsilon}$ and temperature, the constitutive equation of material is simplified as

$$\sigma = \sigma(\varepsilon) \tag{2.3}$$

The combination of Eqs (2.2) and (2.3) gives

$$\frac{\partial v}{\partial t} = \frac{1}{\rho_0}\frac{\partial \sigma}{\partial x} = \frac{1}{\rho_0}\frac{\mathrm{d}\sigma}{\mathrm{d}\varepsilon}\frac{\partial \varepsilon}{\partial x} = c^2\frac{\partial \varepsilon}{\partial x} \tag{2.4}$$

For the material with stress monolithically increasing with strain, the wave speed in the above relation is $c^2 \equiv \dfrac{1}{\rho_0}\dfrac{\mathrm{d}\sigma}{\mathrm{d}\varepsilon} > 0$. Thus, substitution Eq. (2.1) into Eq. (2.4) gives a 1D wave equation:

$$\frac{\partial^2 u}{\partial t^2} = c^2\frac{\partial^2 u}{\partial x^2} \tag{2.5}$$

where c is the wave speed given by

$$c = \sqrt{\frac{1}{\rho_0}\frac{\mathrm{d}\sigma}{\mathrm{d}\varepsilon}} = \begin{cases} c_0, & \sigma \le Y, \mathrm{d}\sigma/\mathrm{d}\varepsilon = E \\ < c_0, & \sigma > Y, \mathrm{d}\sigma/\mathrm{d}\varepsilon < E \end{cases} \tag{2.6}$$

It can be seen from Eq. (2.6) that the stress wave speed depends on the slope of the stress–strain curve. The elastic wave speed is constant, but the plastic stress wave speed

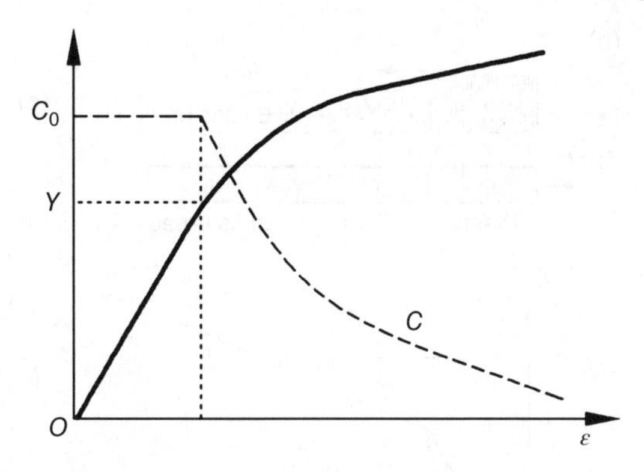

Figure 2.3 The wave speed related to the stress–strain curve.

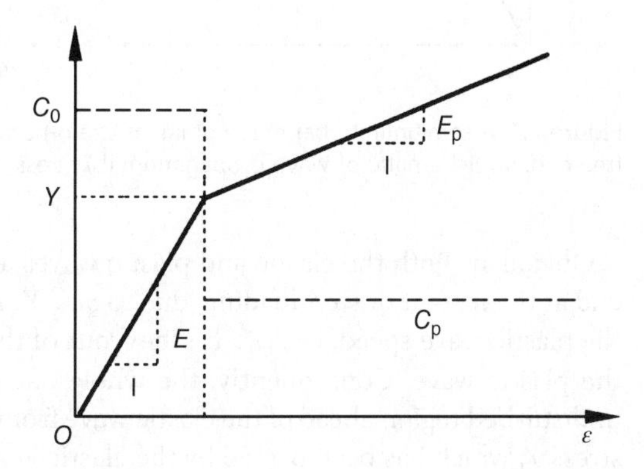

Figure 2.4 Stress–strain curve and the wave speed of the linear strain-hardening material.

is a function of strain or a function of stress. Most engineering materials have a stress–strain curve ($\sigma - \varepsilon$) that is concave downwards, reflecting a decreasingly strain-hardening material, so that the larger the stress or strain, the slower the speed of the stress wave, as shown in Figure 2.3.

As a special case, in some materials, such as a certain kind of aluminum alloy and polymer, the stress–strain relationship can be simplified as a linear strain-hardening material or a bilinear material, as shown in Figure 2.4. That is, in the plastic range, $\sigma > Y$, the plastic wave speed is also a constant, related to the plastic modulus E_p:

$$c_p = \sqrt{\frac{E_p}{\rho_0}} = \text{constant} \tag{2.7}$$

2.1.1 A Semi-Infinite Bar Made of Linear Strain-Hardening Material Subjected to a Step Load at its Free End

As illustrated in Figure 2.5(a) and (b), suppose a semi-infinite uniform bar made of linear strain-hardening material is suddenly subjected to a step-loaded pressure σ_0 at its free end, where the pressure is higher than the yield stress of the material, $\sigma_0 > Y$.

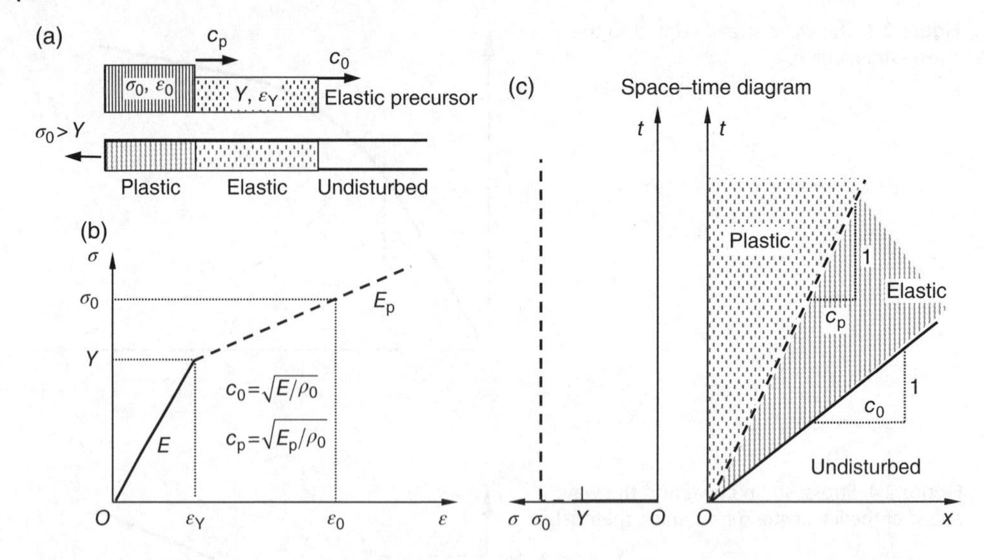

Figure 2.5 A semi-infinite bar made of linear strain-hardening material subjected to a step load at its free end. (a) Schematic of wave propagation; (b) stress–strain curve; (c) space–time diagram.

Obviously, both the elastic and plastic waves are produced simultaneously at the free end at the instant of step-loading, due to $\sigma_0 > Y$. As the elastic wave speed is higher than the plastic wave speed, i.e., $c_0 > c_p$, the front of the elastic wave must be ahead of that of the plastic wave. Consequently, the whole bar can be divided into three regions: the undisturbed region ahead of the elastic wave front with zero stress; the elastic region with stress Y, which has been passed by the elastic wave front but has not reached the plastic wave; and the plastic region with stress at the loading pressure $\sigma_0 > Y$, which has been passed by both the elastic and plastic waves. From the space–time diagram given in Figure 2.5(c), the undisturbed region is shrinking with time, while the elastic region and the plastic region are both expanding with time.

2.1.2 A Semi-Infinite Bar Made of Decreasingly Strain-Hardening Material Subjected to a Monotonically Increasing Load at its Free End

Suppose a semi-infinite uniform bar made of decreasingly strain-hardening material is subjected to a monotonically increasing pressure σ at its free end. As shown in Figure 2.6 (a), in the plastic regime of a decreasingly strain-hardening material, the stress increases with strain, i.e., $\dfrac{d\sigma}{d\varepsilon} > 0$, whilst the slope of the stress–strain curve decreases with the strain, i.e., $\dfrac{d^2\sigma}{d\varepsilon^2} < 0$.

After the application of a monotonically increasing pressure, at the beginning only an elastic wave is propagating along the bar, because the applied pressure is lower than the yield stress, i.e., $\sigma < Y$. In this case, elastic waves with different stress amplitudes are all initiating from the free end and propagating to the right, at the same elastic wave speed, c_0. In the space–time diagram shown in Figure 2.6(b), these elastic waves are represented by a set of parallel straight lines with the same slope, $1/c_0$. Then, when $\sigma > Y$, plastic waves

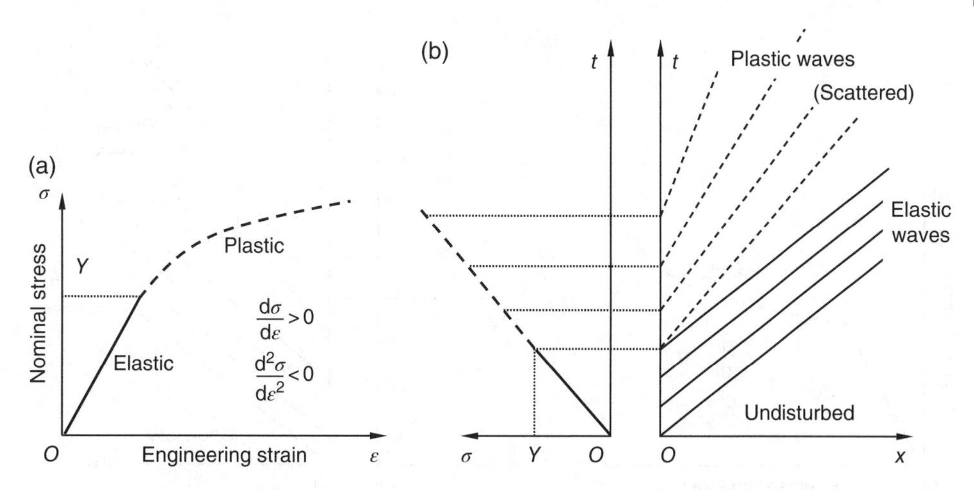

Figure 2.6 A semi-infinite bar made of decreasingly strain-hardening material subjected to a monotonically increasing load at its free end. (a) Stress–strain curve; (b) space–time diagram.

are generated and propagate along the bar. For decreasingly strain-hardening material, the speed of the plastic wave decreases with strain because $\frac{d^2\sigma}{d\varepsilon^2} < 0$; that is, the plastic waves with high stress amplitudes, which are generated later, propagate at slower speeds than the preceding plastic waves with low stress amplitudes. Therefore, in the space–time diagram, plastic waves are represented by a set of straight lines with increasing slopes; i.e., the plastic waves generated earlier have smaller slopes (higher speed), while the plastic waves generated later have bigger slopes (lower speed). Consequently, the lines representing plastic waves are scattered as time increases, i.e., the set of plastic waves is dispersed and changes its wave profile during the propagation process.

2.1.3 A Semi-Infinite Bar Made of Increasingly Strain-Hardening Material Subjected to a Monotonically Increasing Load at its Free End

In contrast, let's consider a semi-infinite uniform bar made of increasingly strain-hardening material subjected to a monotonically increasing pressure σ at its free end. As shown in Figure 2.7(a), in the plastic regime of an increasingly strain-hardening material, the stress increases with the strain, i.e., $\frac{d\sigma}{d\varepsilon} > 0$, while the slope of the stress–strain curve also increases with the strain, i.e., $\frac{d^2\sigma}{d\varepsilon^2} > 0$.

After the application of a monotonically increasing pressure, at the beginning only an elastic wave propagates along the bar, because $\sigma < Y$. In this case, elastic waves with different stress amplitudes all start from the free end and propagate to the right at the same elastic wave speed, c_0. In the space–time diagram shown in Figure 2.7(b), these elastic waves are represented by a set of parallel straight lines with the same slope, $1/c_0$. Then when $\sigma > Y$, plastic waves are generated and propagate along the bar. For increasingly strain-hardening material, the speed of the plastic wave increases with the strain, as

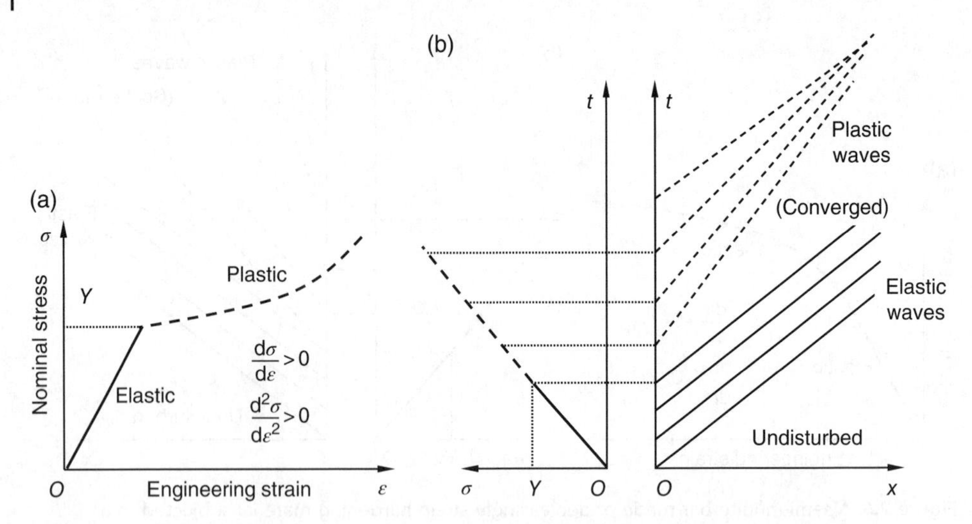

Figure 2.7 A semi-infinite bar made of increasingly strain-hardening material subjected to a monotonically increasing load at the free end. (a) Stress–strain curve; (b) space–time diagram.

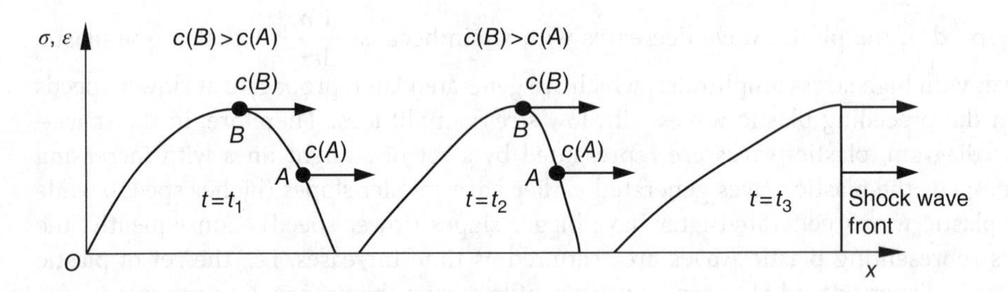

Figure 2.8 Schematic of the formation of a shock wave.

$\frac{d^2\sigma}{d\varepsilon^2} > 0$; that is to say, the plastic waves with high stress amplitudes, which are generated later, propagate with higher speeds compared with the preceding plastic waves with lower stress amplitudes. Therefore, in the space–time diagram, plastic waves are represented by a set of straight lines of decreasing slope – i.e., the plastic waves generated earlier have bigger slopes (lower speed), while the plastic waves generated later have smaller slopes (higher speed). Hence, the lines representing plastic waves converge as time increases.

As the speed of the plastic waves generated later is greater than that of the preceding waves, the later plastic waves will catch up with the preceding ones as time passes. In this case, the wave front will become steeper and steeper, and finally it will form a shock wave. As shown in Figure 2.8, at time $t = t_1$, the plastic wave in a solid material has a fairly smooth profile, and a higher stress point B is located behind a lower stress point A. Because the wave speed at B is higher than at A, $c(B) > c(A)$, point B moves closer to point A as time passes, leading to a steeper wave profile, as seen at $t = t_2$; eventually at time t_3, point B catches up to point A and forms a shock wave front.

From the above discussions, it is clear that a shock wave will be generated in the increasingly strain-hardening material, which is rarely seen in conventionally homogeneous solid materials. However, for cellular materials, such as lattice, honeycomb, and foam, which are increasingly being used in lightweight structures and energy absorption devices, their compressive stress–strain curves do exhibit increasingly strain-hardening behavior, as shown in Figure 2.7(a), due to the densification of cells at large equivalent strain. Therefore, under the impact loading, converged plastic waves or even shock waves are probably generated in these cellular materials.

2.1.4 Unloading Waves

As illustrated earlier, the difference between a non-linear elastic material and an elastic-plastic material is displayed in the unloading process. For an elastic-plastic material, the slope of the unloading curve is the same as that during elastic loading, $\frac{\mathrm{d}\sigma}{\mathrm{d}\varepsilon} = E$, so the disturbance generated by unloading must propagate at the elastic wave speed $c_0 = \sqrt{E/\rho_0}$. As the real cases always involve unloading, the propagation of the unloading wave and its interaction with loading waves are also worth a thorough investigation.

To show a clear physical picture, the interaction between unloading waves and loading waves is analyzed by taking the linear strain-hardening material as an example. Suppose a semi-infinite bar made of linear strain-hardening material is subjected to a rectangular pulse, $\sigma_0 > Y$, at its free end. As shown in Figure 2.9, in the loading phase $0 \le t \le t_d$, the propagations of the elastic loading wave and the plastic loading wave are the same as that described before in the case of step loading (see Figure 2.5). At $t = t_d$, the applied pressure at the free end is suddenly removed, which produces an unloading disturbance transmitted along the bar, i.e., it generates an unloading wave propagating towards the right at elastic wave speed c_0. As the elastic wave travels more quickly than the plastic wave ($c_0 > c_p$), the loading plastic wave will be caught up and then unloaded by the unloading wave. At $t = t_u$, the plastic wave is completely unloaded by the unloading wave.

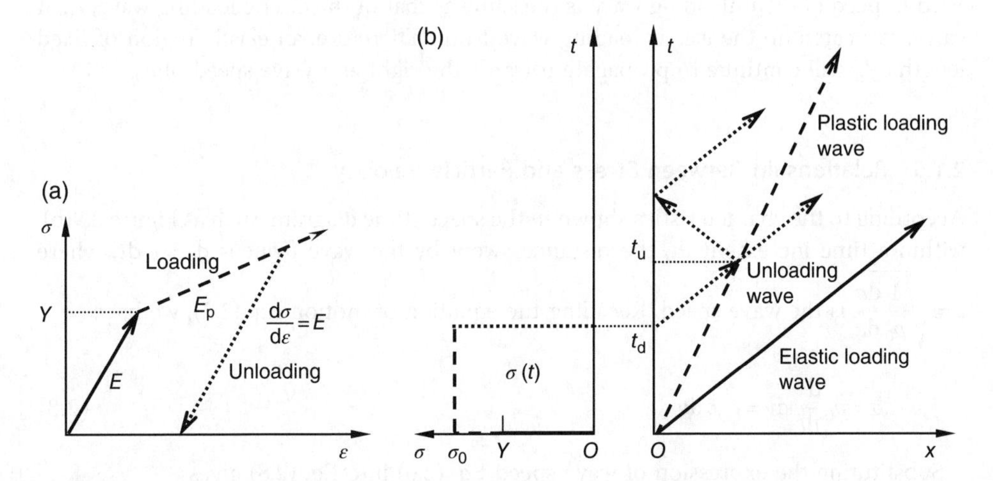

Figure 2.9 Unloading wave in a semi-infinite bar made of linear strain-hardening material. (a) Stress–strain curve; (b) space–time diagram.

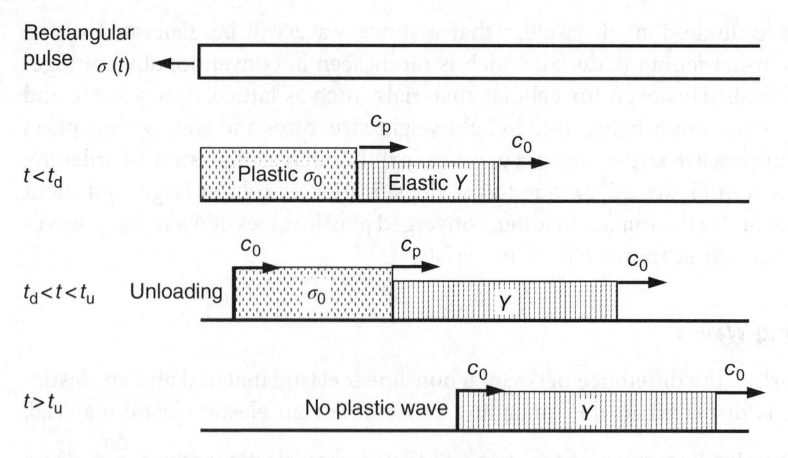

Figure 2.10 Distribution of different wave regions in a semi-infinite bar subjected to a rectangular pulse.

Different regions generated by wave propagation on the space–time plane are shown in Figure 2.10. At $t = 0$, a rectangular pulse $\sigma(t)$ is suddenly applied at the free end of the bar. At time $t < t_d$, the applied pressure remains at $\sigma(t) = \sigma_0$, and thus the bar is under the loading condition, i.e., both the elastic loading wave and the plastic loading wave are propagating towards the right. A plastic region with stress σ_0 and an elastic region with stress Y are both expanding with time, with speeds of c_p and c_0, respectively.

At $t = t_d$, the applied pressure is suddenly removed, reducing to $\sigma(t) = 0$, and an unloading wave departs from the free end, following the loading waves at an elastic wave speed of c_0. The zero stress and a residual strain are left in the region behind the unloading wave. Therefore, in the time period $t_d < t < t_u$, there are three regions in the bar: the elastic region, the plastic region, and the unloading region. At $t = t_u$, the front of plastic wave is caught up by the unloading wave, and so the plastic stress is completely unloaded. As the speed of the unloading wave is the same as that of the elastic loading wave, c_0, it can never catch up the elastic loading wave front. Therefore, an elastic region of fixed length $c_0 t_d$ will continue to propagate towards the right at a wave speed of c_0.

2.1.5 Relationship Between Stress and Particle Velocity

According to the characteristics shown in the space–time diagram, such as Figure 2.9(b), within a time increment dt, the distance swept by the wave front is $dx = c\ dt$, where $c = \sqrt{\frac{1}{\rho_0}\frac{d\sigma}{d\varepsilon}}$ is the wave speed. Recalling the equation of motion Eq. (2.2), we have

$$d\sigma = \rho_0 \frac{dx}{dt}\, dv = \rho_0 c\, dv \tag{2.8}$$

Substituting the expression of wave speed Eq. (2.6) into Eq. (2.8) gives

$$d\sigma = \rho_0 c\, dv = \sqrt{\rho_0 (d\sigma/d\varepsilon)}\, dv \tag{2.9}$$

where $\rho_0 c$ is the wave impedance or *mechanical impedance*, which in general is a function of stress or strain. As a special case, for an elastic wave the speed $c_0 = \sqrt{E/\rho_0}$ is constant; the impedance $\rho_0 c_0$ is therefore also a constant, so that the amplitude of stress is proportional to the particle velocity, $\sigma = \rho_0 c_0 v$.

If a velocity v is applied at the free end of a semi-infinite bar, then within the elastic range, a stress $\sigma = \rho_0 c_0 v$ will result. When this stress reaches the yield stress Y of the material, the critical velocity,

$$v_Y \equiv \frac{Y}{\rho_0 c_0} = \frac{Y}{\sqrt{E\rho_0}} \tag{2.10}$$

is called *yield velocity*. For example, for mild steel with elastic modulus $E = 205$ GPa and density $\rho_0 = 7800\ \mathrm{kg/m^3}$, the mechanical impedance is $\rho_0 c_0 = \sqrt{E\rho_0} = 40 \times 10^6\ \mathrm{kg\ m^{-2}\ s^{-1}}$. If the yield stress of this mild steel is $Y = 400$ MPa, then the corresponding yield velocity is $v_Y = 10$ m/s. This implies that permanent plastic deformation will take place in the mild steel bar under impact, with the velocity > 10 m/s.

For some dynamic experimental devices, such as Hopkinson bars (introduced in Chapter 3), plastic deformation has to be avoided in order to make the device usable time and again. In this case, it is necessary to ensure that the impact velocity is lower than the material's yield velocity. It should also be noted that apart from the yield stress, the magnitude of the yield velocity also depends on the elastic modulus and density of the material. For example, the yield velocity of aluminum could be higher than that of steel, due to its lower material density.

Figure 2.11 presents a typical stress–strain curve for an engineering material under tension. After an elastic deformation stage, the material will yield at yield strain ε_Y, corresponding to yield stress Y. If the applied tensile stress is high, $\sigma_0 > Y$, the material will undergo plastic deformation. Integrating Eq. (2.9) gives

$$v = \int_0^{\sigma_0} \frac{d\sigma}{\sqrt{\rho_0 (d\sigma/d\varepsilon)}} = \int_0^{\varepsilon_0} \sqrt{\frac{d\sigma/d\varepsilon}{\rho_0}}\, d\varepsilon \tag{2.11}$$

This relation provides a means to correlate the impact velocity with the maximum strain generated.

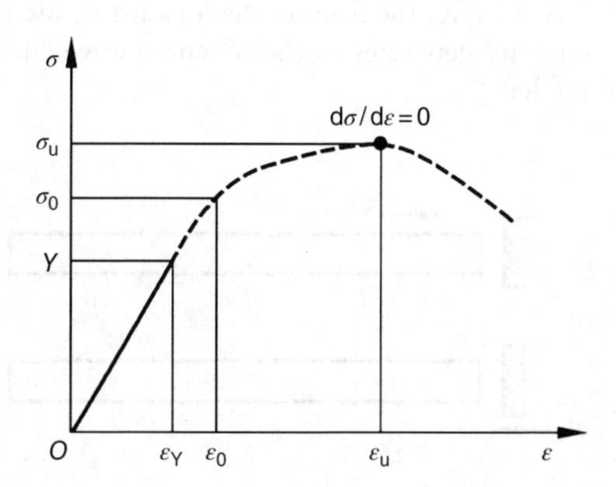

Figure 2.11 Tensile stress–strain curve of a material with ultimate strength.

It can be seen from the stress–strain curve shown in Figure 2.11 that in the plastic deformation stage, the stress will increase with the strain until an ultimate point is reached. The ultimate strength is the maximum value of the stress, $\sigma = \sigma_u$, which occurs at the ultimate strain $\varepsilon = \varepsilon_u$. At the point of ultimate strain, the slope of the stress–strain curves becomes zero, $d\sigma/d\varepsilon = 0$. In the tensile test of a ductile material, the ultimate strength indicates the occurrence of necking, followed by tensile failure. The loading velocity, which generates the ultimate strain ε_u, can be found from Eq. (2.11) as

$$v_c = \int_0^{\varepsilon_u} \sqrt{\frac{d\sigma/d\varepsilon}{\rho_0}}\, d\varepsilon \tag{2.12}$$

This v_c is called the *von Karman critical velocity*, which results in fracture of the bar under high-speed tension. In fact, *von Karman critical velocity* can be regarded as a dynamic fracture criterion, different from the static fracture criterion.

As discussed earlier, there are two critical velocities for each elastic-plastic material: the yield velocity, v_Y, valid for both tensile and compression waves; and the von Karman critical velocity, v_c, valid for tensile waves only.

2.1.6 Impact of a Finite-Length Uniform Bar Made of Elastic-Linear Strain-Hardening Material on a Rigid Flat Anvil

A classical wave interaction problem was solved by Lensky (1949) as follows.

As shown in Figure 2.12, a uniform bar of length l is made of an elastic, linear strain-hardening material. At $t = 0$, the bar impinges onto a rigid flat anvil with initial velocity v^* that is higher than the yield velocity, $v_Y < v^* < 2v_Y$. In fact, this problem is equivalent to that of a rigid flat anvil with initial velocity v^* impinging on a stationary bar of length l.

The space–time diagram of this problem is depicted in Figure 2.13(a). At $t = 0$, the elastic compressive wave and the plastic compressive wave are both initiated from the left end of the bar, and propagate towards the right at elastic wave speed c_0 and plastic wave speed c_p, respectively.

When $0 < t < l/c_0$ (see Figure 2.13b) the elastic wave front has not yet reached the free end on the right. Hence, at this time three regions co-exist in the bar: the undisturbed region O with zero stress and strain; the elastic region I with yield stress Y; and the plastic region II.

At $t = l/c_0$, the front of the forward elastic compressive wave arrives at the right free end; this generates an elastic tensile wave due to the reflection, and this travels towards the left.

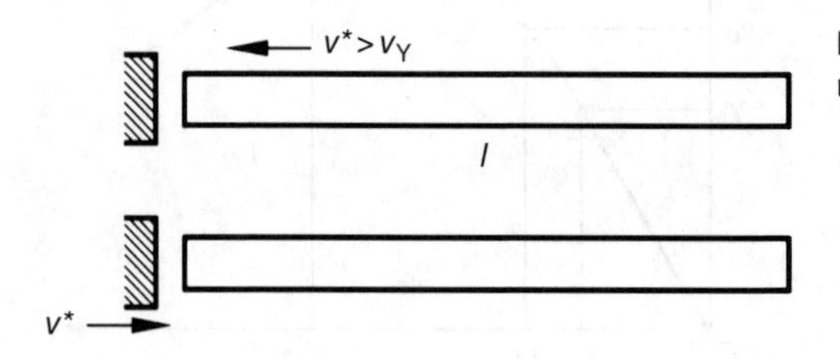

Figure 2.12 A uniform bar impinges on a rigid flat anvil.

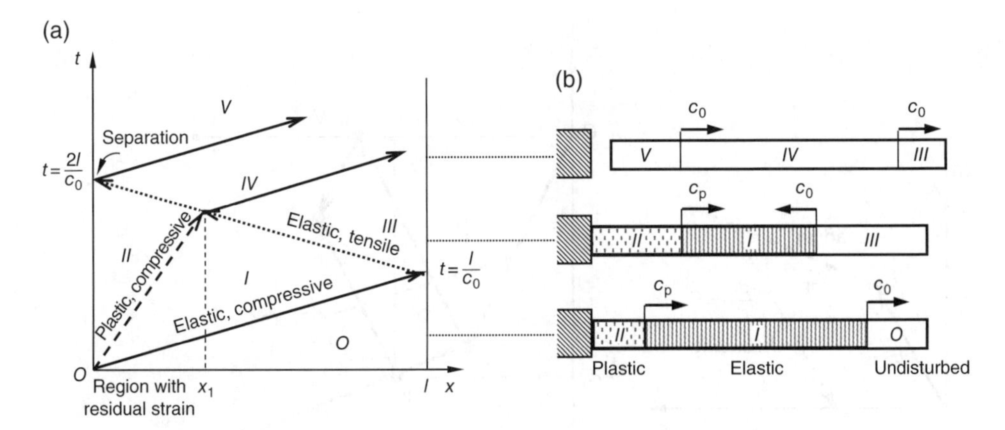

Figure 2.13 A bar made of elastic linear strain-hardening material impacts a rigid flat anvil. (a) Space–time diagram; (b) distribution of wave regions.

For $l/c_0 < t < 2l/c_0$, the backward-traveling elastic tensile wave will unload the forward elastic compressive loading wave first, thus creating an elastic unloading region *III* with zero stress and strain. Then, the elastic unloading wave will meet the plastic loading wave, and, as a result of the unloading, a plastic unloading region *IV* with residual strain is formed. The size of this plastic unloading region can be calculated from the speeds of various waves:

$$\frac{x_1}{c_p} = \frac{2l - x_1}{c_0}, \; x_1 = \frac{2c_p}{c_0 + c_p} l \tag{2.13}$$

In the bar, plastic residual strain exists in the region of $x < x_1$. It can be seen from Eq. (2.13) that the size of the plastic deformation region depends on the elastic wave speed, the plastic wave speed and the original length of the bar, but is independent of the impact velocity. Due to fact that the elastic wave speed is always higher than the plastic one, $c_0 > c_p$, it is clear that $x_1 < l$, i.e., the plastic residual deformation is always limited to part of the bar, rather than the entire bar, no matter how high the impact velocity. It is known from this example (Figure 2.13) and the example discussed earlier (Figure 2.9) that the residual plastic deformation region can be formed in two ways: face to face unloading, or chased from behind unloading by an elastic wave.

At $t = 2l/c_0$, the elastic unloading wave reaches the impinging surface. At this time, the particle velocity of the bar becomes higher than the velocity of the rigid anvil $v*$. Hence, the bar will separate from the rigid anvil and move towards the right. The elastic unloading wave is then reflected, generating a compressive elastic wave at the left free end, which travels towards the right.

For $t > 2l/c_0$, elastic waves will propagate back and forth in the bar, and they will be reflected at the two free ends repeatedly. Apart from the kinetic energy, the bar also possesses deformational energy.

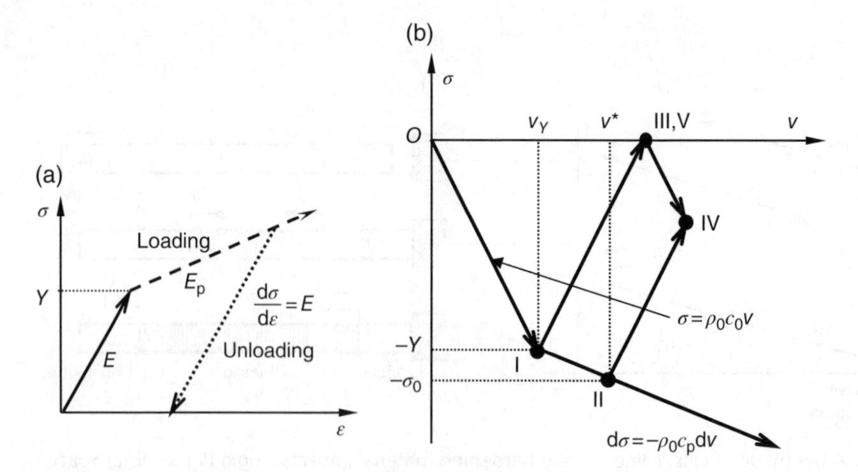

Figure 2.14 A bar made of elastic linear strain-hardening material impacts a rigid flat anvil. (a) Stress–strain curve; (b) stress–particle velocity diagram.

Furthermore, the stress amplitude and particle velocity can be found. For this purpose, the stress–strain curve of the elastic, linear strain-hardening material, and the stress–particle velocity diagram are depicted in Figure 2.14.

The stress amplitude in the elastic loading region I is determined first. Obviously, the stress in this region is equal to the yield stress Y, and the particle velocity must be the yield velocity v_Y, due to the relation $Y = \rho_0 c_0 v_Y$.

In the plastic loading region II, as the impact velocity $v^* > v_Y$, the relation between the stress and the particle velocity increment $d\sigma = \rho_0 c_p \, dv$ leads to $\sigma_0 - Y = \rho_0 c_p (v^* - v_Y)$, i.e.

$$\sigma_0 = Y + \rho_0 c_p (v^* - v_Y) \tag{2.14}$$

represents the stress amplitude in the plastic loading region II. The particle velocity in region II is v^*. From the stress–strain curve shown in Figure 2.14(a), for the plastic regime,

$$\sigma_0 - Y = E_p(\varepsilon_0 - \varepsilon_Y) \tag{2.15}$$

The strain corresponding to stress σ_0 is ε_0, within which the plastic strain component is

$$\varepsilon_p = \varepsilon_0 - \frac{\sigma_0}{E} = \varepsilon_Y + \frac{\sigma_0 - Y}{E_p} - \frac{\sigma_0}{E} \tag{2.16}$$

Another expression of the plastic residual strain is given by substituting Eq. (2.14) into Eq. (2.16) as

$$\varepsilon_p = \rho_0 c_p (v^* - v_Y)\left(\frac{1}{E_p} - \frac{1}{E}\right) \tag{2.17}$$

which indicates that the amplitude of the plastic residual strain increases with the impact velocity v^*.

2.2 High-Speed Impact of a Bar of Finite Length on a Rigid Anvil (Mushrooming)

At the end of Section 2.1, Lensky's analysis of a finite-length bar impinging on a rigid flat anvil is described. By adopting an elastic, linear strain-hardening material model, the size of the plastic deformation region and the amplitude of plastic residual strain are found depending on the impact velocity, whilst the reason for the localization of the plastic deformed zone is well explained. However, the Lensky slender bar model is based on the 1D assumption, i.e., the cross-section of the bar is assumed to be unchanged during the whole impact process. What would happen if the impact velocity were very high, so the bar can't remain as a 1D structure? In this case, the change in cross-section would have to be considered. This problem is related to the mushrooming of bullets or projectiles during the impact process and was first analyzed by G. I. Taylor in 1948.

Consider a short cylindrical bar, which impinges on a rigid anvil at very high initial velocity, so that the mushrooming around the impact end is related to large plastic deformation. Suppose the impact velocity is at the level of $\rho_0 v_0^2 \approx Y$, which is about $v_0 \approx 300$ m/s for steel. The following assumptions are adopted in Taylor's theoretical model:

- Rigid-plastic idealization of the material – i.e., since plastic deformation is dominant for the bar under very high-impact velocity, the effect of elastic deformation could be ignored.
- The bar is short, so that no buckling occurs after impact.
- The bar is cylindrical with initial cross-sectional area A_0, and the dynamic deformation of the bar remains axis-symmetric.

2.2.1 Taylor's Approach

Taylor's model is depicted in Figure 2.15 for a short cylindrical bar impinging on a rigid anvil. Assume that the material of the bar (projectile) is rigid plastic; then the elastic wave is disregarded. After the impact, plastic deformation occurs and the plastic

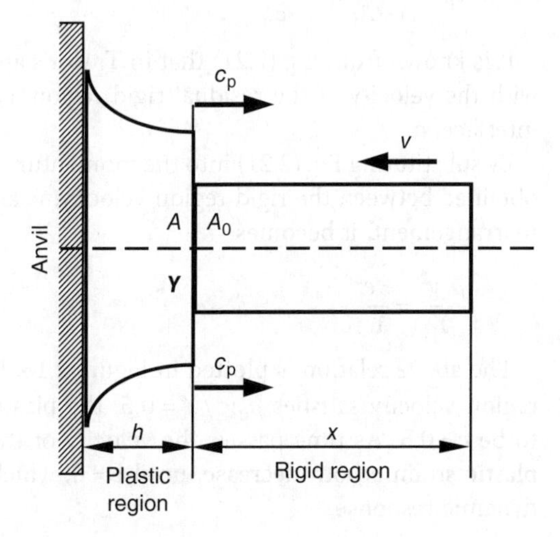

Figure 2.15 A short cylindrical bar impinges a rigid anvil: Taylor's model.

wave propagates towards the free end on the right at a speed of c_p, dividing the bar into two regions, a plastic region and a rigid region. It should be noted that the particle velocity in the rigid section of the bar will also change, so it drops from the initial impact velocity v_0 to v at time t. At the same time, at the interface between the plastic region and the rigid region, a jump in the cross-sectional area takes place, i.e., the cross-sectional area increases from A_0 in the rigid region to A in the plastic region.

The continuity equation at the interface of these two regions requires that

$$A_0\left(v+c_p\right)=Ac_p \tag{2.18}$$

During a small time increment dt, the reduction in the length of the rigid region is dx, where $dx=\left(v+c_p\right)dt$. By using the conservation of momentum for the increment dx, $\rho_0 A_0\, dx\, v = Y(A-A_0)\, dt$, we have

$$\rho_0 A_0 v\left(v+c_p\right)=Y(A-A_0) \tag{2.19}$$

We now need to define the strain at the interface between the plastic region and the rigid region. For a small segment of initial length dl_0 and volume dV, its length is changed to dl after passing through the interface. Therefore, the engineering strain related to this axial compression is $e=(dl_0-dl)/dl_0$. Owing to the axis-symmetric assumption, the length change of this small segment is related to the cross-sectional area change:

$$e=\frac{dl_0-dl}{dl_0}=\frac{dV/A_0-dV/A}{dV/A_0}=1-\frac{A_0}{A} \tag{2.20}$$

It is evident from Eq. (2.20) that $\frac{A_0}{A}=1-e$, or $\frac{A}{A_0}=\frac{1}{1-e}$. Here it should be noted that both the engineering strain e and the cross-sectional area A are variables defined at the interface of the rigid region and plastic region, rather than variables that apply to the whole plastic region.

Substitution of the strain at the interface (Eq. 2.20) into the continuity equation (Eq. 2.18) leads to the speed of the plastic wave front:

$$c_p=\frac{A_0}{A-A_0}v=\frac{1-e}{e}v \tag{2.21}$$

It is known from Eq. (2.21) that in Taylor's approach, the plastic wave speed c_p varies with the velocity of the residual rigid region, v, as well as the plastic strain across the interface, e.

By substituting Eq. (2.21) into the momentum conservation (Eq. 2.19), a relationship is obtained between the rigid region velocity, v, and the plastic strain at interface, e; after rearrangement, it becomes

$$\frac{\rho_0 v^2}{Y}=\frac{e^2}{1-e} \tag{2.22}$$

The above relation is plotted in Figure 2.16. For example, at the time when the rigid region velocity satisfies $\rho_0 v^2/Y=0.5$, the plastic strain at the interface is determined to be $e=0.5$. As time passes, the velocity of the residual rigid region v, as well as the plastic strain e both decrease, until $e=0$, which corresponds to the end of the whole dynamic response.

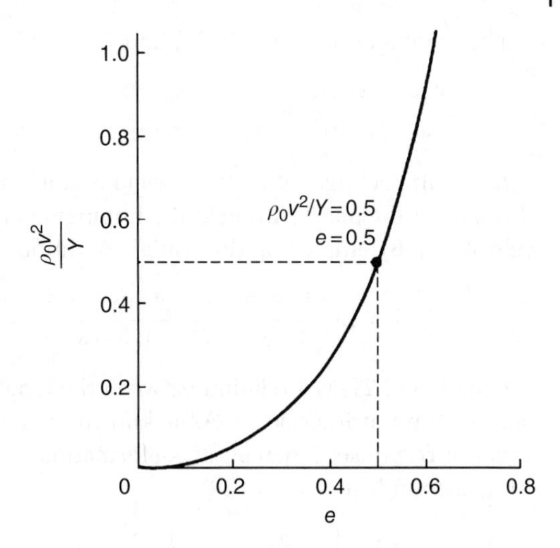

Figure 2.16 Velocity of the residual rigid region *v* versus the plastic strain at interface *e*.

We will also determine the final length of the bar after impact. Refer to Figure 2.15: if the length of the plastic region is denoted by h, and that of the rigid region by x, then their rates of change are

$$\frac{dh}{dt} = c_p, \frac{dx}{dt} = -(v + c_p) \tag{2.23}$$

The equation of motion of the rigid region can be written as $YA_0 = -\rho_0 A_0 x \dfrac{d}{dt}(v + c_p)$. If the plastic wave speed c_p is treated as constant, as an approximation, this relation is simplified to $YA_0 \approx -\rho_0 A_0 x \dfrac{dv}{dt}$, which leads to

$$\frac{dv}{dt} = -\frac{Y}{\rho_0 x} \tag{2.24}$$

Dividing the second equation of Eq. (2.23) by Eq. (2.24), we find

$$\frac{dx}{dv} = \frac{1}{Y}(v + c_p)\rho_0 x \tag{2.25}$$

By using the expression of c_p given by Eq. (2.21), it is found that $v + c_p = \left(1 + \dfrac{1-e}{e}\right)v = \dfrac{v}{e}$. Then, substituting this relation into Eq. (2.25) gives

$$\frac{dx}{x} = \frac{\rho_0 v}{Ye} dv \tag{2.26}$$

At the same time, differentiating Eq. (2.22) gives

$$\frac{2\rho_0 v}{Y} dv = \frac{2e - e^2}{(1-e)^2} de \tag{2.27}$$

The combination of Eq. (2.26) and Eq. (2.27) leads to

$$2\frac{\mathrm{d}x}{x} = \frac{2-e}{(1-e)^2}\,\mathrm{d}e \tag{2.28}$$

At the impact instant $t = 0$, the rigid region covers the whole bar, so $x = L$. Suppose at this instant the plastic strain at the impinging end is $e = e_0$, which depends on the initial velocity v_0. By employing this initial condition, the integration of Eq. (2.28) results in

$$\ln\left(\frac{x}{L}\right)^2 = \ln\left(\frac{1-e_0}{1-e}\right) + \frac{e-e_0}{(1-e)(1-e_0)} \tag{2.29}$$

Equation (2.29) is a relation between the length of the residual rigid region x and the plastic strain at interface e. As is known, e and x both decrease with time during the dynamic response. When plastic deformation ceases, $e = 0$, the final length of rigid region $x = x_f$ is found from Eq. (2.29) as

$$\ln(L/x_f) = \frac{1}{2}\left(\frac{e_0}{1-e_0} + \ln\frac{1}{1-e_0}\right) \tag{2.30}$$

To summarize, the following calculation procedure is adopted in Taylor's approach:

1) When v_0, ρ_0, and Y are specified, use a special case of Eq. (2.22), $\frac{\rho_0 v_0^2}{Y} = \frac{e_0^2}{1-e_0}$, to determine the plastic strain at the impinging end e_0.
2) For every value of $e < e_0$,
   a) determine A/A_0 from Eq. (2.20), i.e., $A/A_0 = 1/(1-e)$;
   b) determine the velocity of the rigid region from Eq. (2.22), i.e., $\frac{\rho_0 v^2}{Y} = \frac{e^2}{1-e}$;
   c) and determine the length of the rigid region from Eq. (2.29).
3) When the plastic strain reduces to zero, $e = 0$, determine the final length of rigid region x_f from Eq. (2.30).

Here we may note that in Taylor's approach, the plastic strain on the interface e, which is gradually decreasing from e_0 to 0, serves as a process parameter instead of the real time t. The formulation by employing e makes the calculation much simpler.

Finally, the deformation shape, i.e. mushrooming, can be calculated. By using Eq. (2.23) and the expression of plastic wave speed in Eq. (2.21), i.e., $c_p = \frac{1-e}{e}v$, we obtain

$$\frac{\mathrm{d}h}{\mathrm{d}x} = -\frac{c_p}{v+c_p} = e-1 \tag{2.31}$$

And the integration of Eq. (2.31) gives

$$\frac{h}{L} = \int_{x/L}^{1} (1-e)\,\mathrm{d}\left(\frac{x}{L}\right) \tag{2.32}$$

When the final length of the rigid region, $x = x_f$, is applied, Eq. (2.32) provides the final length of the plastic region h_f, as well as the final total length of the Taylor bar, $x_f + h_f$.

Table 2.1 A numerical example of Taylor's approach (Johnson, 1972)

e	0	0.1	0.2	0.3	0.4	0.5
Time	t_f					0
x/L	0.43	0.48	0.54	0.635	0.7	1.0
h/L	0.38	0.34	0.28	0.21	0.12	0.0
d/d_0	1.00	1.05	1.12	1.20	1.29	1.41
v_0t/L	0.34	0.29	0.24	0.18	0.10	0.0

For a cylindrical Taylor bar with initial diameter d_0, the deformed diameter d after impact, which provides the profile of the "mushroom", is given by the cross-sectional area or by the plastic strain on the interface:

$$\frac{d}{d_0} = \sqrt{\frac{A}{A_0}} = \frac{1}{\sqrt{1-e}} \tag{2.33}$$

In Table 2.1, a set of numerical results produced by Taylor's approach is listed. This example pertains to the case of $e_0 = 0.5$, and the impact velocity satisfies $\frac{\rho_0 v_0^2}{Y} = \frac{e_0^2}{1-e_0} = 0.5$. The final profiles of the Taylor bar under impact are shown in Figure 2.17 with different initial velocities.

Taylor's model was developed during the Second World War. The development of weapons required scientists to determine the dynamic yield stress of relevant materials under the velocity range of weapon attack. At that time the dynamic experiment facility and measuring technology were under-developed; with no electronic device and no computer, it was very difficult to measure the strain rate and the dynamic yield stress under high-speed loading. Taylor's approach suggested a useful experimental method to overcome these difficulties. The dynamic yield stresses can be estimated by measuring the final profiles of Taylor bars under various impact velocities. Based on this theoretical analysis, valuable experimental procedures were carried out to determine the dynamic yield stress of various materials, which exhibited a strain rate effect and a temperature effect, such as those reported by Whiffin in 1948.

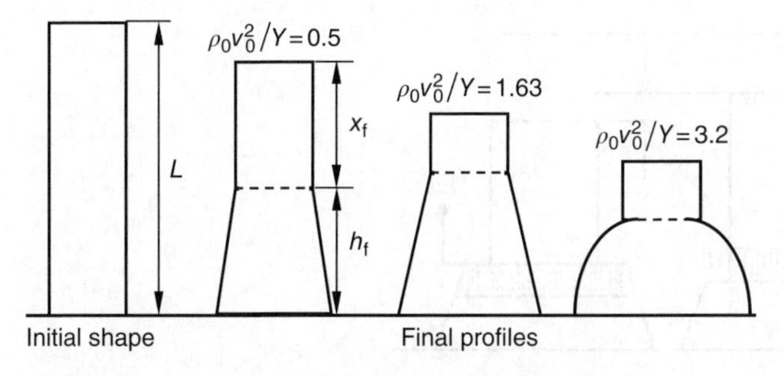

Figure 2.17 Final profiles given by Taylor's approach.

However, compared with experimental results, Taylor's prediction on mushrooming showed notable discrepancies for specimens under very high-impact velocity – some experiments resulted in the final shapes of bars having concave profiles, which differed significantly from those given by Taylor's model, especially for large values of $\rho_0 v_0^2/Y$.

2.2.2 Hawkyard's Energy Approach

Hawkyard (1969) adopted the same deformation mode as that proposed by Taylor (1948), but he employed the *global energy balance* instead of the *local momentum balance* at the plastic front as used by Taylor (1948).

In Hawkyard's approach, the continuity equation (Eq. 2.18), the definition of plastic strain (Eq. 2.20), and the plastic wave speed (Eq. 2.21) in Taylor's approach all remain unchanged, but the equation of motion (Eq. 2.19) is abandoned. According to the illustrations in Figure 2.18

$$\frac{dy}{dt} = c_p,\ \frac{ds}{dt} = v,\ \frac{dx}{dt} = -\left(\frac{ds}{dt} + \frac{dy}{dt}\right) = -\left(v + c_p\right) \tag{2.34}$$

The energy dissipation rate for crossing the plastic front is

$$\frac{dw}{dt} = \frac{1}{dt}\left(A_0\ |dx|\ Y\ \ln\frac{A}{A_0}\right) = A_0\left(v + c_p\right)Y\ \ln\frac{A}{A_0} \tag{2.35}$$

In Hawkyard's approach, the true strain, $\ln(A/A_0)$, is adopted to replace the engineering strain used in Taylor's approach, which produces a reasonable improvement in large deformation cases.

The rate of loss in the kinetic energy of the projectile is

$$\frac{dE}{dt} = \frac{1}{dt}\left(\frac{1}{2}A_0\rho_0\ |dx|\ v^2 + A_0 Y\ ds\right) = A_0 v\left[\frac{1}{2}\rho_0 v\left(v + c_p\right) + Y\right] \tag{2.36}$$

where $A_0\rho_0|dx|v^2/2$ is the kinetic energy of the *deformed element* dx, while $A_0 Y\ ds$ is the work done by the force in the distance equal to the length change of the rigid segment.

From the energy conservation, at any time

$$\frac{dw}{dt} = \frac{dE}{dt} \tag{2.37}$$

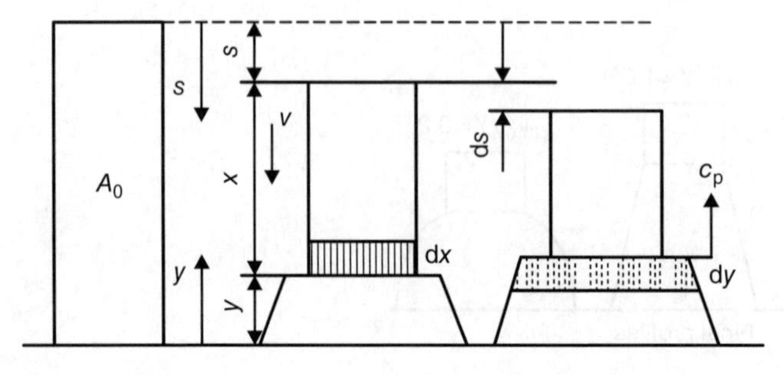

Figure 2.18 Hawkyard's energy approach to the Taylor bar.

Combination of Eqs (2.35)–(2.37) gives

$$\left(v+c_{\mathrm{p}}\right)Y\ln\frac{A}{A_0}=\frac{1}{2}\rho_0 v^2\left(v+c_{\mathrm{p}}\right)+vY \tag{2.38}$$

From Eq. (2.38) and the continuity equation (Eq. 2.18), $A_0\left(v+c_{\mathrm{p}}\right)=Ac_{\mathrm{p}}$, we have

$$\frac{\rho_0 v^2}{Y}=2\left[\ln\frac{A}{A_0}-\left(1-\frac{A_0}{A}\right)\right]=2\left(\ln\frac{1}{1-e}-e\right) \tag{2.39}$$

Obviously, the relation between the impact velocity and the plastic strain given by Hawkyard's energy approach is different from that by Taylor's approach (Eq. 2.22).

A deformation analysis can now be carried out. Similar to that in Taylor's approach (Eq. 2.24), the equation of motion for the rigid segment is

$$YA_0=-\rho_0 A_0 x\left(v\,\frac{\mathrm{d}v}{\mathrm{d}s}\right) \tag{2.40}$$

A relationship between $\mathrm{d}v$ and $\mathrm{d}e$ is found by the differentiating Eq. (2.39). Substitution of this relationship into Eq. (2.40) gives

$$-\frac{\mathrm{d}s}{x}=\frac{e}{1-e}\,\mathrm{d}e \tag{2.41}$$

From Eq. (2.34), $\mathrm{d}s=-e\,\mathrm{d}x$, so that Eq. (2.41) can be rewritten as

$$\frac{\mathrm{d}x}{x}=\frac{\mathrm{d}e}{1-e} \tag{2.42}$$

Using the initial condition, $x=L$, $e=e_0$, the integration of Eq. (2.42) leads to

$$x=L\left(\frac{1-e_0}{1-e}\right) \tag{2.43}$$

The final length of the rigid region is obtained by taking $e=0$ in Eq. (2.43), i.e.

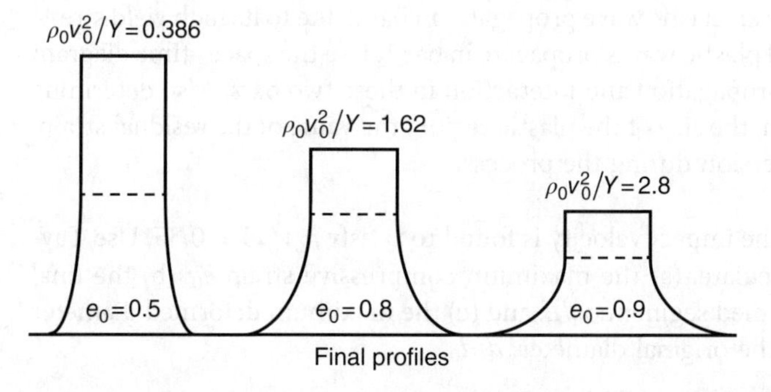

Figure 2.19 Final profiles of the Taylor bar using Hawkyard's approach.

$$x_f = L(1 - e_0) \tag{2.44}$$

As discussed earlier, Hawkyard's approach employed the *global energy balance* to replace the *local momentum balance* at the plastic front used by Taylor. The loss in the kinetic energy is equal to the increase of the deformation energy. Besides, the true strain is adopted instead of the engineering strain. The final profiles of Taylor bar predicted by Hawkyard's approach are shown in Figure 2.19. As can be seen, Hawkyard's approach results in the final profiles of the Taylor bars having a concave shape, which is in better agreement with the experiments.

Questions 2

1 If the applied pressure is a triangular pulse with the peak higher than the material yield stress, what will happen to the wave propagation in a semi-infinite bar?

Problems 2

2.1 If the stress-strain ($\sigma - \varepsilon$) relation of a material over an elastic-plastic range (until its failure) can be approximated by $\sigma = 2(2\varepsilon - 5\varepsilon^2)$ GPa, and the density of the material is $\rho_0 = 8000$ kg/m^3, please determine the von Karman's critical velocity, v_c, for the material.

2.2 A bar of length L is made of an elastic, linear strain-hardening material with $E_p = E/16$ and density ρ_0, with yield stress Y. The right end of the bar is fixed, while the left end is free. At $t = 0$, a constant compressive stress $\sigma = -Y$ is suddenly applied at the free end. What will happen when this disturbance reaches the fixed end?

2.3 Two bars of length L and cross-sectional area A are both made of elastic, linear strain-hardening material of density ρ_0 and modulus $E_p = E/16$. The yield stress of bar 1 is Y, while the yield stress of bar 2 is much higher than that of bar 1. Collinear collision occurs when the two bars move towards to each other, both with initial velocity v. Only an elastic wave propagates in bar 2 due to its high yield stress, while both elastic and plastic waves propagate in bar 1. Use the space–time diagram to analyze the wave propagation and interaction in these two bars. Also determine the time of separation, the size of the plastic region, the value of the residual strain, and the energy conversion during the process.

2.4 In a Taylor bar test, the impact velocity is found to satisfy $\rho_0 v_0^2/Y = 0.35$. Use Taylor's approach to calculate: (a) the maximum compressive strain e_0; (b) the final length of the undeformed segment, x_f/L; and (c) the maximum deformed diameter in comparison with the original diameter, d/d_0.

(Answers: $e_0 = 0.4419$, $x_f/L = 0.5028$, $d/d_0 = 1.339$)

Part 2

Dynamic Behavior of Materials under High Strain Rate

3

Rate-Dependent Behavior of Materials

In this chapter, the strain rate-dependent behavior of engineering materials will be investigated and an experimental technique to determine the material properties at different levels of strain rate will be described.

3.1 Materials' Behavior under High Strain Rates

Under dynamic loading, the mechanical behavior of most materials depends on the strain rate. It should be noted that in dynamic experiments on materials, the critical parameter is the *strain rate*, not the *velocity of deformation*. In this chapter, we present some experimental techniques for calculating the strain rates of specimens. The majority of the content is taken from the monograph by Meyers (1994), *Dynamic Behavior of Materials*.

Strain rate is the rate of change of strain with time. Its unit is 1/second (s^{-1}):

$$\dot{\varepsilon} = \frac{d\varepsilon}{dt} \tag{3.1}$$

Below are two examples showing the calculation of the strain rate.

Example 1 Consider Figure 3.1(a), in which a specimen is under dynamic tension, and the length along which the deformation is measured and evaluated is $l_0 = 0.1$ m. If the velocity of the tensile machine is $v_0 = 1$ m/s, the strain rate of this specimen is

$$\dot{\varepsilon} = \frac{\Delta\varepsilon}{\Delta t} = \frac{\Delta l/l_0}{\Delta l/v_0} = \frac{v_0}{l_0} = \frac{1\ \mathrm{m/s}}{0.1\ \mathrm{m}} \tag{3.2}$$

i.e.,

$$\dot{\varepsilon} = 10\ s^{-1} \tag{3.3}$$

which means that the strain rate of this specimen is only 10 s^{-1} even if it experiences a tensile deformation at a very fast speed, 1 m/s. In a conventional tensile test, the loading speed is about 6 mm/min, so the corresponding strain rate of the same specimen is only 10^{-3} s^{-1}. This estimation shows that the magnitude of strain rate depends not only on

Introduction to Impact Dynamics, First Edition. T.X. Yu and XinMing Qiu.

Published 2018 by John Wiley & Sons Singapore Pte. Ltd.

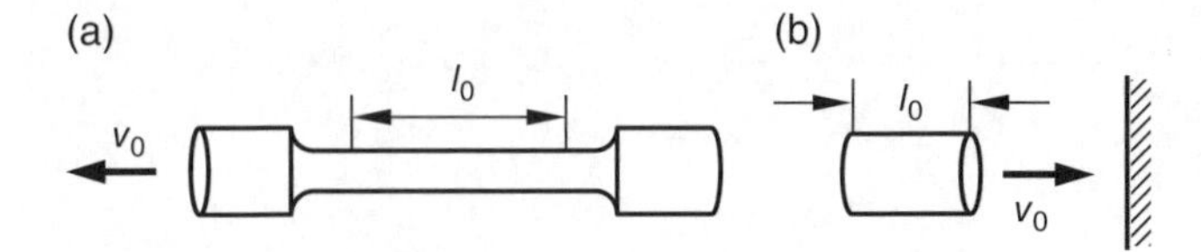

Figure 3.1 Examples of strain rate calculations. (a) A specimen under dynamic tension; (b) a projectile hits a rigid wall.

the loading velocity, but also on the size of the specimen – i.e., under the same loading velocity, the shorter the specimen, the larger the strain rate.

Example 2 Consider Figure 3.1(b), in which a cylindrical projectile of length $l_0 = 50$ mm impinges on a rigid target at a velocity of 1000 m/s. Suppose the projectile decelerates linearly to rest, becoming a truncated cone, and the length is reduced to 25 mm. Then the average strain rate of the projectile during this process can be estimated as,

$$\dot{\varepsilon} = \frac{\Delta l}{l_0 t} = \frac{25}{50t} \tag{3.4}$$

The response time t can be calculated based on the assumption of linear deceleration:

$$\Delta l = 25 \text{ mm}, \frac{1}{2} v_0 t = \Delta l, t = \frac{2\Delta l}{v_0} = 5 \times 10^{-5} \text{s} \tag{3.5}$$

Substituting Eq. (3.5) into Eq. (3.4) gives

$$\dot{\varepsilon} = \frac{25}{50 \times 5 \times 10^{-5} \text{s}} = 10^4 \text{s}^{-1} \tag{3.6}$$

Obviously, high strain rates will be a factor in weapons-related applications. It is difficult to produce a strain rate of this magnitude with a conventional static testing device, and hence other dynamic experimental techniques have to be adopted in order to determine the relevant material properties.

The strain rates found in these two examples, which are $< 10^4$ s^{-1}, are at the low to medium end of strain rate values. At the higher end, a shock wave produced by a nuclear explosion has a very sharp wave front, the thickness of which decreases as the explosive pressure increases, so the strain rate in the wave front could be as high as $10^8 - 10^9 \text{s}^{-1}$, which is much higher than those in weapons-related applications. At the other end of the spectrum, creep tests produce strain rates that can be lower than 10^{-7}s^{-1}. Similar to the creep of engineering materials, the geological orogenic movement is also a mechanical phenomenon with very low strain rates.

Figure 3.2 gives a list of testing techniques and mechanical considerations for a wide spectrum of strain rates. The strain rates can be classified into two groups: $< 5\,\text{s}^{-1}$ and $> 5\,\text{s}^{-1}$. For lower strain rates, the inertial forces (illustrated in Chapter 5) are generally negligible, while equilibrium is always maintained. Under higher strain rates, on the

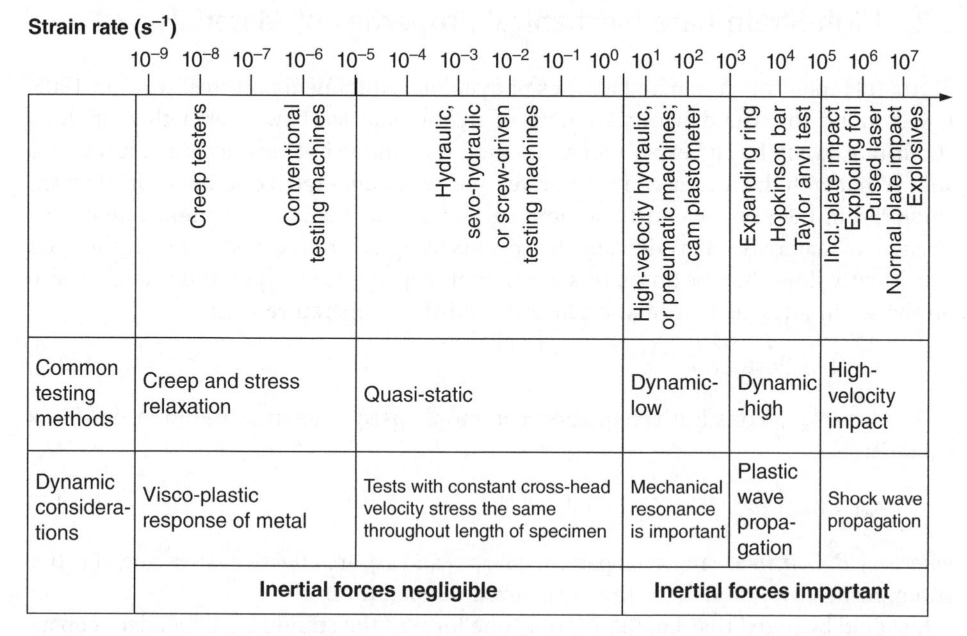

Figure 3.2 Schematic classification of testing techniques according to strain rate. *Source*: Meyer (1994). Reproduced with permission of Wiley.

other hand, the influence of inertial forces and the effect of wave propagations will increase with the strain rate.

Very low strain rates appear along with the phenomena of creep and stress relaxation, which involve the visco-plastic responses of metal or polymer materials. Creep testing machines with strain rates of 10^{-9}–10^{-7} s^{-1}, or conventional testing machines with strain rates of 10^{-7}–10^{-5} s^{-1} are usually adopted.

Quasi-static loadings result in relatively low strain rates in the range 10^{-5}–10 s^{-1}. In this range, the elastic wave propagations in specimens and testing machines become important. The testing machines used are hydraulic, servo-hydraulic, and pneumatic. In these tests, the stress and strain should ideally be uniform along the length of the entire specimen. Also, the cross-head speed of the machine should remain constant.

The tests with low-speed dynamic deformation, i.e., in the strain rate range 10–10^{3} s^{-1}, are usually most difficult. Loading in this range often causes resonance with the testing machine.

For high-speed dynamic deformation, i.e., in the strain rate range10^{3}–10^{5} s^{-1}, the testing techniques are relatively mature, and include the Hopkinson bar, the Taylor bar, and the expanding ring.

In the strain rate range 10^{5}–10^{8} s^{-1}, shear wave and shock wave propagations are involved, and various means allowing rapid deposition of energy at the surface of the material are adopted. This can be accomplished by impact (normal or inclined), by the detonation of explosives in contact with the material, or by pulsed laser or other sources of radiation.

Various experimental techniques are discussed in the following sections. But first the behavior of materials at high strain rates is briefly described.

3.2 High-Strain-Rate Mechanical Properties of Materials

Bertram Hopkinson conducted a series of dynamic experiments on steel wires in 1905. It was found that the dynamic strength of steel was at least twice as high as its low-strain-rate strength. The experimental method was simple but very accurate. It was also shown that with the increase of strain rate, steel wires undergo a ductile-to-brittle transition. From then on, scientists became curious about the effect of strain rate on the strength of materials. Many dynamic test results using different experimental techniques consistently show that the flow stress of materials depends not only on the strain, but also on the strain rate, the deformation history, and the temperature. That is,

$$\sigma = f(\varepsilon, \dot{\varepsilon}, \text{history}, T) \tag{3.7}$$

For example, a constitutive equation commonly used, known as the Johnson–Cook equation, is

$$\sigma_{\text{eff}} = \left(\sigma_0 + B\varepsilon_{\text{eff}}^n\right)\left(1 + C\ln\dot{\varepsilon}^*\right)\left(1 - T^{*m}\right) \tag{3.8}$$

where σ_0, B, C, n, m are material parameters, σ_{eff} and ε_{eff} are effective stress and effective strain, respectively, and T^* is the normalized temperature.

It should be noted that Eq. (3.8) is only one form of the strain rate-dependent constitutive relationship. There have been many attempts to predict the parameters from first principles, but so far none have been successful. Thus, people are forced to obtain the material parameters experimentally. In the following sections of this chapter, the experimental techniques for determining the dynamic material properties, i.e. the parameters in the constitutive equation taking the strain rate effect into account, will be described.

3.2.1 Strain Rate Effect of Materials under Compression

In Xu *et al.* (2015a), in order to evaluate the dynamic recrystallization behavior of as-cast Mg-9.15Li-3.23Al-1.18Nd duplex magnesium alloy, a series of compression tests with a strain of 0.7 were performed on a Gleeble 3500 thermo-mechanical simulator in the temperature range of 473–673 K and the strain rate range of 0.001–1 s^{-1}. The typical true stress–strain flow curves of as-cast LA93-1.0Nd alloy are presented in Figure 3.3. It can be seen that the peak flow stress and steady flow stress increase significantly with increasing strain rate and decreasing deformation temperature. Clearly, the materials are strain rate-sensitive under compression.

3.2.2 Strain Rate Effect of Materials under Tension

In Figure 3.4, the stress–strain curves obtained from dynamic uniaxial tensile tests on mild steel are compared with those of the quasi-static test (0.001 s^{-1}). Quasi-static experiments are conducted on a universal testing machine, while a hydropneumatic machine and a modified Hopkinson bar are used to investigate the dynamic tensile behavior of mild steel specimens at medium and high strain rates, respectively.

The engineering stress versus engineering strain and true stress versus true strain curves of mild steel specimens at various rates (0.001–750 s^{-1}) of tensile loading are

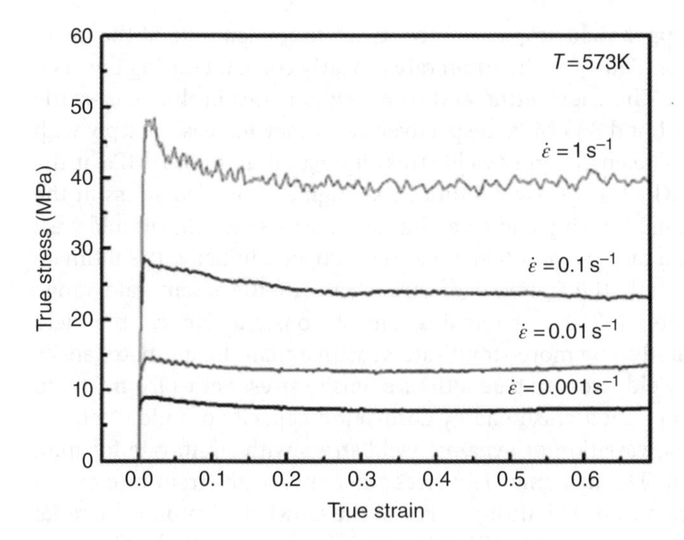

Figure 3.3 Effect of strain rate on the stress–stain response under compression. *Source*: Xu *et al.* (2015a). Reproduced with permission of Elsevier.

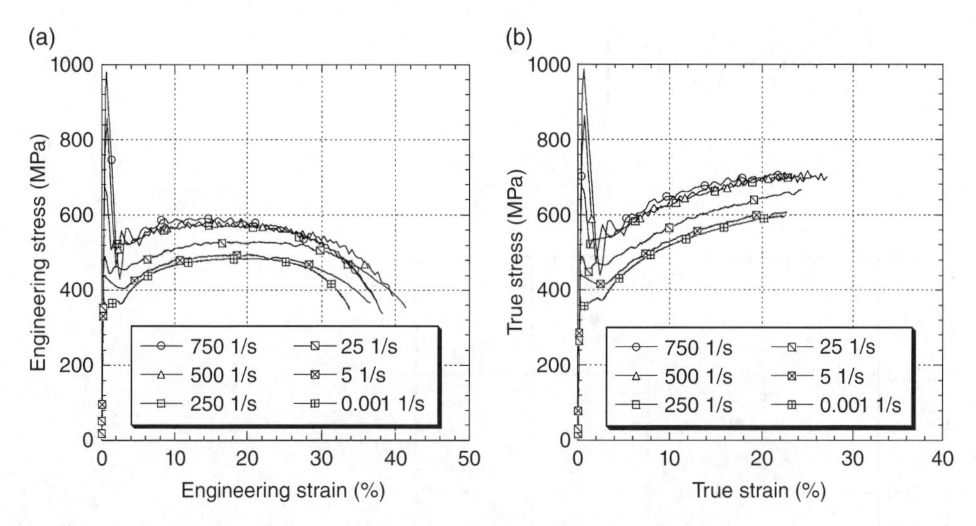

Figure 3.4 Stress–strain curves of steel under uniaxial tension: (a) engineering stress–strain; (b) true stress–strain. *Source*: Singh *et al.* (2013). Reproduced with permission of ASCE.

compared in Figure 3.4. The strain hardening in the material increases up to the strain rate 250s^{-1} and then changes marginally in the strain rate range 250–750 s^{-1}. The fracture energy and toughness values at high strain rates are greater than those at quasi-static and medium strain rates because the area under the stress–strain curve is greater at high strain rates. The yielding instability of upper and lower stress peaks increases with the increasing strain rate. The ratio of upper yield point stress to lower yield point stress varies from 1.25 to 2.02 as the strain rate increases from

250 to 750 s^{-1}. Thus, the upper yield stress is almost twice the magnitude of the lower yield stress at high strain rate (750 s^{-1}). The strain rate is nearly constant during the plastic deformation of material. The engineering and true yield stresses in the quasi-static (0.001 s^{-1}) condition are 361 and 363 MPa, respectively, and they increase sharply with the increase in strain rate. The engineering yield stress increased up to 891 MPa at the high strain rate (750 s^{-1}), which is nearly 2.5 times the engineering yield stress in the quasi-static condition. The engineering and true ultimate tensile strengths of mild steel under quasi-static condition are 483 and 596 MPa, respectively. Initially, the ultimate tensile strength increases with the increase of strain rate in the strain rate range 0.001–250 s^{-1}, and thereafter (>250 s^{-1}) remains almost constant. Hence, the yield strength of mild steel is found to be more strain rate-sensitive than the ultimate tensile strength. The ratio of true yield stress to true ultimate tensile stress varies from 0.60 to 1.40, which indicates the very good crash energy absorption capacity of mild steel.

Figure 3.5 summarizes the variation of dynamic yield stress with strain rate for mild steel under uniaxial tension. The dynamic yield stress increases with strain rate over a relatively wide range of strain rate. The fitting curve of the Cowper–Symonds material model (introduced in Chapter 4) is also plotted in Figure 3.5. According to the Cowper–Symonds model, the dynamic yield stress of the mild steel tested under tension can be approximately given as,

$$\sigma_d = (363)\left[1 + \left(\frac{\dot{\varepsilon}}{301}\right)^{\frac{1}{2.4639}}\right] \tag{3.9}$$

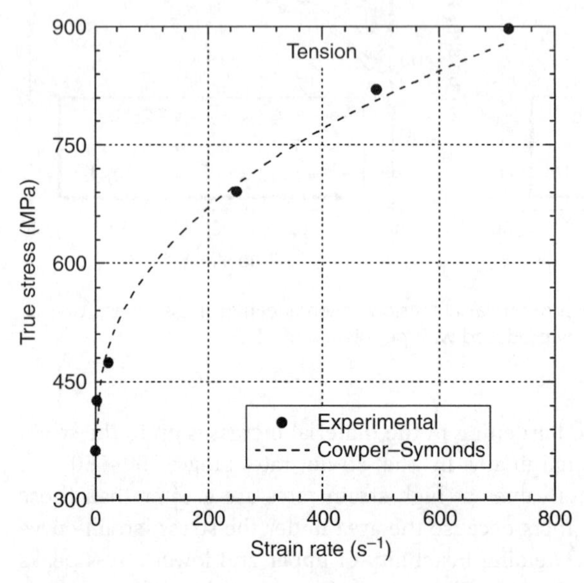

Figure 3.5 Variation of dynamic yield stress with strain rate for mild steel under uniaxial tension. *Source:* Singh *et al.* (2013). Reproduced with permission of ASCE.

Therefore, it is concluded that mild steel is a strain rate-dependent material (also called a rate-sensitive material), and has the following characteristics: yield stress and ultimate stress increase with strain rate; and ductility decreases with strain rate.

3.2.3 Strain Rate Effect of Materials under Shear

Clifton (1983) reported that dramatic increases in flow stress were attained at strain rates of the order of 10^5 s^{-1} in experiments on 1100-0 aluminum involving the inclined impact of flat plate (Figure 3.6). For lower stain rates, i.e., $\dot{\gamma} < 10^5$ s^{-1}, the shear stress, τ, of aluminum increased slightly with shear strain rate, whereas dramatic "hardening" was observed as the strain rate reached the order of 10^5 s^{-1}. This rise in strength is not accommodated by the conventional constitutive equations and requires special modification of the equation. This rise leads some scientists to believe that there is, indeed, a limiting strain rate at which the strength of the material approaches infinity.

It should be emphasized that when the strain rate increases, the deformation process changes gradually from fully isothermal to fully adiabatic. In the conventional experimental process, heat exchange between the specimen and the environment is sufficient due to the slow loading velocity, so that the temperature of the specimen is always close to room temperature and the deformation process could be regarded as fully isothermal. By contrast, in the case of a high-speed loading experiment, the temperature of the specimen will increase as there is not enough time for the heat generated by the plastic deformation to escape from the body. The stress obtained at room temperature (e.g., 20° C) could be the stress of the material at 200° C, so that the effect of the temperature rise on the material properties is very significant. This give rise, in some cases, to adiabatic shear instabilities that have a profound effect on the mechanical response of the material. The deformation of the specimen will be concentrated on a narrow-band region and therefore the equivalent strain calculated by uniform deformation will be largely under-estimated. Hereafter, the constitutive

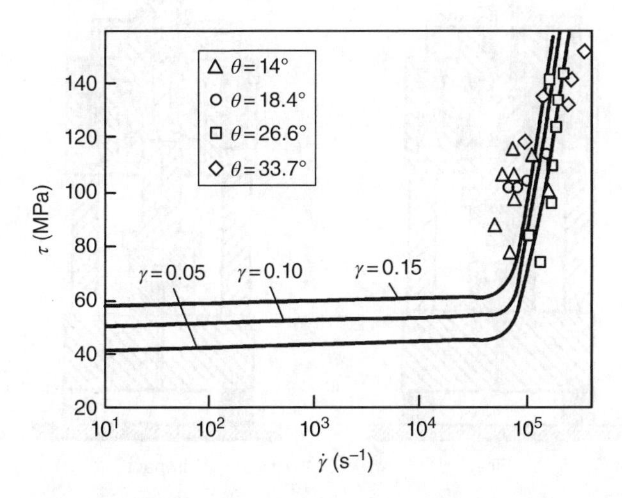

Figure 3.6 Dependence of flow stress (at different plastic strains) on the shear strain rate for 1100-0 aluminum. *Source*: Meyer (1994). Reproduced with permission of Wiley.

relation of a material should be analyzed stage by stage according to the relevant deformation mechanisms.

3.3 High-Strain-Rate Mechanical Testing

3.3.1 Intermediate-Strain-Rate Machines

As discussed in Section 3.1, high-strain-rate mechanical testing cannot be done with conventional testing machines, because the loading apparatus cannot be accelerated fast enough. In order to achieve an intermediate strain rate, the most feasible approach is to use dynamic loading equipment with stored potential energy, e.g., a compressed-gas machine, a drop weight tester, or a rotary flywheel.

Compressed-Gas Machine

A compressed-gas machine is shown in Figure 3.7. The compression specimen is placed on the anvil of the machine. A movable piston passes through a reservoir. Support jacks keep the piston "cocked" (see Figure 3.7a). When the auxiliary reservoir is pressurized after the support jacks are lowered, the piston is accelerated downward, impacting the anvil (Figure 3.7b). This results in the compression of the specimen, as shown in Figure 3.7(c). This machine, a DYNAPAK, is capable of velocities up to 15 m/s. The diagnostics are required to record stress and strain (not shown in the figure). A simpler version of this machine is the "drop hammer", in which a large mass is dropped freely.

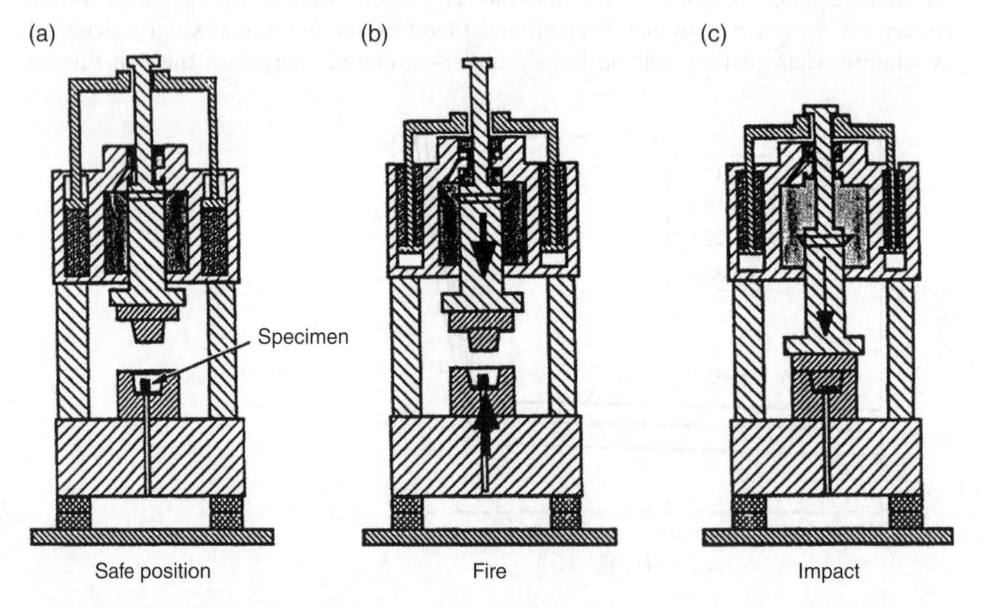

Figure 3.7 Schematic of intermediate-strain-rate compressed-gas machines. *Source*: Meyer (1994). Reproduced with permission of Wiley.

Drop Weight Tester

The principle of a drop weight tester is to impact the specimen by dropping a large weight freely. Different impact velocity and impact kinetic energy can be achieved by adjusting the releasing height and the mass of the drop weight. In addition to acceleration due to gravity, the weight can be accelerated further using the pneumatic auxiliary device, in order to obtain a higher impact velocity. To measure the dynamic response of materials or structural models, a drop weight tester is usually associated with a high-speed camera and dynamic force sensor. The drop weight tester is an important experimental facility to get the response of materials at intermediate strain rates, i.e., 10^{-1}–10^{3} s^{-1}.

For example, the CEAST 9350, as shown in Figure 3.8, is a floor-standing impact system designed to deliver 0.59–757 J or up to 1800 J with an optional high-energy system. It works with an impact software and data acquisition system to make analysis simpler. This versatile instrument can be used to test anything from composites to finished products and is suitable for a range of impact applications.

In addition to the commercial machines available, many research institutions design and make drop weight testers according to their own experimental requirements. Similar to the compressed-gas machine, the drop weight tester also has the shortcoming

Figure 3.8 CEAST 9350 Drop Tower Impact System. *Source*: http://www.instron.us/en-us/products/testing-systems

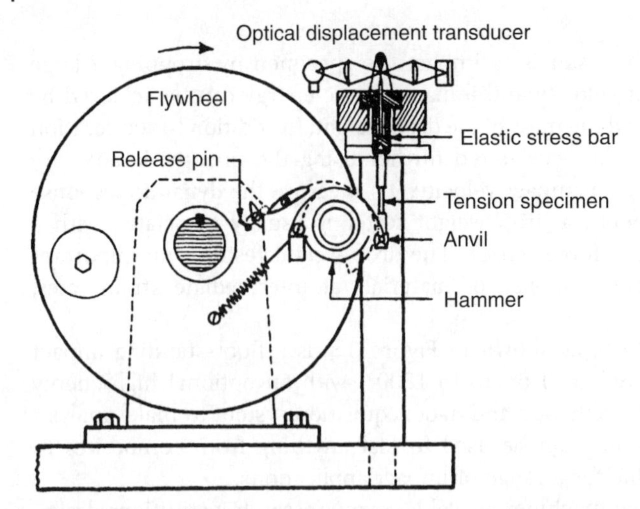

Figure 3.9 High-speed tension machine using rotary flywheel. *Source*: Meyers (1994). Reproduced with permission of Wiley.

that the impact velocity gradually decreases during the impact process, and hence the strain rate is not a constant. The drop weight is also generally assumed to be a rigid body with no deformation, so the impact force can be directly obtained by using Newton's second law.

Rotary Flywheel Machine

In a rotary flywheel machine (Figure 3.9), an electric motor drives a large flywheel in the clockwise direction. When the desired flywheel velocity is reached, a release pin triggers the hammer (shown in Figure 3.9 in the two positions), which then impacts the bottom of a tensile specimen. The mass of the flywheel is sufficiently large to ensure a constant velocity. The release of the hammer is synchronized with the position of the specimen so that it occurs prior to impact with the specimen. The displacements of an elastic stress bar and the bottom part of the specimen are monitored by optical methods. The displacement of the stress bar provides the stress, whereas that of the bottom of the specimen provides the strain. Thus, it is possible to obtain a continuous stress–strain curve. Usually specimens with a gauge length of 25 mm are used. Strain rates between 0.1 and 10^3 s^{-1} can be achieved by this method.

Servo-hydraulic Dynamic Testing Machine

Utilizing servo-hydraulic and the supporting control technologies, some companies offer fully integrated testing solutions designed for the high strain rate and high-speed test requirements. For example, Figure 3.10 shows a typical servo-hydraulic dynamic testing machine, the INSTRON VHS8800 High Strain Rate System. The rigid beam at the top of the machine is fixed, and when the high-stiffness load frame below the beam moves rapidly down or up, the specimen in the middle will be stretched or compressed dynamically. The maximum tensile velocity of the specimen is 25 m/s, and the maximum compressive velocity is 10 m/s. According to Eq. (3.2), if the lengths of a tensile and a compression specimen are 100 and 50 mm, respectively, the corresponding maximum

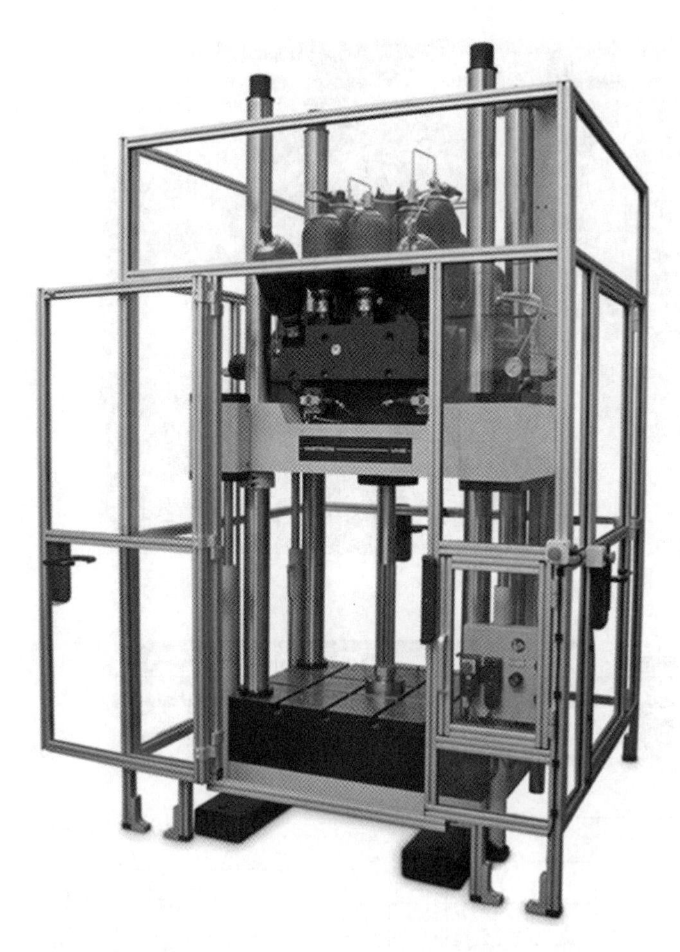

Figure 3.10 INSTRON VHS8800 High Strain Rate System. *Source*: http://www.instron.cn/zh-cn/products/testing-systems/dynamic-and-fatigue-systems/8800-high-strain-rate

tensile strain rate and compression strain rate will be 250 and 200 s^{-1}, respectively. These are typical "intermediate strain rate" values which are difficult to obtain with other testing techniques. The capacity of the applied force of this machine is between 40 and 100 kN, which is measured by the specialized measurement transducers.

The key to achieving rapid loading lies in using a hydraulic system with 28 MPa pressure supplied, which results in very high acceleration and load performance, so as to overcome the inertia of the moving part of the machine. The specimen will be clamped until the loading chuck reaches a predetermined velocity, and then will be stretched or compressed. The maximum upward displacement of the frame is 300 mm, which is much larger than the moving distance during the acceleration stage. Of course, in the design of the experiment the length of the specimen itself should also be taken into account. In addition, this testing machine is very sensitive to the lateral load, i.e., even very small lateral loads may cause damage to the components. Therefore, a transverse load protection system needs to be employed in order to protect the machine.

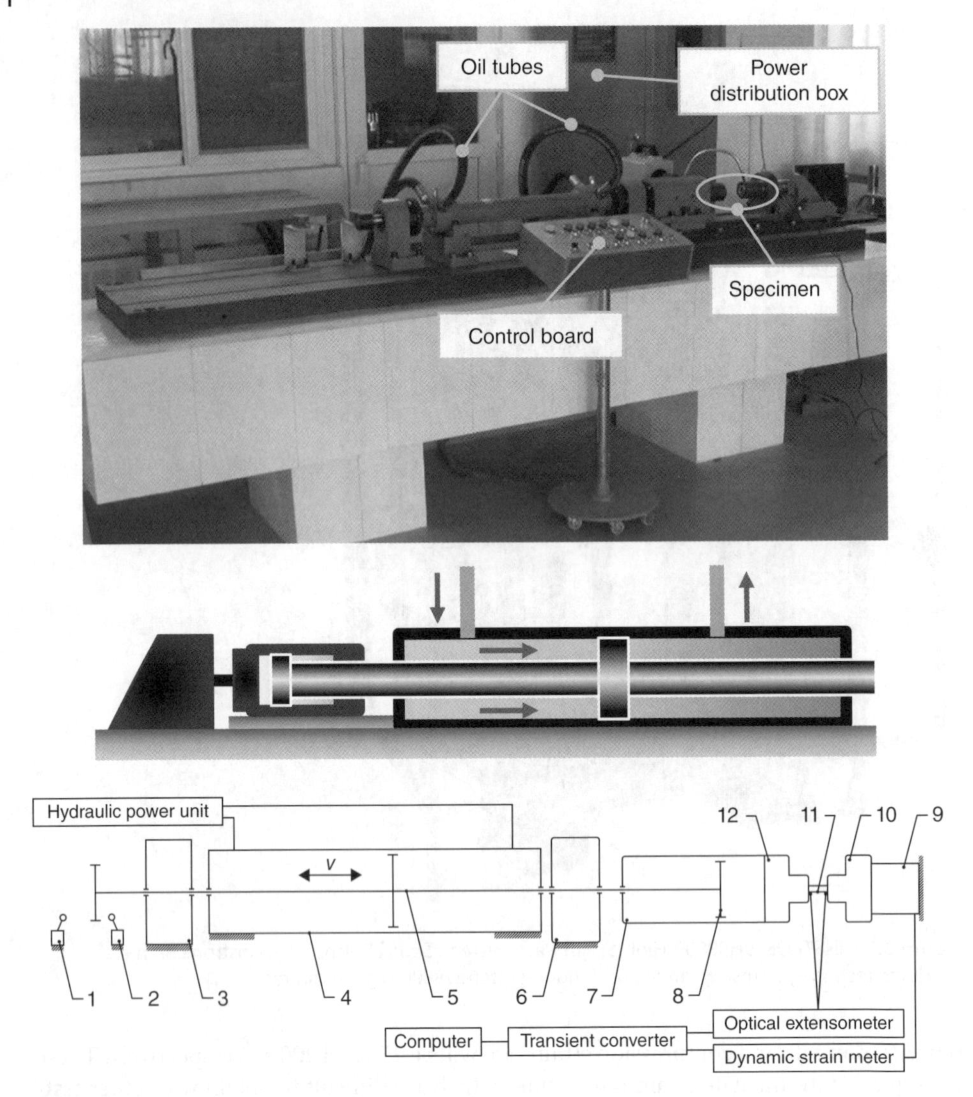

Figure 3.11 Intermediate-strain-rate materials testing apparatus. Top: photograph of the apparatus; middle and bottom: schematic of the apparatus. 1, front travel switch; 2, rear travel switch; 3, front retainer; 4, hydraulic cylinder; 5, piston rod; 6, rear retainer; 7, connector; 8, impact hammer; 9, load cell; 10, rear grip; 11, specimen; 12, front grip. *Source*: Wu *et al.* (2005).

Intermediate-Strain-Rate Machine

A self-developed material testing apparatus is shown in Figure 3.11. The machines work in the strain rate range 0.1–50 s^{-1} and is suitable for tension/compression loading/loading–unloading experiments on columns or flat specimens. By using the technique of fluid drive, multi-level speed regulation and cushioning impact, stable loading pulse with steep rising edge is generated to carry out the intermediate strain rate experiment. The self-centering fixture and hydraulic cylinder are employed to assure the precision of the experiment. The strain and stress are measured by using the optical tensimeter, the force sensor and self-developed dynamic strain indicator.

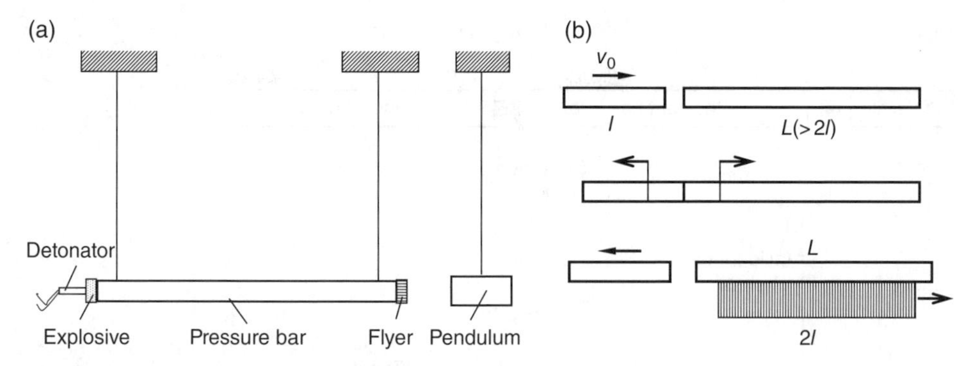

Figure 3.12 Bertram Hopkinson experimental device: (a) schematic diagram of the original Hopkinson pressure bar; (b) stress wave with finite length.

3.3.2 Split Hopkinson Pressure Bar (SHPB)

The Compression Split Hopkinson Bar

The original Hopkinson pressure bar was designed by B. Hopkinson in 1914 to produce a pressure pulse of controlled length and to accelerate a flyer. About 100 years ago, there was no instrument (e.g., strain gauge, laser and optical technology, high-speed photography) to measure the dynamic deformation of a specimen. It was very challenging to measure the stress wave propagation generated by an explosion in experiments. In 1914, Bertram Hopkinson invented a pressure bar to measure the pressure produced by high explosives or the high-speed impact of bullets (Figure 3.12a). The device could not only accelerate the flyer, but also measure the pulse of the stress wave. A cylindrical pressure bar hanging from light wires was impinged by the detonation of the explosive attached to its left end. After explosion, a compressional pressure pulse propagated in the pressure bar towards the far end. When the compressional pulse was reflected at the free end of the pressure bar, a light flyer originally stuck with oil would fly off, due to the reflected tensile pulse. A pendulum hanging separately was then impinged by the flyer. By measuring the final swing angle of the pendulum, its initial velocity could be estimated, as well as the initial momentum of the flyer. In this way, the average stress of the pulse captured by the flyer, the length of which was twice its thickness, was calculated.

As shown in Figure 3.12(b), if a cylindrical bullet of length l is adopted instead of the original explosive as the power source, after impact, the pressure pulse generated in the Hopkinson bar is a rectangular pulse of length $2l$. After separation of the bullet from the pressure bar, the rectangular pulse will propagate along the latter.

Kolsky (1949) was the first person to extend the Hopkinson bar technique to measure the stress–strain response of materials under impact loading conditions. Figure 3.13(a) shows a compressional SHPB, in which the specimen is sandwiched between two bars. A projectile of length l impinges on the incident bar of length L_i at initial velocity v_0. An elastic pressure pulse is generated by the projectile and propagates from the impact end of the incident bar. This elastic pulse is a rectangular pulse of length $2l$, and the pulse amplitude is proportional to the impact velocity v_0. The length and amplitude of the rectangular pulse can be adjusted by varying the length and initial velocity, respectively,

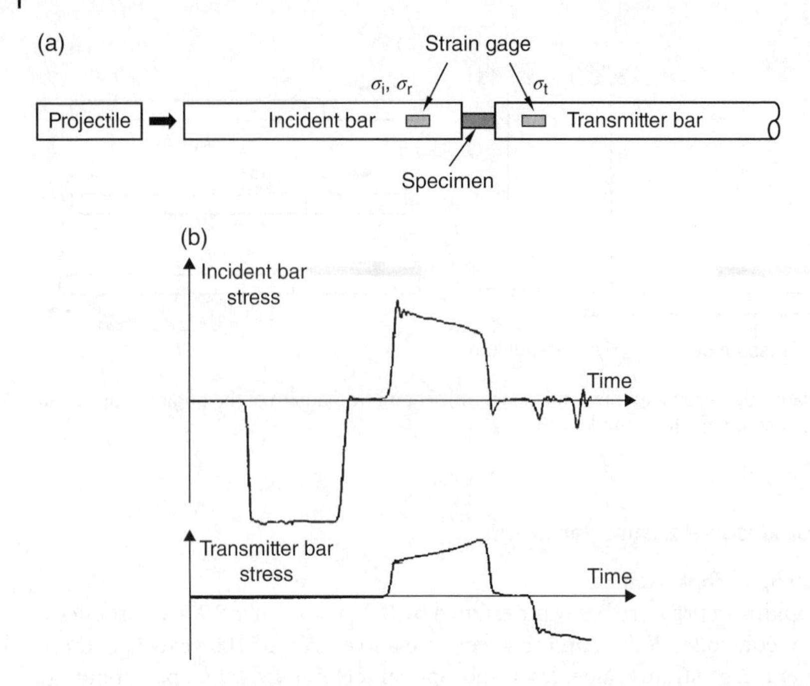

Figure 3.13 Split Hopkinson pressure bar. (a) Schematic of experimental device; (b) stress waves.

of the incident bar. Once the elastic rectangular pulse reaches the specimen between two long bars (incident and transmitter bars), a complex pattern of reflections is produced, but the pulse passes through the specimen (deforming it plastically), and continues its course in the transmitter bar.

The compressional SHPB has found wide acceptance as the standard instrument for intermediate strain rate testing (10^2–10^4 s^{-1}) since the 1970s. Typically, an incident bar (input bar), a transmission bar (output bar) and an optional extension bar are fabricated from the same material and are of the same diameter. As the stress waves inside the bars are measured by surface strains, the bar material should be linearly elastic with a high yield strength. To ensure one-dimensional wave propagation in the bars, they must be physically straight and free to move on their supports with minimized friction. The whole bar system must be perfectly aligned along a common straight axis, which is the loading axis of the system.

The amplitude of the rectangular pulse should be high enough to produce plastic deformation in the specimen. When the elastic compressive wave in the incident bar reaches the specimen, part of the wave passes through the specimen and enters the transmitter bar, and part of the wave is reflected back to the incident bar. From the strain gauges located in the appropriate positions on the incident bar and the transmitter bar, the incident and transmission stress waves varying with time can be measured, as shown in Figure 3.13(b).

In order to obtain reliable stress and strain values from the signal of the elastic wave, the design of an SHPB should satisfy the following requirements:

1) The lengths of incident bar L_i and the transmitter bar L_t should be much longer than the length of the projectile l, i.e., $L_i \gg l$ and $L_t \gg l$;

2) The length of the specimen L_s should be much shorter than the length of the rectangular compressional pulse $2l$, i.e., $l \gg L_s$.

The former ensures that the strain signals in the incident bar and the transmitter bar are "clean", i.e., are not disturbed by reflected signals from the far ends. The latter ensures the state of stress and strain in the specimen is almost uniform after many wave reflections between its two ends. The recordings of the strain gauges on two long bars can then be converted into the stress in the specimen. Finally, a stress–strain curve can be constructed from the data gathered in an SHPB test. As the (average) strain rate in the test can be also calculated, this curve is useful to show the dynamic behavior of the material at high strain rate, and to provide necessary information for mathematical modeling and numerical simulations via constitutive equations.

Analysis of SHPB

The standard SHPB analysis is based on two assumptions. First, it is assumed that the wave profiles are known not only at the measuring points but also everywhere in the bar because an elastic wave can be shifted to any locations without dispersion according to the uniaxial elastic wave propagation theory. Thus, the transmitted wave can be shifted to the transmitter bar–specimen interface to obtain the transmitter force and velocity, whereas the incident force and velocity can be determined by the incident and reflected waves shifted to the incident bar–specimen interface.

If the strains carried by the incident pulse, the transmitted pulse and the reflected pulse are ε_i, ε_t, and ε_r, respectively, then the forces and velocities at both ends of the specimen are given by the following equations:

$$\begin{aligned} F_{\text{input}}(t) &= S_B E[\varepsilon_i(t) + \varepsilon_r(t)] \\ F_{\text{output}}(t) &= S_B E \varepsilon_t(t) \\ v_{\text{input}}(t) &= c_0[\varepsilon_i(t) - \varepsilon_r(t)] \\ v_{\text{output}}(t) &= c_0 \varepsilon_t(t) \end{aligned} \tag{3.10}$$

where F_{input}, F_{output}, v_{input}, v_{output} are forces and particle velocities at the interfaces; S_B, E, and c_0 are cross-sectional area of the bars, Young's modulus, and the longitudinal wave speed, respectively, and $\varepsilon_i(t)$, $\varepsilon_r(t)$, $\varepsilon_t(t)$ are the strains known at the bar–specimen interfaces.

Secondly, from the forces and velocities at both bar–specimen interfaces, the standard analysis assumes the axial uniformity of the stress and strain fields in the specimen, and thus the stress–strain curve can be obtained (like those obtained from a quasi-static test). The compressional strain rate $\dot{\varepsilon}_s(t)$ is calculated first by the difference in velocities at both ends of the specimen;

$$\dot{\varepsilon}_s(t) = \frac{v_{\text{output}}(t) - v_{\text{input}}(t)}{L_s} \tag{3.11}$$

By the uniform deformation assumption, the specimen is supposed to be in the equilibrium state, i.e., $F_{\text{input}}(t) = F_{\text{output}}(t)$. Hence, the stress in the specimen is,

$$\sigma_s(t) = \frac{F_{\text{output}}(t)}{S_s} \tag{3.12}$$

where L_s and S_s are the length and cross-sectional area of the specimen, respectively.

Again by the uniform deformation assumption, the strains in the specimen satisfy $\varepsilon_i(t) + \varepsilon_r(t) = \varepsilon_t(t)$. By substituting Eq. (3.10) into Eq. (3.11), the strain rate of the specimen is

$$\dot{\varepsilon}_s(t) = \frac{2c_0}{L_s}\varepsilon_r(t) \tag{3.13}$$

To integrate the strain rate over time, the strain of the specimen is found as

$$\varepsilon_s(t) = \frac{2c_0}{L_s}\int_0^t \varepsilon_r(t)\,\mathrm{d}\tau \tag{3.14}$$

The stress of the specimen can be obtained from the force on the interface between the specimen and the transmitter bar:

$$\sigma_s(t) = \frac{S_B E}{S_s}\varepsilon_t(t) \tag{3.15}$$

It should be noted that the strain rate, strain and stress of the specimen are obtained from Eqs (3.13)–(3.15), only utilizing the strain signals of the reflected and transmitted pulses, i.e., $\varepsilon_r(t)$ and $\varepsilon_t(t)$. Hence, these formulas are termed the two-wave formula of the SHPB. It is easy to see that combining the stress in Eq. (3.15) and the strain in Eq. (3.14) gives the stress–strain curve under a certain strain rate, Eq. (3.13).

As such an homogeneous assumption does not always hold under dynamic loading, at least in the early stages of the test because of the transient effects, the loading starts at one end of the specimen, whereas the other end remains at rest. Therefore, a three-wave analysis has been proposed to use the average of the two forces to calculate the stress instead of Eq. (3.12):

$$\sigma_s(t) = \frac{F_{\text{input}}(t) + F_{\text{output}}(t)}{2S_s} \tag{3.16}$$

Accordingly, without requiring the uniform deformation in the specimen, which is $\varepsilon_i(t) + \varepsilon_r(t) = \varepsilon_t(t)$, Eq. (3.16) leads to the following three-wave SHPB formulas:

$$\sigma_s(t) = \frac{S_B E}{2S_s}[\varepsilon_t(t) + \varepsilon_r(t) + \varepsilon_i(t)] \tag{3.17}$$

$$\dot{\varepsilon}_s(t) = \frac{c_0}{L_s}[\varepsilon_t(t) + \varepsilon_r(t) - \varepsilon_i(t)] \tag{3.18}$$

$$\varepsilon_s(t) = \frac{c_0}{L_s}\int_0^t [\varepsilon_t(t) + \varepsilon_r(t) - \varepsilon_i(t)]\,\mathrm{d}\tau \tag{3.19}$$

Both the two-wave formula and the three-wave formula of SHPB are derived from non-dispersive wave propagation and homogeneous stress–strain field assumptions. However, it should be emphasized that the homogeneous stress–strain field assumption is not necessary for the tests using SHPB. If one simply uses Eq. (3.10) in SHPB analysis, such a test provides an accurate measurement of forces and velocities at both sides of the specimen.

From the stress, strain and strain rate of the specimen given by the two-wave formula or the three-wave formula of SHPB, the stress–strain curve under a certain average strain rate can be constructed for the tested material. As the average strain rate is directly related to the impact velocity, the stress–strain curves under different strain rates can be obtained by varying the impact velocity. Thus, the parameters employed in the strain rate-dependent constitutive equation of the material will be determined, so as to provide necessary information for mathematical modeling and numerical simulations.

When using SHPB to determine the $\sigma - \varepsilon - \dot{\varepsilon}$ relations, we have in fact made three assumptions:

1) The state of stress and strain in the specimen is uniform under the high-strain-rate dynamic loading.
2) The bars and the specimen are in a one-dimensional stress state. The bars should be thin and long, and the specimen should have dimensions of $L_s/d_s = \sqrt{3}/2$, which minimizes the effect of transverse inertia.
3) The friction between the specimen and the ends of the bars is negligible. Lubrication at the interfaces can be applied to effectively reduce the friction.

In the design of an SHPB experiment, the cylindrical specimen dimensions should be determined first. In order to avoid stress concentration, the homogeneity of the material should be satisfied. The dimension of the specimen should be at least one order larger than the microstructure of the material. For conventional metals, the microstructure contains crystal grains and therefore a small specimen will satisfy the homogenous assumption, i.e., at least 10 times greater than the grain size. However, nowadays the SHPB technique is also widely employed in testing the brittle materials, such as ceramics, glass, ice, rocks, concrete, bricks, cortical bones, and some composites. These materials are extremely sensitive to stress concentration caused by bar misalignment or non-parallel loading surfaces. SHPB technique is also adopted in testing the equivalent mechanical properties of porous materials or cellular materials, such as honeycombs, lattices, and foams. For these materials, satisfying the above three basic assumptions for SHPB is not as easy as it is for metals. Take the concrete as an example. The aggregate (broken stones) in concrete usually has the dimension of centimeters. To ensure the homogeneity and avoid stress concentration, the specimen size should be 10 times of the aggregate, i.e., typically in the order of 100 mm. But the big specimen will bring a series of difficulties: (a) the size of incident bar and transmitter bar should be even larger, in order to satisfy the one-dimensional stress state; (b) it is difficult to achieve high strain rates because the strain rate of specimen is inversely proportional to the specimen size; and (c) the effect of transverse inertia will be very significant. Thereby, the strain rate-dependent properties of concrete measured by the SHPB technique are still controversial.

Honeycomb, foam, and lattice are classified as generalized cellular materials. Although they are made of conventional uniform material, i.e., metal, they have microstructures typically in the order of 1–10 mm. In the static state, the cellular materials are usually regarded as uniform materials, which means that the detailed structural response at the micro level is neglected, while the equivalent properties at the macro level are of interest, e.g. equivalent elastic modulus and yield strength, and so on. Similar to the case of concrete discussed earlier, in dynamic experiments, the size of the specimen of cellular material should also be large (at least 10 cells included) to ensure the homogeneity. Therefore, larger incident and transmitter bars are required in an SHPB experiment design. In addition, the density, yield strength, stiffness, wave speed, and mechanical impedance should all be small. Using a metal bar that has a large mechanical impedance will produce a weak transmitted signal, while using a nylon/PMMA bar will introduce stress wave dispersion in the viscoelastic material. The most challenging issue is to facilitate uniform deformation in the specimen. With the SHPB technique, it is assumed that the specimen deforms uniformly so that the measurement of average deformation over the specimen length can represent the deformation at any point of the specimen. However, this assumption may not be satisfied automatically in cellular materials under intense dynamic loading. Under rapid compression, shock waves will be generated and localized deformation will take place in the cellular specimens. Consequently, only a portion of specimen in contact with the incident bar will be compressed, so that no uniform stress–strain fields can be measured to construct the desired constitutive relation.

In conclusion, in employing the SHPB technique to characterize the above-mentioned materials, careful design is required to ensure that the basic assumptions are not violated, rendering the results meaningless or misleading.

Split Hopkinson Tensile Bar

The tensile and compressive properties of some materials are different, such as cast iron and unidirectional fiber-reinforced composite. Furthermore, some materials, such as fibers, can only sustain tension. Exploring the high-rate response of these materials requires a tensile version of the split Hopkinson technique. The tensile version of the SHPB started to emerge in the 1960s. The working principles of these bars are similar to those of compressive bars. However, the loading mechanisms are more complicated than the simple bar-to-bar impact seen in compression experiments. The specimens in both tension and torsion experiments must be attached to the ends of the bar, which introduces the complication of gauge-section identification in strain rate calculations (Chen & Song, 2011).

A number of experimental fixtures have been developed, and some of these are shown in Figure 3.14 for the split Hopkinson tensile bar.

The simplest modification of the SHPB is to use a hat-shape specimen, as shown in Figure 3.14(a). A "top hat" specimen is sandwiched between the incident bar and a hollow transmission tube. The compressive stress waves in the incident bar strike the inside of the specimen hat, causing a tensile load on the specimen gauge section along the side. To increase the stress amplitude in the specimen, sometimes the specimen gauge section is not entirely solid. For example, it can be split into several arms.

In Figure 3.14(b), a specimen is threaded onto the ends of the incident bar and the transmitter bar in the test section. A rigid collar is placed over the specimen to allow the compressive wave to pass through the collar and leave the specimen virtually untouched by the initial compressive wave. The cross-sectional area of the collar is much larger than that of the specimen. Most of the compressive energy in the incident bar due to the impact of striker is transferred into the transmitter bar. When the compressive stress wave travels to the free end of the transmitter bar, it is reflected back as a tensile wave propagating towards the specimen. When this tensile wave arrives at the specimen,

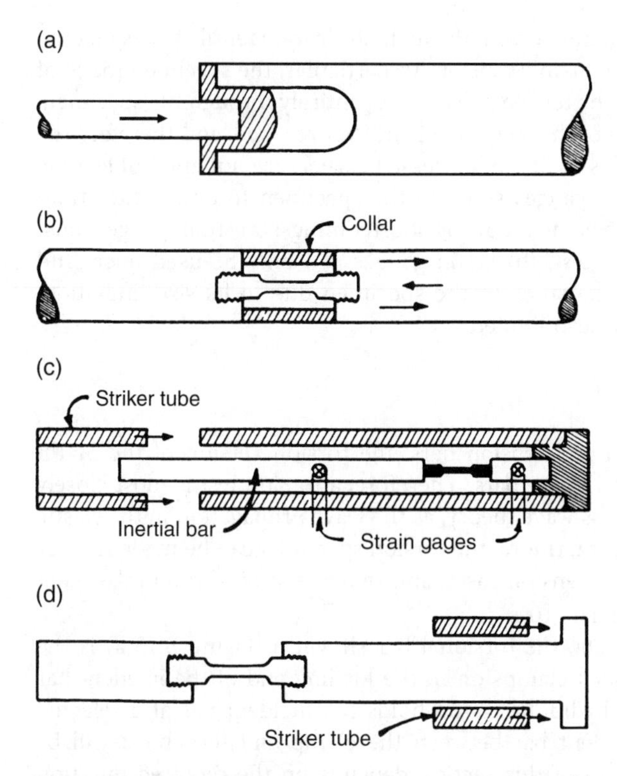

Figure 3.14 Setups for testing specimens in a Hopkinson bar in tension mode: (a) hat-shaped specimen; (b) collared specimen; (c) inertial-bar configuration; (d) collar impacting bolt head. *Source*: Meyers (1994). Reproduced with permission of Wiley.

the rigid collar cannot support the tensile wave and the specimen is subjected to a dynamic tensile pulse.

In the design of Figure 3.14(c), the inertial bar configuration is indicated. A hollow tube is adopted instead of the original projectile, see the striker tube on the left. Inside the main tube, an elastic inertial bar is connected to the specimen as a transmitter bar; the specimen is clamped to an incident bar which is connected to the yoke at the end of the main tube. The striker tube impinges the main tube, generating a compressive pulse. This compressive pulse acts on the specimen, pulling it to the right. The inertial bar counteracts this. The result is tension acting on the specimen.

The most commonly used loading method in an SHPB is direct tension. There are two main types of direction methods. One is to strike a flange at the end of the incident bar with a form of kinetic energy, as shown in Figure 3.14(d). This design requires the collar as striker to slide a long distance without friction and thus it is difficult to achieve. The other approach is to generate direction tension at the far end (from the specimen) of the incident bar by releasing the stored energy. The energy could be stored in a rotating disk, a press-stressed bar, a pendulum, or a hammer. In such cases, the released tensile wave may be difficult to control and a pulse shaper should be considered. With the proper use of the material and geometry of the pulse shaper, such as a prefixed metal bar that could be stretched to fracture, thus generating a tensile pulse in the incident bar, the shape of the incident pulse can be controlled to some extent.

One of the problems often encountered with the high-strain-rate tensile test is that the specimen may be broken during dynamic tension. In particular, the specimen made of brittle material may break before the tensile pulse passes entirely through the specimen. In this case, the equilibrium state of the specimen is not yet reached, and therefore the two-wave and three-wave formulas of SHPB, discussed earlier, are not applicable. One possible approach is to use strain gauges stuck to the specimen to obtain the strain history directly. But this method has the following shortcomings: (a) strain gauges must be calibrated before the dynamic test; (b) strain gauges can only be used once; and (c) only a few strain gauges can be stuck to the specimen due to its size limitation, and thus the strain information is also limited.

Split Hopkinson Torsion Bar

Compared with the compression and tension bars, the torsion version of the SHPB eliminates the radial inertia effects in the bars. Therefore, a torsion test is most closely described by one-dimensional stress wave theory, as the wave propagation in the elastic bars should be non-dispersive. Furthermore, the torsion specimen can be made smaller than those in the compression and tension cases, and thus the stress–strain relation at higher strain rate can be reached, i.e., $10^5 s^{-1}$.

For shear stress–shear strain data, the torsional bar shown in Figure 3.15(a) is the proper instrument. The lathe chuck clamps on to the loading end of the incident bar on the left side of the figure. A hydraulic clamp holds the incident bar at a selected location. The section of the incident bar between the clamp and the chuck will be pre-stressed in torsion. The length of this section depends on the required duration of the loading pulse. The specimen is a thin tube which is brazed to the incident and transmitter bars. During an experiment, the pre-stressed section is twisted by the chuck within the elastic range of the incident bar. The clamp is then released suddenly, allowing

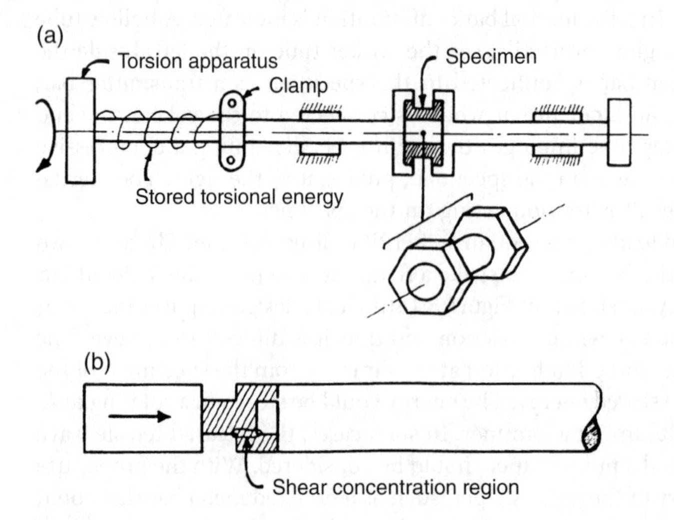

Figure 3.15 (a) Torsion Hopkinson bar for pure shear configuration; (b) hat-shaped specimen for shear stresses in compressive Hopkinson bar. *Source*: Meyers (1994). Reproduced with permission of Wiley.

the torsional strain energy to propagate towards the specimen in the form of one-dimensional shear wave.

One can also create shear stresses in the compressive Hopkinson bar by using a hat-shaped specimen, shown in Figure 3.15(b). For this special designed specimen installed on the compressive Hopkinson bar, the connection part of the specimen is under dynamic shear when the compressive stress wave passes the system.

3.3.3 Expanding-Ring Technique

The expanding ring was introduced by Johnson in 1963 as an alternative technique in high-strain-rate testing. As shown in Figure 3.16, the steel cylinder has a core, in which an explosive is detonated. A shock wave travels outward and enters the metal ring, propelling it in a trajectory with an expanding radius. By measuring the velocity history of the expanding ring using laser interferometry, one can determine the stress–strain curve of this ring at the imposed strain rate.

The mathematical derivation of the governing equations of the expanding ring is straightforward, as given below. Let us consider a small segment of a circular ring with radius r and thickness h (see Figure 3.16). During flight the ring obeys Newton's second law. The stress acting on the ends of this segment is σ. If the instantaneous cross-sectional area is A, the force is σ_A and the density of the material is ρ, we have

$$\mathbf{F} = m\mathbf{a} \tag{3.20}$$

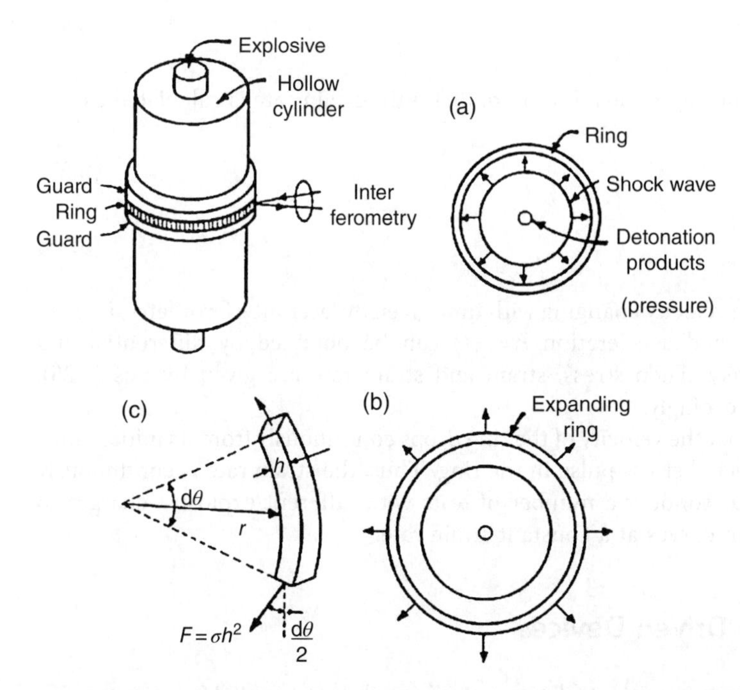

Figure 3.16 Expanding-ring technique: (a) steel block with explosive in core; (b) sections of cylinder just after detonation and during flight of ring; (c) section of ring. *Source*: Meyers (1994). Reproduced with permission of Wiley.

In the radial direction, the above vector relation gives

$$F_{\mathrm{r}} = ma_{\mathrm{r}} \tag{3.21}$$

The circumferential force on the cross-section is F, for a small segment corresponding to angle $\mathrm{d}\theta$,

$$2F\sin(\mathrm{d}\theta/2) = ma_{\mathrm{r}} \tag{3.22}$$

Hence, for small $\mathrm{d}\theta$

$$F\,\mathrm{d}\theta = ma_{\mathrm{r}} \tag{3.23}$$

Thus

$$\sigma h^2\,\mathrm{d}\theta = \rho r\,\mathrm{d}\theta\,h^2\ddot{r} \tag{3.24}$$

which leads to

$$\sigma = \rho r\ddot{r} \tag{3.25}$$

In experiments, the velocity laser interferometry gives us the velocity as a function of time, i.e., $\dot{r} = \dot{r}(t)$, while Eq. (3.25) tells us that if we know $r(t)$ and the deceleration history $\ddot{r}(t)$ of the ring as well as its density, we can determine the stress in it. For the true strain, by assuming a uniaxial strain, we have

$$\varepsilon = \ln\left(\frac{r}{r_0}\right) \tag{3.26}$$

where r_0 is the original ring radius. From Eq. (3.26) the strain rate is calculated as

$$\mathrm{d}\varepsilon = \frac{\mathrm{d}r}{r} \tag{3.27}$$

$$\dot{\varepsilon} = \frac{\mathrm{d}\varepsilon}{\mathrm{d}t} = \frac{1}{r}\frac{\mathrm{d}r}{\mathrm{d}t} = \frac{\dot{r}}{r} \tag{3.28}$$

Therefore, using the velocity changing with time given by laser interferometry, $\dot{r} = \dot{r}(t)$, displacement $r = r(t)$ and acceleration $\ddot{r} = \ddot{r}(t)$ can be obtained by differential and integration, respectively. Then stress, strain and strain rate are given by Eqs (3.25), (3.26), and (3.28), accordingly.

It should be noted that the velocity of the ring drops continuously from its initial value, as a result of the reflected stress pulse in the ring. Thus, the strain rate is continuously varying, so we have to conduct a number of tests with different explosive charges to construct stress–strain curves at a constant strain rate.

3.4 Explosively Driven Devices

Explosively driven systems are the technique requiring the least capital investment, and hence are best suited to the start-up of a shock-loading program. It is assumed that the user is familiar with the safety requirements for the use of explosives. It is obvious that

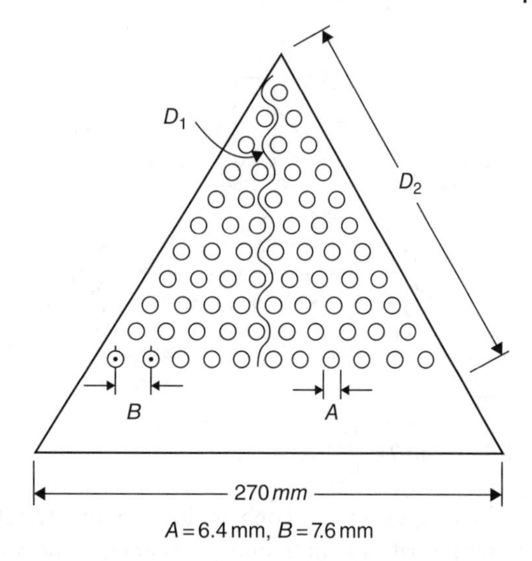

Figure 3.17 Triangular line-wave generator. *Source*: Meyers (1994). Reproduced with permission of Wiley.

great care should be exercised in the handling of explosives; this is especially true when new systems are being tested. Different experimental arrangements have been developed throughout the years to transform a point detonation produced by a detonator into a plane detonation. Several of these are reviewed here, and their design is described.

3.4.1 Line-Wave and Plane-Wave Generators

Line-Wave Generator

Among the various systems designed, the perforated triangle is the most common. As shown in Figure 3.17, the perforated triangle is initiated at one of the vertices by a cap and a small booster charge. The detonation front has to pass between the holes, and the curved trajectory D_1 is the same as the trajectory at the edges D_2. Hence, the wave front is linear. The diameter and spacing of the circles are chosen in such a way that the above conditions are met. Line-wave generators are commercially available but can be easily fabricated by making steel dies with the appropriate hole pattern; the sheet explosive placed between the dies and a punch is used to perforate the explosive.

Plane-Wave Generator

In order to transport a planar shock front to the flyer plate or to the system, or to transform a point detonation into a desired surface configuration, special experimental configurations have to be used.

Explosive Lens

Figure 3.18 shows one of the possible designs of an explosive lens. The detonator transmits the front to two explosives that have different detonation velocities. The insider explosive has a detonation velocity V_{d2} which is lower than the outside one V_{d1}. The angle θ is such that

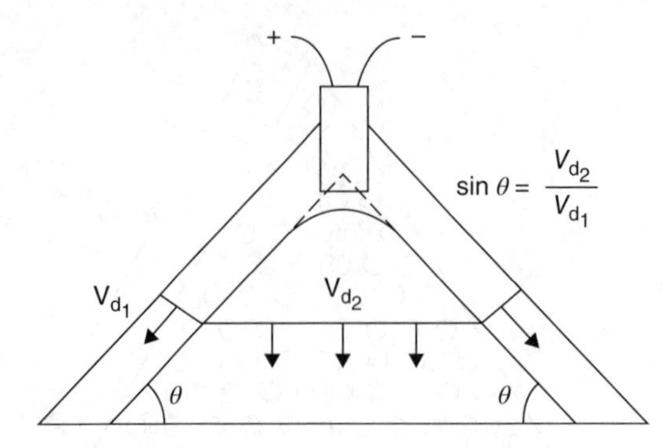

Figure 3.18 Conical explosive lens. *Source*: Meyers (1994). Reproduced with permission of Wiley.

$$\sin\theta = \frac{V_{\mathrm{d2}}}{V_{\mathrm{d1}}} \tag{3.29}$$

The apex of the cone is not exactly straight; a certain curvature is introduced to compensate for initiation phenomena (the steady detonation velocity is not achieved instantaneously) and for the fact that the initiation source is not an infinitesimal point source. Because of these and other complications, such as the requirement for precision casting of explosives, fabrication of explosive lenses is best left to specialists.

Mousetrap Plane-Wave Generator

The mousetrap assembly is frequently used in metallurgical work. Figure 3.19 shows a common setup. Two layers of explosive (2 mm thick each) are placed on top of a 3-mm-thick glass plate titled an angle α to the main charge. The detonation of the explosive will propel the glass into the main charge; all glass fragments should simultaneously hit the top surface of the main charge, resulting in plane detonation. The angle α is calculated from the velocities of detonation and fragments, V_{d} and V_{f}, respectively.

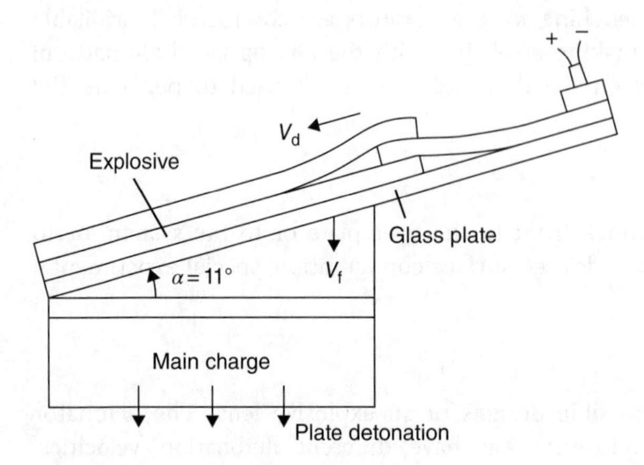

Figure 3.19 Mousetrap-type plane-wave generator. *Source*: Meyers (1994). Reproduced with permission of Wiley.

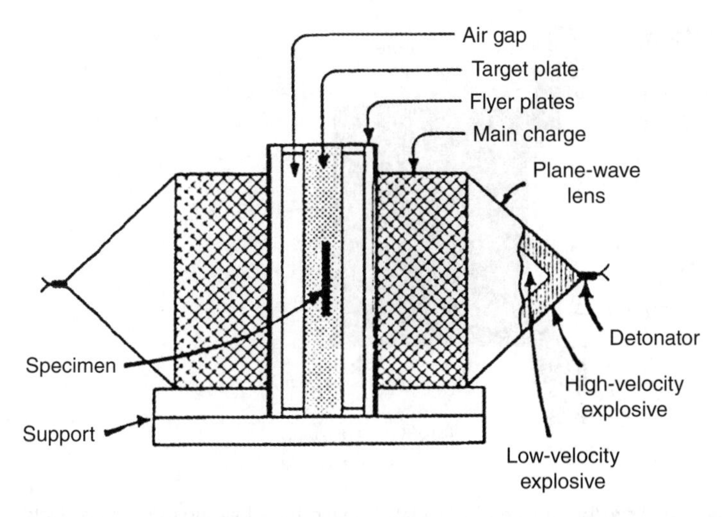

Figure 3.20 Yoshida system using shock superposition. *Source*: Meyers (1994). Reproduced with permission of Wiley.

3.4.2 Flyer Plate Accelerating

The plane-wave generators are used to simultaneously initiate the entire surface of a main charge. The detonation of this main charge, in turn, produces the energy that propels a flyer plate at velocities as high as 3 km/s for simple planar systems. An explosively driven system consisting of two plane-wave lenses simultaneously initiated and driving two flyer plates is shown in Figure 3.20. These flyer plates impact opposite sides of a main target simultaneously and generate two shock waves, which superimpose at the center, producing a corresponding pressure increase. The specimen to be studied is placed in the shock superposition region.

Figure 3.21 shows another system using a gun barrel to launch a disk-shaped flyer plate impacting on the target at high speed. Similar to the expanding-ring technique illustrated in Figure 3.16, the velocity on the backing material is measured using laser interference technology, and then the displacement, acceleration and deformation history of the target are obtained from data analysis.

In plate impact, the planar impact of a disk of material onto a target specimen (Figure 3.21) produces shock waves in both target and impactor materials. The strain rate across a shock front is given by $u_p/U_s\tau$ where u_p is the particle velocity, U_s is the shock velocity and τ is the rise time of shock. Measured values of these parameters (u_p, U_s, τ) range from 0.1 km/s, 2.6 km/s, and 50 ns for polymers to 1 km/s, 10 km/s, and 1 ns for aluminas. These values give a strain rate for materials swept by a shock wave in the range 10^6–10^8 s^{-1}. These are the highest rates of deformation that can be achieved in the laboratory by mechanical means. It should be pointed out that the deformation that takes place at these strain rates is under one-dimensional strain. This is because the inertia of the material involved in the collision acts (for a period of a few microseconds) to rigidly constrain the material in the center of the colliding disks. This state of affairs lasts until lateral release waves reach the center of the disks, i.e. for a time given by r/c_s, where r is the radius of the disc and c_s is the appropriate wave speed in the

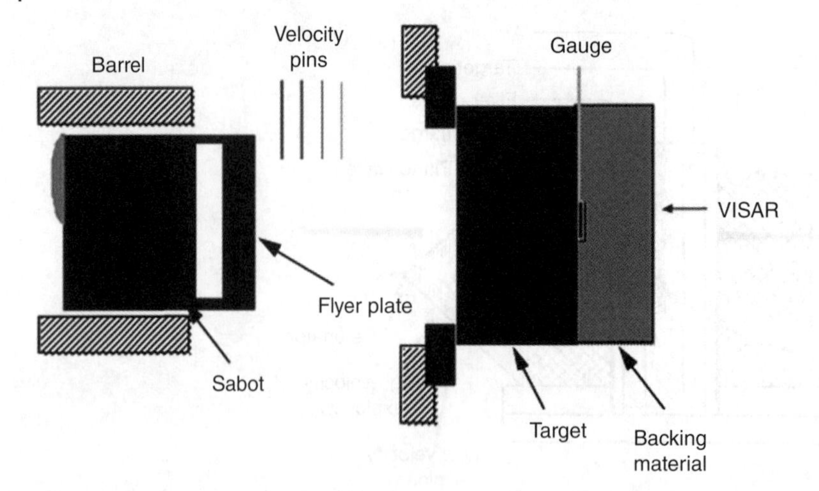

Figure 3.21 Schematic diagram of the 'business end' of a plate impact shock loading gun. *Source*: Field *et al.* (2004). Reproduced with permission of Elsevier.

shocked (and hence densified) material. Hence, the larger the diameter of the impactor/target, the longer the state of one-dimensional shock strain lasts. However, the costs of manufacture and operation of a laboratory gun increase rapidly with the bore size, so most plate impact facilities use guns with bores in the range 50–75 mm. Single-stage guns operated with compressed gas have a typical upper impact speed of around 1.2 km/s if helium is used as the propellant. Higher velocities can be achieved with single-stage guns using solid propellants, but this has the disadvantage of producing a great deal of residue which has to be cleaned out each time the gun is fired. To achieve impact speeds typical, say, of the impact of space debris on an orbiting satellite requires two- or even three-stage guns. One disadvantage is that each successive stage is of smaller diameter than the previous one. Hence, the final projectile is typically only a few millimeters in diameter. For the very highest speeds in such systems, hydrogen is used as the propellant.

3.4.3 Pressure-Shear Impact Configuration

The inclined impact of a flat plate could be achieved by the pressure-shear impact configuration shown in Figure 3.22, with the maximum strain rate of $10^5 s^{-1}$, which is higher than the strain rate of SHPB. As indicated in the figure, in the fiberglass tube with an oblique free end on its right side, an intensive compressive stress wave propagates towards the right. When this stress wave reaches the free end, the reflected tensile wave will drive a flyer to fly off and impact the specimen placed on an anvil. Before the impact, the anvil must be adjusted to be absolutely parallel to the flyer. Therefore, the specimen is subjected to the combination of compressive and shear loading due to the inclined impact. As mentioned in Section 3.1, the strain rate of a specimen is dependent on the loading velocity and also on the size of the specimen. The smaller the specimen, the higher the strain rate. In this design, the thinnest specimen could be a nanofilm, which will lead to a very high strain rate. This device is quite expensive, due to its high

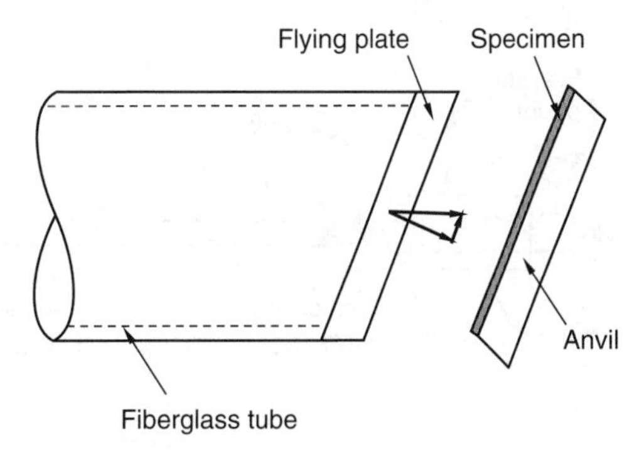

Figure 3.22 Schematic of the inclined impact of a flat plate.

accuracy and applicability to micro- and nanoscale analysis. It is usually only used for special high-grade materials.

3.5 Gun Systems

Guns have been used for many years and will continue to be used as tools to generate impacts of between 100 and 8000 m/s. The principal advantages of guns over other techniques are the reproducibility of the results, the excellent planarity and parallelism at impact, and the relative ease of use of sophisticated instrumentation and diagnostics:

1) For low-velocity impacts, one-stage gas guns with maximum velocity 1100 m/s or propellant guns with maximum velocity 1100 m/s can be chosen.
2) For velocities up to 8 km/s, one uses the two-stage gas gun.
3) Even higher velocities can be reached by means of an electromagnetic gun known commonly as a “rail gun”. Velocities of 15 km/s have been reported with this gun.
4) Super high velocity is given by a plasma accelerator, with claims of velocities of 25 km/s being reached.

3.5.1 One-Stage Gas Gun

The one-stage gas gun is very simple in design. It consists of a breech, a barrel, and a recovery chamber (on target chamber). The high-pressure gas is loaded into the high-pressure chamber and the projectile, mounted in a sabot, is placed in the barrel. A valve is released (or a diaphragm is burst) and the high-pressure gas drives the projectile. One uses light gases to reduce the inertia of gas and hence get the maximum velocity; hydrogen and helium are favorites, although air can also be used. The maximum velocity can be calculated from the maximum rate of expansion of the gas. Figure 3.23 shows a one-stage gas gun that can accelerate a projectile to a velocity of 1200 m/s using helium. It contains a fast-acting valve that is pulled back by a pressure differential and allows the high-pressure chamber to vent into the barrel. The projectile is accelerated in the barrel. Its velocity is measured by the interruption of two laser beams. The recovery chamber is used to capture the specimens with minimum post-impact damage.

A one-stage gun can be used as the driving device of the projectile in the SHPB experiment, and it can also be used to impact material specimens or structural models directly.

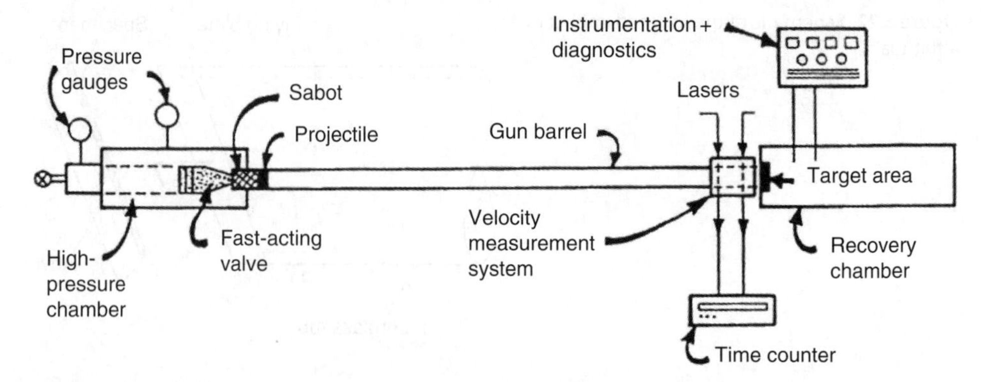

Figure 3.23 Schematic representation of a one-stage light gas gun. *Source*: Meyers (1994). Reproduced with permission of Wiley.

3.5.2 Two-Stage Gas Gun

Figure 3.24 shows a two-stage gas gun. It uses the first stage to drive a fairly massive piston. This piston is usually accelerated by means of the deflagration of a gunpowder load. In Figure 3.24 the four phases of operation of the gun are shown. The powder is placed in

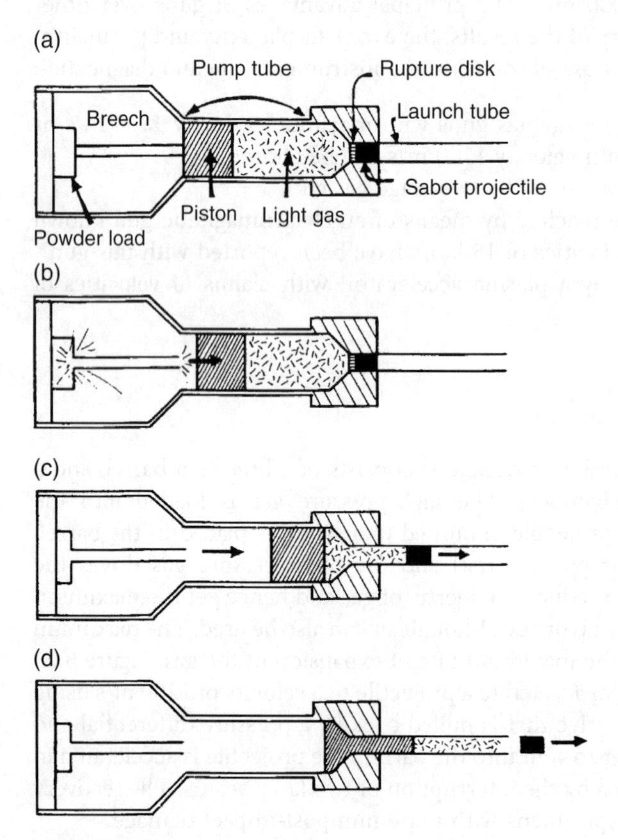

Figure 3.24 Two-stage gas gun: (a) phase 1; (b) phase 2; (c) phase 3; (d) phase 4. *Source*: Meyers (1994). Reproduced with permission of Wiley.

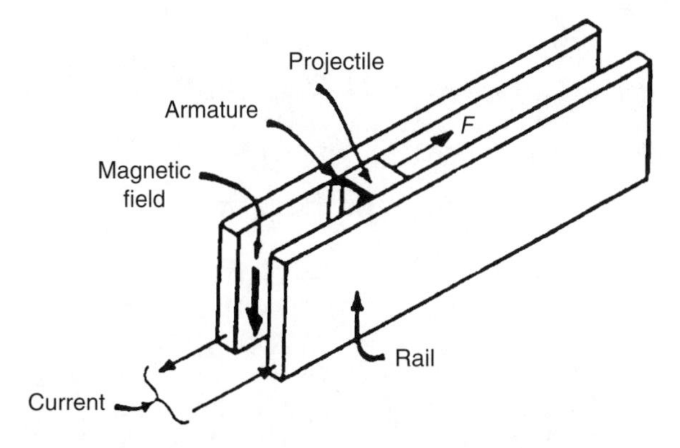

Figure 3.25 Basic operation of a direct-current electric rail gun. *Source*: Meyers (1994). Reproduced with permission of Wiley.

the breech. The first stage has a considerably larger diameter than the second stage. The first stage is filled (between projectile and piston) with a light gas (He or H). The deflagration of the gunpowder accelerates the piston, compressing the light gas. When the pressure in the first stage reaches a critical level, a diaphragm (rupture disk) opens and starts driving the projectile. The piston continues to move and to compress the gas that drives the projectile. The piston eventually partly penetrates into the launch tube. Very high pressures in the light gas are generated that can drive projectiles to velocities that routinely reach 7 km/s. In some cases, higher velocities can be achieved.

3.5.3 Electric Rail Gun

Even higher velocities can be reached with the electric rail gun. This gun uses electromagnetic forces that create a mechanical force (Lorentz force) to drive the projectile. Figure 3.25 shows the operation of a rail gun. Two parallel rails of a conductor (copper) bracket a projectile. This projectile is usually made of a very low-density material (polycarbonate) with a metallic tail (armature). It can be seen that the velocity increases with the force F, which is proportional to the square of the current. High currents are achieved by capacitor discharge.

Problems 3

3.1 A steel bar (E = 200 GPa) is 25 mm long. At one of its ends it is subjected to an impact from a large rigid body of velocity 2.5 m/s. Assuming that the deformation of the bar remains elastic, calculate the average strain rate of the bar after impact.

3.2 A rotary flywheel machine is designed for high-strain-rate testing of materials. A tensile specimen of size 6 mm × 6 mm × 40 mm is supposed to be tested at strain rate $10^3 s^{-1}$, while the flow stress will reach a level of about 800 MPa. Estimate the power required for the rotary flywheel machine.

3.3 Briefly illustrate the basic assumptions and working principles of the split Hopkinson pressure bar.

4

Constitutive Equations at High Strain Rates

In this chapter, the constitutive relations of materials, taking the strain rate effect into account, will be illustrated. The majority of the content is taken from the monograph by Meyers (1994), *Dynamic Behavior of Materials.*

4.1 Introduction to Constitutive Relations

High-strain-rate plastic deformation of materials is often described by constitutive equations that relate stress σ with strain ε, strain rate $\dot{\varepsilon}$ and temperature T. This is expressed by

$$\sigma = f(\varepsilon, \dot{\varepsilon}, T) \tag{4.1}$$

As plastic deformation is an irreversible and thus path-dependent process, the response of the material at a certain point depends not only on the current stress–strain (σ, ε) status, but also on the substructure resulting from its deformation history. Thus, we have to add the general term "deformation history" to Eq. (4.1):

$$\sigma = f(\varepsilon, \dot{\varepsilon}, T, \text{deformation history}) \tag{4.2}$$

From the theory of elasticity, we know that stress and strain are both second-order symmetric tensors, each having six independent components. In order to express the constitutive equations in scalar form rather than tensor form, the effective stress σ_{eff} and effective strain ε_{eff} can be introduced as follows:

$$\sigma_{\text{eff}} = \frac{\sqrt{2}}{2}\sqrt{(\sigma_1 - \sigma_2)^2 + (\sigma_2 - \sigma_3)^2 + (\sigma_3 - \sigma_1)^2} \tag{4.3}$$

$$\varepsilon_{\text{eff}} = \frac{\sqrt{2}}{3}\sqrt{(\varepsilon_1 - \varepsilon_2)^2 + (\varepsilon_2 - \varepsilon_3)^2 + (\varepsilon_3 - \varepsilon_1)^2} \tag{4.4}$$

Similar to Eq. (4.4), the effective strain rate $\dot{\varepsilon}_{\text{eff}}$ can also be defined. Thereby, rather than dealing with tensors, we deal only with the scalar quantities σ_{eff}, ε_{eff} and $\dot{\varepsilon}_{\text{eff}}$. Alternatively, we may use the effective shear stress τ_{eff}, the effective shear strain γ_{eff} and the effective shear strain rate to construct constitutive equations. In fact, shear

Introduction to Impact Dynamics, First Edition. T.X. Yu and XinMing Qiu.

Published 2018 by John Wiley & Sons Singapore Pte. Ltd.

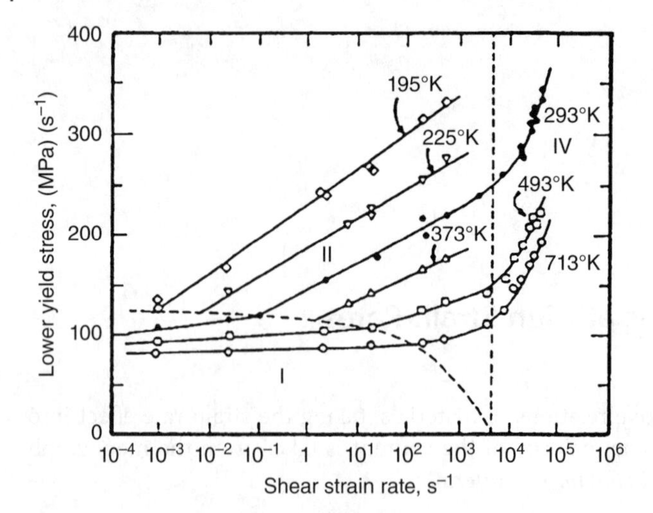

Figure 4.1 Lower yield stress versus strain rate of mild steel. *Source*: Meyers (1994). Reproduced with permission of Wiley.

stresses are dominant components in plastic deformation, i.e., metals and polymers deform plastically by shear.

Figure 4.1 shows the classic plot by Campbell and Ferguson (1970). For mild steel, the lower yield stress against the logarithm of the strain rate is given at different temperatures. Clearly, the yield stress decreases with the increasing temperature. Furthermore, for strain rate $< 10^3\ \mathrm{s}^{-1}$, yield stress changes with strain rate rather slowly, whereas for strain rate $> 10^3\ \mathrm{s}^{-1}$, the yield stress changes rapidly. Two observations can be made from Figure 4.1:

- The yield stress increases with strain rate.
- The increase of yield stress is more remarkable at lower temperatures.

Decreasing the temperature and increasing the strain rate have a similar effect to increasing the yield stress and reducing the ductility of the material. Inspired by this feature, the test of creep can usually be accelerated by increasing temperature.

A successful constitutive equation should be capable of representing all the data displayed in this figure and being interpolated or extrapolated.

4.2 Empirical Constitutive Equations

There are a number of equations that have been proposed and successfully used to describe the plastic behavior of materials as a function of strain, strain rate, and temperature. At low and constant strain rates, most metals are known to approximately exhibit so-called power hardening:

$$\sigma = \sigma_0 + k\varepsilon^n \tag{4.5}$$

where σ_0 is the yield stress, n is the strain-hardening coefficient, and k is the pre-exponential factor. If $n = 0$, Eq. (4.5) represents a rigid, perfectly plastic material; $n = 1$ refers to a linear strain-hardening material. For most metals, the value of n is in the range of 0.2–0.3. Note that although the equation is one-dimensional, it can be used for a complex loading situation by using the effective stress and effective strain given in Eqs (4.3) and (4.4).

From the data for mild steel shown in Figure 4.1, the effect of temperature on the flow stress can be represented by

$$\sigma = \sigma_r \left[1 - \left(\frac{T - T_r}{T_m - T_r} \right)^m \right] \tag{4.6}$$

where, T_m is the melting point; T_r is a reference temperature, at which a reference stress σ_r is measured; and T is the temperature at which σ is calculated. Eq. (4.6) is a simple curve-fitting equation, where m is obtained by fitting the experimental data.

The effect of strain rate can be simply approximated by

$$\sigma \propto \ln \dot{\varepsilon} \tag{4.7}$$

which is observed very often at strain rates that are not too high, i.e., $\dot{\varepsilon} \leq 10^2\ \text{s}^{-1}$ (see data in Figure 4.2).

Johnson and Cook (1983) combined these ingredients and proposed the following equation:

$$\sigma = (\sigma_0 + B\varepsilon^n) \left(1 + c \ln \frac{\dot{\varepsilon}}{\dot{\varepsilon}_0} \right) [1 - T^{*m}] \tag{4.8}$$

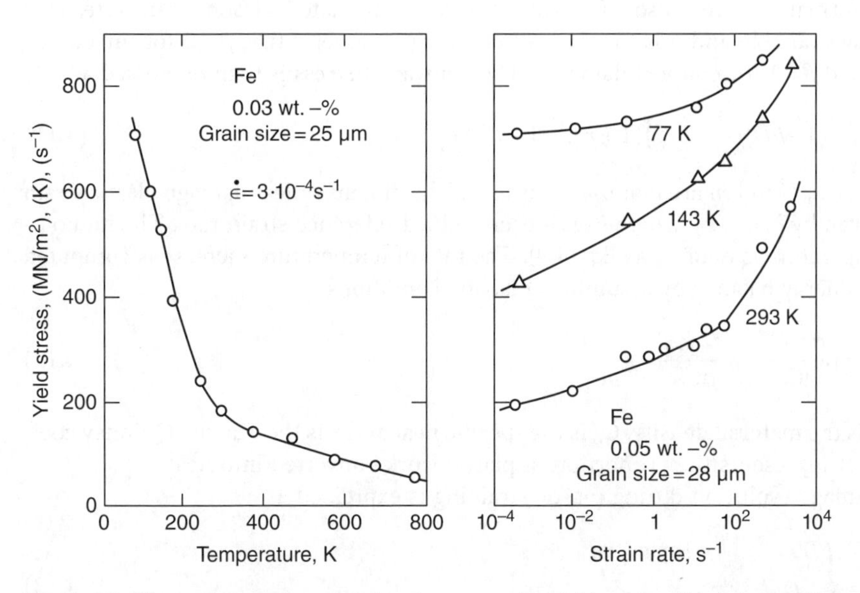

Figure 4.2 Effects of temperature and strain rate on the yield stress of iron. *Source*: Meyers (1994). Reproduced with permission of Wiley.

where T^* denotes a normalized temperature,

$$T^* = \frac{T - T_r}{T_m - T_r} \tag{4.9}$$

T_r is the reference temperature at which σ_0 is measured, and $\dot{\varepsilon}_0$ is the reference strain rate (e.g. taken to be $1\,s^{-1}$).

The Johnson–Cook equation (Eq. 4.8) contains five parameters (σ_0, B, C, n, m), all of which can be determined by experiments. The Johnson–Cook equation is a highly useful and successful constitutive model. One of the problems with this equation is that all parameters are coupled as they are multiplied by each other.

In addition to the effects of large plastic strains, high strain rates, and softening due to adiabatic heating, the damage softening is another important effect in the problem of ballistic impacts on ductile materials. The damage softening may either be uncoupled or coupled with the constitutive equation. In the uncoupled approach, the yield condition, the plastic flow, and the strain hardening are assumed to be unaffected by the damage evolution (i.e., the nucleation and growth of voids in ductile materials). By contrast, damage affects the plastic deformation and may lead to softening in the final stage before fracture in the coupled approach (Børvik *et al.*, 2009). A modified version of the well-known Johnson–Cook constitutive relation (or the MJC model, in short) was given in Børvik *et al* (2001). By assuming an isotropic (von Mises) material, and by adopting the effective stress concept and the principle of strain equivalence, the equivalent stress σ_{eq} and the accumulated plastic strain rate $\dot{\varepsilon}_{eq}$ are defined as

$$\sigma_{eq} = \sqrt{\frac{2}{3}\boldsymbol{\sigma}' : \boldsymbol{\sigma}'} = (1 - \beta D)\tilde{\sigma}_{eq}, \quad \dot{\varepsilon}_{eq} = \sqrt{\frac{2}{3}\boldsymbol{d}^{p} : \boldsymbol{d}^{p}} = \frac{\dot{r}}{(1 - \beta D)} \tag{4.10}$$

where $\boldsymbol{\sigma}'$ is the deviatoric stress tensor, $\tilde{\sigma}_{eq}$ is the damage equivalent stress, $\boldsymbol{d}^p$ is the plasticdeformation rate tensor, $\dot{r}$ is the damage accumulated plastic strain rate, D is the damage variable and β is the damage coupling parameter (i.e., $\beta = 0$ for uncoupled damage and $\beta = 1$ for coupled damage). The equivalent stress is then expressed as

$$\sigma_{eq} = (1 - \beta D)(A + Br^n)(1 + \dot{r}^*)^C(1 - T^{*m}) \tag{4.11}$$

where A, B, n, C and m are material constants. The dimensionless damage plastic strain rate is given by $\dot{r}^* = \dot{r}/\dot{\varepsilon}_0$, where $\dot{\varepsilon}_0$ is a user-defined reference strain rate. The homologous temperature is defined as Eq. (4.9). The rate of temperature increase is computed from the energy balance by assuming adiabatic conditions

$$\dot{T} = \chi\frac{\boldsymbol{\sigma} : \boldsymbol{d}^p}{\rho C_p} = \chi\frac{\sigma_{eq}\dot{\varepsilon}_{eq}}{\rho C_p} = \chi\frac{\tilde{\sigma}_{eq}\dot{r}}{\rho C_p} \tag{4.12}$$

where ρ is the material density, C_p is the specific heat and χ is the Taylor–Quinney coefficient that represents the proportion of plastic work converted into heat.

The damage evolution during plastic straining is expressed as

$$\dot{D} = \begin{cases} 0 & \text{for } \varepsilon_{eq} < \varepsilon_d \\ \dfrac{D_C}{\varepsilon_f - \varepsilon_d}\dot{\varepsilon}_{eq} & \text{for } \varepsilon_{eq} > \varepsilon_d \end{cases} \tag{4.13}$$

where D_C is the critical damage and ε_d is a damage threshold. The fracture strain is given as

$$\varepsilon_f = (D_1 + D_2 \exp(D_3\sigma^*))\left(1 + \dot{\varepsilon}^*_{eq}\right)^{D_4}(1 + D_5 T^*) \quad (4.14)$$

where $D_1, \cdots, D_5$ are material constants determined from material tests, $\sigma^* = \sigma_H / \sigma_{eq}$ is the stress triaxiality ratio where σ_H is the hydrostatic stress, and $\dot{\varepsilon}^*_{eq} = \dot{\varepsilon}_{eq}/\dot{\varepsilon}_0$ is the dimensionless strain rate. Fracture occurs by element erosion (i.e., the stresses in the integration points are set to zero) when the damage of a material element is equal to the critical damage D_C (≤ 1). Note that if the damage coupling parameter β in Eq. (4.10) is set to zero, $\tilde{\sigma}_{eq} \rightarrow \sigma_{eq}$, $\dot{r} \rightarrow \dot{\varepsilon}_{eq}$ and $D_C \rightarrow 1$, and the uncoupled damage formulation (as used in the original Johnson–Cook model) reappears.

Alternatively, for uncoupled damage (i.e., $\beta = 0$ in Eq. 4.11) failure can be modeled using a fracture criterion proposed by Cockcroft and Latham (CL), where it is assumed that fracture depends on the stresses imposed as well as on the strains developed by Deya *et al.* (2007). The model can be expressed as

$$D = \frac{W}{W_{cr}} = \frac{1}{W_{cr}} \int_0^{\varepsilon_f} \langle \sigma_1 \rangle d\varepsilon_{eq} \quad (4.15)$$

where W is the CL integral, and σ_1 is the major principal stress, $\langle \sigma_1 \rangle = \sigma_1$ when $\sigma_1 \ge 0$ and $\langle \sigma_1 \rangle = 0$ when $\sigma_1 < 0$. It is seen that fracture cannot occur in this model when there is no tensile stress operating, which implies that the effect of stress triaxiality on the failure strain is implicitly taken into account. The advantage with the CL failure criterion is that the critical value W_{cr} can be determined from one uniaxial tensile test. Moreover, the model captures some main experimental observation for many steels exposed to impact.

Other empirical equations have been developed by different research groups around the world. They are reviewed by Meyer (1992). For instance, Klopp (1985) used the following equation:

$$\tau = \tau_0 \gamma^n T^{-\nu} \dot{\gamma}_p^m \quad (4.16)$$

where τ and γ denote the shear stress and shear strain, respectively, ν is a temperature-softening parameter, and n and m represent strain hardening and strain rate sensitivity, respectively.

Some other investigators favor the equation

$$\tau = \tau_0 \left(1 + \frac{\gamma}{\gamma_0}\right)^n \left(\frac{\dot{\gamma}}{\dot{\gamma}_r}\right)^m e^{-\lambda \Delta T} \quad (4.17)$$

where τ_0 is the yield stress of the material ($\gamma = 0$) at the reference strain rate $\dot{\gamma}_r$, and ΔT is the change in temperature from the reference value to current value ($T - T_0$). This equation possesses an exponential thermal softening. This expression is often expressed explicitly by $\dot{\gamma}$. By making

$$\sigma_r = \tau_0 e^{-\lambda \Delta T} \left(1 + \frac{\gamma}{\gamma_0}\right)^n \quad (4.18)$$

the simplified form of Eq. (4.17) is

$$\dot{\gamma} = \dot{\gamma}_r \left(\frac{\tau}{\sigma_r}\right)^{1/m} \quad (4.19)$$

Two additional examples are given below:

$$\tau = \tau_0 \gamma^n \left(\frac{\dot{\gamma}}{\dot{\gamma}_0}\right)^m \exp\left(\frac{W}{T}\right) \tag{4.20}$$

where τ_0, W, n, and m are the parameters that have to be determined experimentally, and τ, γ, and $\dot{\gamma}$ are shear stress, strain, and strain rate, respectively. Equation (4.20) was proposed by Vinh *et al.* (1979) and is essentially identical to Eq. (4.17). Campbell *et al.* (1977) used the following equation successfully for copper:

$$\tau = A\gamma^n[1 + m\ln(1 + \dot{\gamma}/B)] \tag{4.21}$$

In structural impact problems where the rigid, perfectly plastic idealization is adopted, a widely employed constitutive equation for rate-dependent materials is the Cowper–Symonds equation:

$$\dot{\varepsilon} = D\left(\frac{\sigma_0^d}{\sigma_0} - 1\right)^q \tag{4.22}$$

which can be written in the following equivalent form:

$$\frac{\sigma_0^{\mathrm{d}}}{\sigma_0} = 1 + \left(\frac{\dot{\varepsilon}}{D}\right)^{1/q} \tag{4.23}$$

where σ_0^{d} is the dynamic flow stress at a uniaxial strain rate $\dot{\varepsilon}$, σ_0 is the associated static flow stress, and D and q are material constants. Clearly, the Cowper–Symonds equation (Eq. 4.23) provides a relation between flow stress and strain rate, without consideration of temperature. Based on experimental results, $D = 40\ \mathrm{s}^{-1}$ and $q = 5$ for mild steel; and $D = 6500\ \mathrm{s}^{-1}$ and $q = 4$ for aluminum alloy. For example, if the strain rate is $\dot{\varepsilon} = 100\ \mathrm{s}^{-1}$, the dynamic flow stress σ_0^{d} calculated by Eq. (4.23) is $\sigma_0^{\mathrm{d}}/\sigma_0 = 2.20$ for mild steel and $\sigma_0^{\mathrm{d}}/\sigma_0 = 1.35$ for aluminum alloy. Obviously, the mild steel is more sensitive to strain rate than is the aluminum alloy.

Thus, the Cowper–Symonds equation is capable of estimating the dynamic flow stress σ_0^{d} at given strain rate, so σ_0^{d} could be adopted to replace the static flow stress σ_0 in analysis and calculation. This procedure is simple and practical, and also easy to be incorporated into finite element simulations. Because of this, its use is popular among engineers.

4.3 Relationship between Dislocation Velocity and Applied Stress

The deformation mechanism of metal plasticity at the micro-scale, i.e., the motion of dislocations varying with the strain rate and temperature at the macro-scale, is discussed briefly in this section.

4.3.1 Dislocation Dynamics

Plastic deformation of materials may be related to various physical mechanisms, depending on the material, loading condition and temperature, etc., e.g., dislocation motion or dislocation glide, mechanical twinning, and phase transformation. We will only consider dislocation movement as the agent for plastic deformation in this section.

In materials science, a dislocation is a crystallographic defect, or irregularity, within a crystal structure. The presence of dislocations strongly influences many of the properties of materials. Some types of dislocation can be visualized as being caused by the termination of a plane of atoms in the middle of a crystal. It suffices here to recall that dislocations produce shear strain by movement and that they move under the action of shear stresses.

There are two primary types: edge dislocations and screw dislocations. Mixed dislocations are intermediate between these. Figure 4.3 shows an edge dislocation moving under the action of a shear stress τ. The force on the dislocation per unit length, under this ideal orientation arrangement, is given by

$$F = \tau b \tag{4.24}$$

where b is the Burger vector, which describes the magnitude and direction of distortion to the lattice. In an edge dislocation, the Burger vector is perpendicular to the line direction. There are frictional forces resisting the movement of a dislocation; thus a force is required to make it move. As shown in Figure 4.3(a), a single edge dislocation will reach the new equilibrium position if it moves the distance of a Burger vector. The movement of arrays of dislocations will produce a shear strain γ, as shown in Figure 4.3(b). Here it is assumed that the dislocations do not interact. This shear strain can be directly related to the number of dislocations, N, per unit area (or their density ρ) by

$$\gamma = \tan\theta = \frac{Nb}{l} = \frac{Nbl}{l^2} \tag{4.25}$$

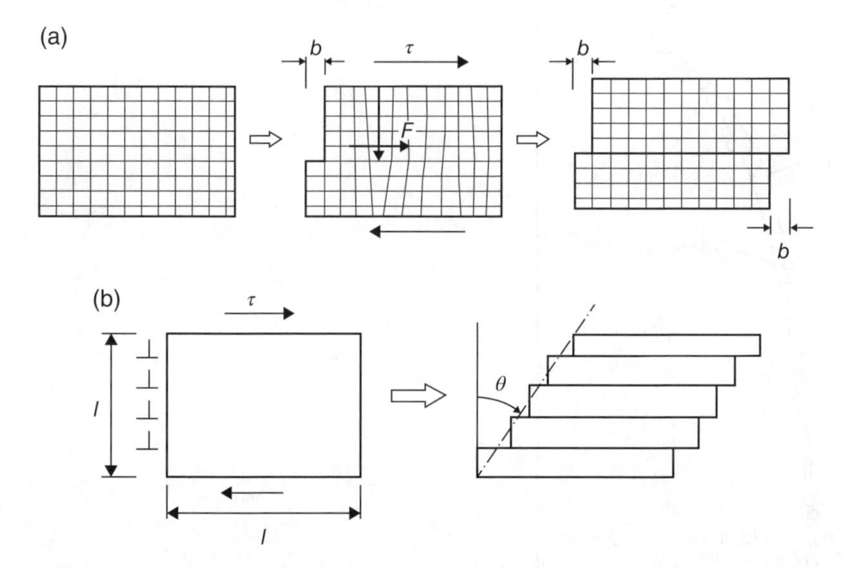

Figure 4.3 Dislocations in metal crystal. (a) Movement of one dislocation causing displacement b; (b) movement of an array of dislocations causing shear strain $\gamma = \tan\theta$. *Source*: Meyers (1994). Reproduced with permission of Wiley.

But $\rho = N/l^2$, so we have

$$\gamma = \rho b l \tag{4.26}$$

The shear strain rate is given by,

$$\dot{\gamma} = \rho b v \tag{4.27}$$

where v is the dislocation velocity. Here Eq. (4.27) gives a relation between the shear strain rate and the dislocation velocity. It should be pointed out that, by considering many slip systems with various orientations in a three dimensional space, the general relationship between strain rate and dislocation velocity should take a vector form. The shear strain can be converted into a longitudinal strain by adding an orientation factor M (Meyers & Chawla, 1984). This orientation factor is often taken as 3.1 for face-centered cubic (FCC) crystals and 2.75 for body-centered cubic (BCC) crystals:

$$\frac{d\varepsilon}{dt} = \dot{\varepsilon} = \frac{1}{M}\rho b v \tag{4.28}$$

The effect of applied stress on dislocation velocity was first studied by Gilman and Johnston (1957) in their classic experiments, in which the velocities of dislocation in lithium fluoride (LiF) single crystals as a function of stress were measured, as shown in Figure 4.4. Their work was then followed by many others, who measured the velocities of dislocations for a number of materials.

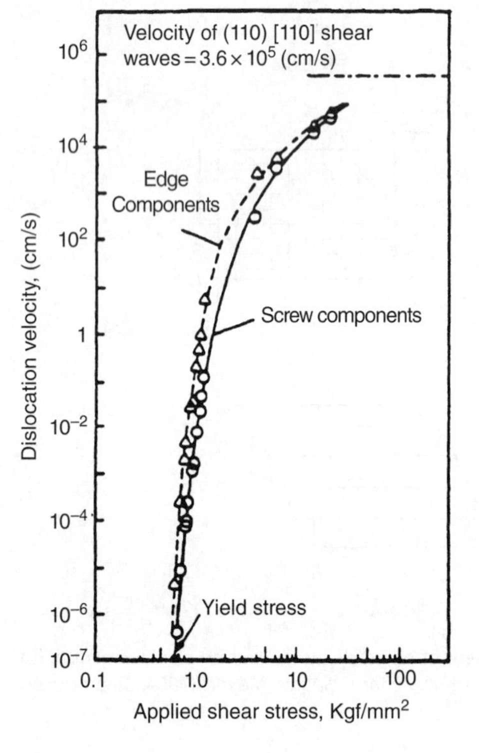

Figure 4.4 Shear stress changing with dislocation velocity. *Source*: Meyers (1994). Reproduced with permission of Wiley.

The log-log coordinate is taken in Figure 4.4. The curves indicate that the dislocations stand still if the applied shear stress is lower than the yield stress of the material; and the dislocation velocity gradually increases with the increasing applied shear stress, if the latter is higher than the yield stress.

These experimental results were then fitted into different types of equations, some empirical, some with theoretical backing; e.g., Johnston and Gilman (1959) expressed the dislocation velocity as

$$v \propto \sigma^m e^{-E/kT} \tag{4.29}$$

where m varies between 15 and 25 when velocity v is expressed in cm/s and stress σ is in kgf/mm^2. At a constant temperature, Eq. (4.29) becomes

$$v = K\sigma^m \tag{4.30}$$

Stein and Low (1960) found that their data fitted more closely the expression

$$v = A\exp\left(-\frac{A}{\tau}\right) \tag{4.31}$$

where τ is the resolved shear stress.

For copper, Greenman et al. (1967) found the following relationship:

$$v = v_0\left(\frac{\tau}{\tau_0}\right)^m \tag{4.32}$$

where $m = 0.7$, v_0 is unit velocity, and τ_0 is a material constant, $\tau_0 = 0.25 \times 10^3$ Pa. But the data could also satisfactorily fit the equation with $m = 1$ and $\tau_0 = 2.7 \times 10^3$ Pa; in this case the dislocation velocity and shear stress have a simple linear relation:

$$v = K\tau \tag{4.33}$$

where K is a constant.

Some other researchers have also obtained a linear relationship between dislocation velocity v and applied shear stress τ for various materials. The approximation in Eq. (4.33) is valid at moderate stresses.

It is very confusing to have so many experimental data and empirical equations, which relate velocity of dislocation to shear stress. A good way to summarize them is to plot all the data in a log-log diagram, as shown in Figure 4.5. Thus, the general tendency is that the slope decreases as the velocity increases.

The schematic representation of the stress–velocity behavior of nickel, shown in Figure 4.6, indicates that there may exist three regimes (I, II and III). For the coefficients of Eq. (4.30), $m_{\mathrm{I}} > m_{\mathrm{II}} > m_{\mathrm{III}}$ and $m_{\mathrm{II}} = 1$. The limiting velocity of the dislocation is set as the propagation speed of an elastic shear wave. Three regimes of behavior, shown in Figure 4.6, are generally accepted as defining three different mechanisms governing plastic deformation. If a diagram with the logarithm of the dislocation velocity as the abscissa and the applied stress as the ordinate is drawn, one can obtain a plot of the same general shape as Figure 4.1. In the next three subsections, the three governing mechanisms will be reviewed. They are thermally active dislocation motion in region I, $m_{\mathrm{I}} > 1$; phonon drag in region II, $m_{\mathrm{II}} = 1$; and relativistic effects in region III, $m_{\mathrm{III}} < 1$, respectively.

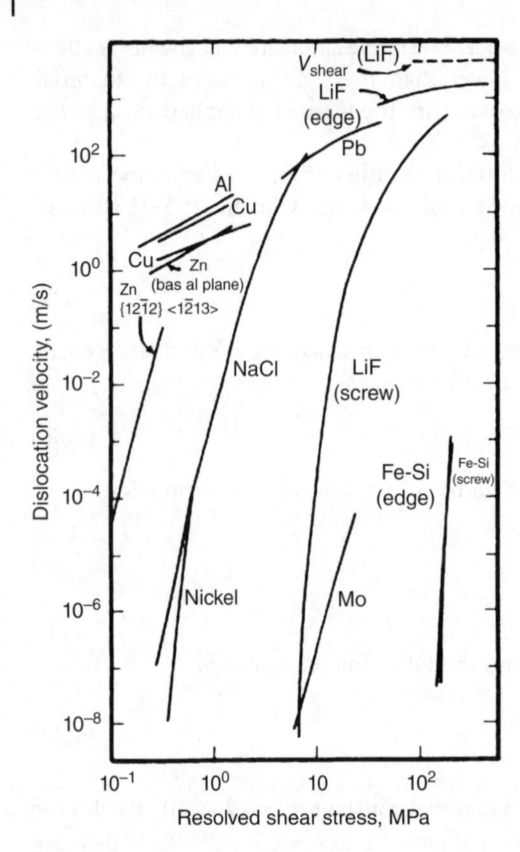

Figure 4.5 Compilation of results in the literature relating dislocation velocities and stress. *Source*: Meyers (1994). Reproduced with permission of Wiley.

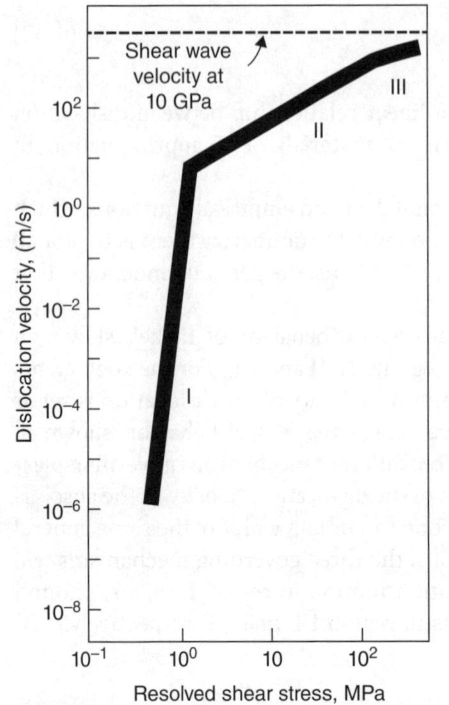

Figure 4.6 Schematic representation of the stress–velocity behavior of nickel. *Source*: Meyers (1994). Reproduced with permission of Wiley.

4.3.2 Thermally Activated Dislocation Motion

A dislocation continuously encounters obstacles as it moves through the lattice. Some of the typical obstacles are shown in Figure 4.7, including solute atoms (interstitial and substitutional), vacancies, small-angle grain boundaries, vacancy clusters, inclusions, precipitates, and so on. Other dislocations can also block dislocation motion. Various types of obstacle make the movement of dislocations more difficult.

At the atomic level, when a dislocation moves from one equilibrium atomic position to the next, it has to overcome an energy barrier termed the Peierls–Nabarro barrier (see Figure 4.8), i.e., a force has to be applied to it. The stress required to move the

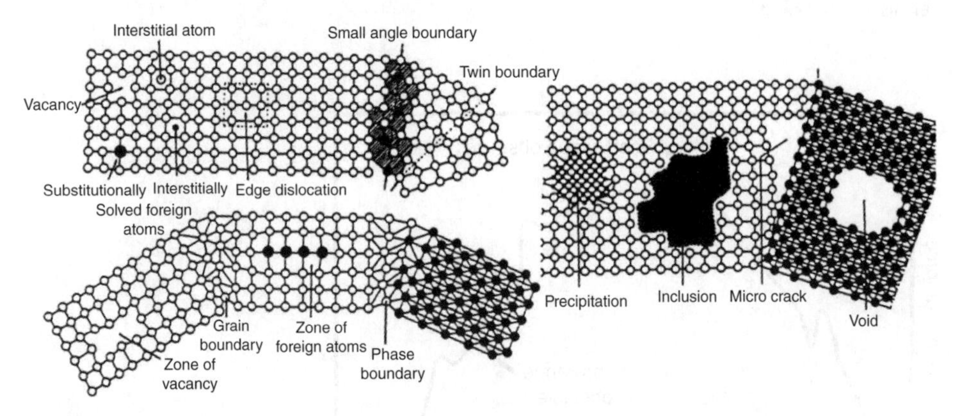

Figure 4.7 Different defects in crystalline materials. *Source*: Meyers (1994). Reproduced with permission of Wiley.

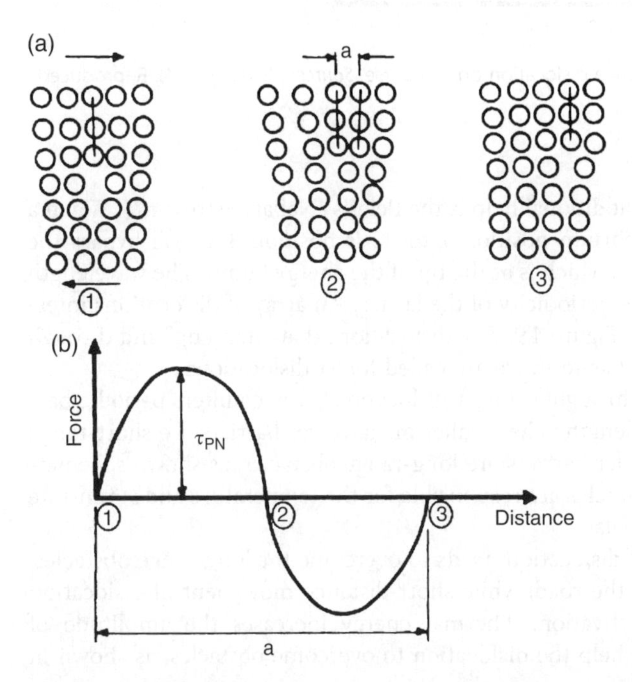

Figure 4.8 Peierls–Nabarro force. (a) Movement of dislocation from one equilibrium position to next; (b) applied stress versus distance. *Source*: Meyers (1994). Reproduced with permission of Wiley.

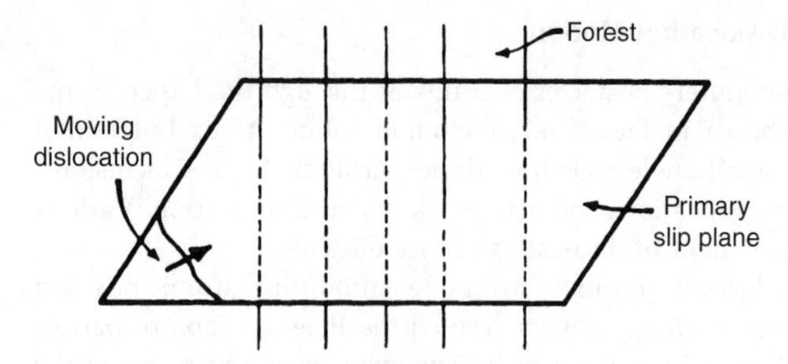

Figure 4.9 Dislocation cutting through a dislocation forest. *Source*: Meyers (1994). Reproduced with permission of Wiley.

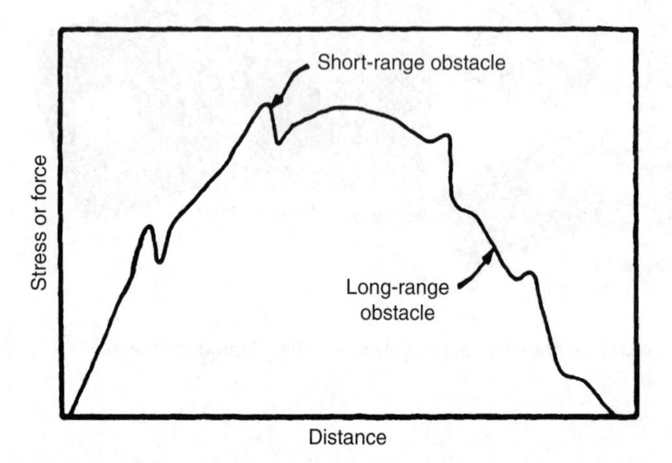

Figure 4.10 Barriers encountered by a dislocation on its course. *Source*: Meyers (1994). Reproduced with permission of Wiley.

dislocation without any other additional help is the Peierls–Nabarro stress τ_{PN}. When a dislocation moves from equilibrium position 1 towards position 3, it will bypass the unstable equilibrium position 2, which is at the tip of the energy hump. The wavelength of these barriers is equal to the periodicity of the lattice. An array of dislocations intersecting a slip plane is shown in Figure 4.9. The dislocations that "stand up" and through which the moving dislocation has to move are called forest dislocations.

When a dislocation moves through a "forest dislocation", it encounters periodic barriers of different spacing and length. The smaller or narrower barriers are short-range obstacles, and the larger or wider barriers are long-range obstacles, as shown schematically in Figure 4.10. These obstacles are responsible for the temperature and strain-rate sensitivity of crystalline materials.

Long-distance movement of dislocation needs to overcome the long-range obstacles, such as other dislocations on the road, while short-distance movement of dislocation can be driven by thermal activation. Thermal energy increases the amplitude of vibration of atoms, and it can help the dislocation to overcome obstacles, as shown in

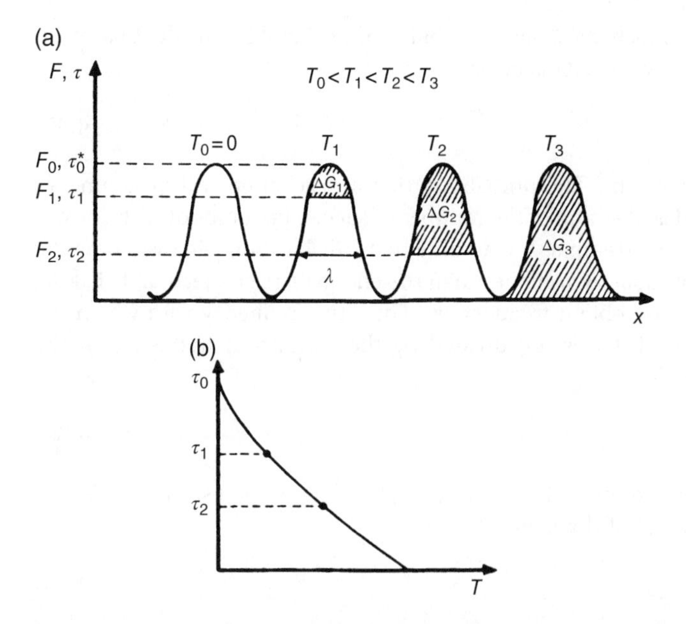

Figure 4.11 (a) Overcoming of barriers by thermal energy; (b) stress required to overcome an obstacle as a function of temperature. *Source*: Meyers (1994). Reproduced with permission of Wiley.

Figure 4.11. The barrier is shown at four temperatures, T_0, T_1, T_2, T_3, where $T_0 = 0$ and $T_0 < T_1 < T_2 < T_3$. The thermal energies $\Delta G_1, \Delta G_2, \Delta G_3$ have been shown by hatching. The area under a force–distance curve is the energy term. The effect of the thermal energy is to decrease the height of the barrier by successively increasing amounts as the temperature increases. Figure 4.11(b) shows the stress required to move the dislocation past that specific obstacle as a function of temperature. In fact, an atom is always in the state of random motion. When temperature rises, the atom will be more active and thus its ability to overcome the barrier is increased.

On the other hand, as the strain rate is increased, there is less time available to overcome the barrier and the thermal energy will be less effective. Thus, larger shear stress is required to move dislocations at a higher strain rate.

It should be noted that the obstacles can be classified into short-range (thermal activated) and long-range obstacles that cannot be overcome by thermal energy (non-thermally activated). Therefore, the flow stress of a material is expressed as

$$\sigma = \sigma_G(\text{structure}) + \sigma^*(T, \dot{\varepsilon}, \text{structure}) \tag{4.34}$$

The term σ_G is due to the athermal barriers (long range) determined by the structure of the materials. The term σ^* is due to thermally activated barriers (i.e., the barriers that can be overcome by thermal energy), which are short-range ones, typically, the Peierls–Nabarro stress, which is particularly important for BCC metals. In ceramics, the strong ionic and covalent bonds as well as their high directionality (bond angles fixed by electronic structure) are responsible for extremely high Peierls–Nabarro stresses. As a result, ceramics fail by alternative mechanism, e.g., crack nucleation and growth.

According to statistical mechanics, the possibility of an equilibrium fluctuation in energy greater than a given value ΔG is given by

$$p_{\mathrm{B}} = \exp\left(-\frac{\Delta G}{\kappa T}\right) \tag{4.35}$$

where κ is Boltzmann's constant. The probability that a dislocation will overcome an obstacle can be considered as the ratio of the number of successful jumps over the obstacle divided by the number of attempts. We assume that a dislocation will overcome the obstacle if it has an energy equal to or higher than the energy barrier ahead of it. Taking these values per unit time, we obtain frequencies. Thus, the frequency with which the dislocation overcomes the obstacles, ν_1, divided by the vibrational frequency of the dislocation ν_0, is equal to p_{B}:

$$\nu_1 = \nu_0 \exp\left(-\frac{\Delta G}{\kappa T}\right) \tag{4.36}$$

Frequency is proportional to the reciprocal of time; the definition of strain rate is also proportional to the reciprocal of time, and thus

$$\dot{\varepsilon} = \dot{\varepsilon}_0 \exp\left(-\frac{\Delta G}{\kappa T}\right) \tag{4.37}$$

where

$$\dot{\varepsilon}_0 = \nu_0 \rho b \frac{\Delta l}{M} \tag{4.38}$$

where Δl is the distance between dislocation barriers, M is the orientation factor in Eq. (4.28), ρ is the density of dislocation and b is the Burger vector.

Expressing Eq. (4.37) explicitly in terms of ΔG, it can be rewritten as

$$\Delta G = \kappa T \ln\left(\frac{\dot{\varepsilon}_0}{\dot{\varepsilon}}\right) \tag{4.39}$$

This is a very important expression, indicating that the thermally activated energy ΔG increases with T, as shown schematically in Figure 4.11, and decreases with increasing strain rate $\dot{\varepsilon}$. In Figure 4.12, we plot the combined effects of strain rate and temperature

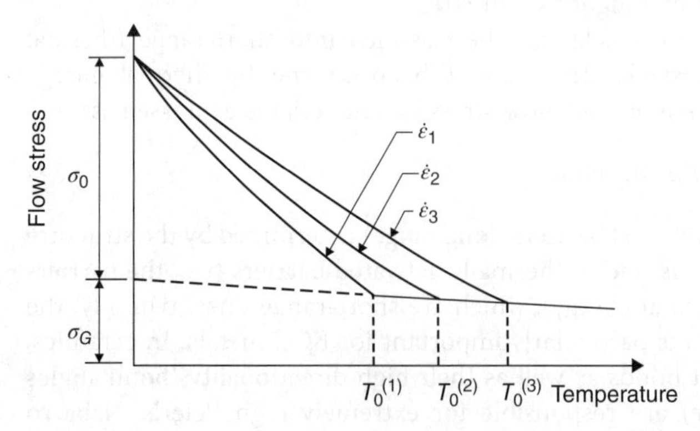

Figure 4.12 Flow stress as a function of temperature for different strain rates; thermal and athermal components of stress indicated. *Source*: Meyers (1994). Reproduced with permission of Wiley.

on the flow stress of a metal obeying a relationship of the type described above. The thermal and athermal components of flow stress are marked.

4.3.3 Dislocation Drag Mechanisms

In regime II of Figure 4.6, we observe that the exponent in Eq. (4.30) is $m = 1$, i.e.

$$v = K\tau \quad \text{or} \quad v = K\sigma \tag{4.40}$$

Hence, we have to explain why the velocity (rather than acceleration) of dislocations is proportional to the applied stress.

It is known that dislocation motion causes a temperature increase. It is also known that only a small part (typically 5–20%) of the input energy is spent in deforming a material (relevant to changing its substructure), and that most of the input energy is dissipated by forces opposing the applied stress. To incorporate these observations into a model, as a first approximation, the solid can be assumed to act as Newtonian viscous material with respect to the dislocation:

$$f_v = Bv \tag{4.41}$$

where B is the viscous damping coefficient, which is independent of v for a Newtonian fluid. Under the application of a certain external stress, the dislocation will accelerate until it reaches a steady-state velocity. Using the force on the dislocation per unit length Eq. (4.24)

$$\tau b = Bv \tag{4.42}$$

and applying Eq. (4.28), one obtains

$$\tau b = \frac{BM\dot{\varepsilon}}{\rho b} \tag{4.43}$$

Making $\tau = \sigma/2$, we get

$$\sigma = \frac{2BM\dot{\varepsilon}}{\rho b^2} \tag{4.44}$$

Thus, the flow stress should be proportional to the strain rate if dislocation drag mechanisms are operative.

Many theoretical analyses and experiments have been conducted to study the drag mechanisms, e.g., on the interaction of the dislocation with thermal vibration (phonon drag – a phonon is an elastic vibration propagating through the crystal) and with electrons (electron viscosity). In particular, many studies were aimed to determine constant B in Eqs (4.35)–(4.38). For instance, Kumar and Clifton (1979) used the inclined-plate impact technique to measure B for LiF at room temperature and obtained a value of approximately $3 \times 10^{-5}\,\text{Ns/m}^2$ for an applied stress in the range 10–30 MPa, generating a dislocation velocity of about 100 m/s.

4.3.4 Relativistic Effects on Dislocation Motion

As shown in Figure 4.6, the regime III in dislocation motion is the one at which the velocity asymptotically approaches the shear wave velocity. In experiments, no supersonic

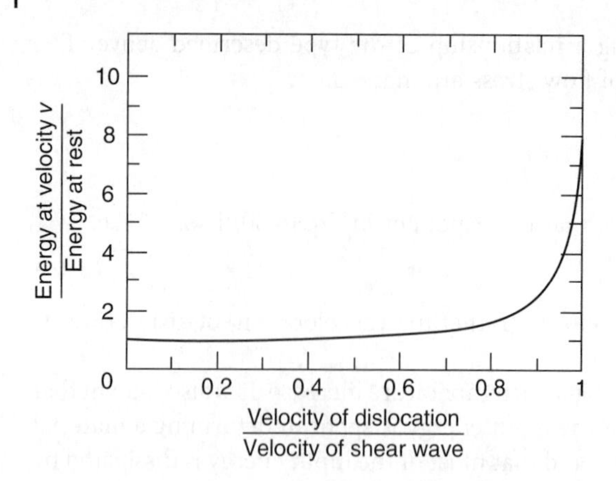

Figure 4.13 Effect of velocity on self-energy of a dislocation. *Source*: Meyers (1994). Reproduced with permission of Wiley.

dislocation has been observed, although it has been theoretically postulated by Eshelby (1949).

When the velocity of a dislocation approaches the shear wave velocity c_s, the total energy will approach infinity. Figure 4.13 presents the effect of velocity on self-energy of a dislocation for nickel. It can be seen that beyond $0.5c_s$, the energy changes very rapidly.

Obviously, when the dislocation velocity is increased to close to the shear wave velocity, the energy needed tends to infinity. This explains why any disturbance cannot propagate at the velocity higher than the shear wave velocity. The analogy to Einstein's relativity theory can be employed here. In fact, Einstein's famous equation $E = mC^2$ states that the energy is dependent on the mass, i.e.

$$m = \frac{m_0}{\sqrt{1 - v^2/C^2}} \tag{4.45}$$

where m is the mass at velocity v, m_0 is the mass at rest, and C is the velocity of light.

Now for dislocations in regime III, similarly we have

$$m = \frac{m_0}{\sqrt{1 - v^2/c_s^2}} \tag{4.46}$$

Thus, it is seen that the moving mass of a body would increase when it is moving at a very high velocity. In other words, the acceleration caused by a certain applied stress would decrease at higher velocity.

4.3.5 Synopsis

The three mechanisms illustrated earlier, i.e., thermally activated glide, drag-controlled dislocation motion, and relativistic motion, seem to realistically describe the motion of dislocations from $v = 0$ to $v = c_s$. The schematic plot is shown in Figure 4.14 based on Regazzoni *et al.* (1987).

When the applied stress is lower than the threshold stress σ_0 (height of activation barrier), thermal activation controls the velocity of propagation; when the applied stress is higher than the short-range barrier height, drag controls the resistance to dislocation

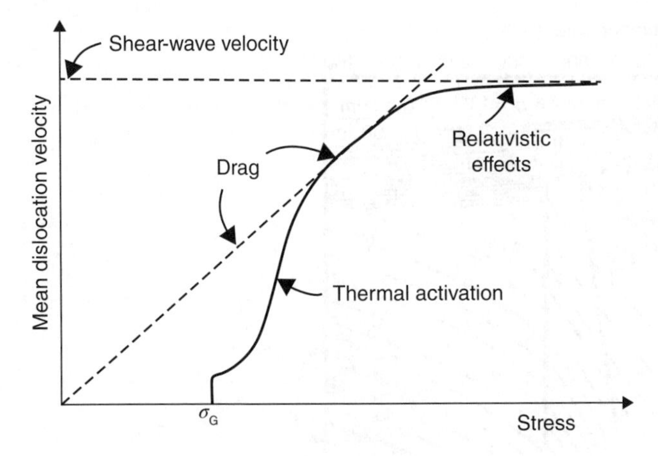

Figure 4.14 Three regimes of dislocation response. *Source*: Meyers (1994). Reproduced with permission of Wiley.

motion and the relationship $F \propto v$ or $\tau \propto v$ applies; and finally, at even higher velocities, in the range of $0.8c_s$, relativistic effects start becoming important.

Example

When copper ($\rho = 10^7 \text{cm}^{-2}$) is deformed, assume that all dislocations are mobile. Calculate the strain rate required to produce a dislocation velocity of $0.8c_s$, where relativistic effect starts gaining importance.

Solution

We apply equation $\dot{\varepsilon} = \rho bv/M$. For copper, the shear modulus is $G = 110$ GPa, and the material density is $\rho_0 = 8.9 \text{ g/cm}^3$. Therefore, the shear wave velocity is

$$c_s = \sqrt{\frac{G}{\rho_0}} = 4000 \text{ m/s}$$

The Burger vector of copper is $b \approx 0.3$ nm, and the Taylor factor for FCC materials is $M = 3.1$. By substituting the velocity of dislocation $v = 0.8c_s$ into the above relations, the corresponding strain rate is $\dot{\varepsilon} = 4 \times 10^4 \text{ s}^{-1}$. This calculation indicates that relativistic effects will become important when the strain rate is above the order of 10^4 s^{-1}.

4.4 Physically Based Constitutive Relations

As we have seen, the deformation mechanisms in metals depend on temperature and strain rate. Frost and Ashby (1982) have represented this dependence by *deformation maps*, as shown in Figure 4.15. The plot covers the strain rate ranging from 10^{-2} to 10^6 s^{-1} and temperature ranging from -200 to 1600°C. Normalized stress level, indicated as σ/G, varies from 10^{-6} to 10^{-2}.

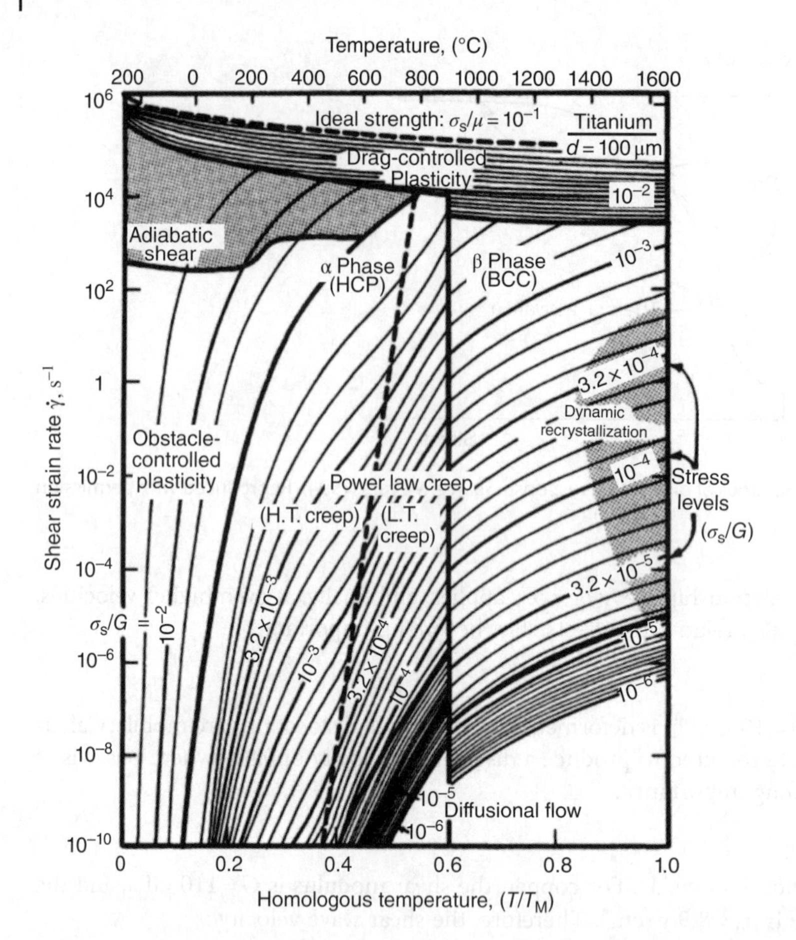

Figure 4.15 Strain rate/homologous temperature deformation map for titanium. BCC, body-centered cubic. H.T., high temperature; L.T., low temperature. *Source*: Meyers (1994). Reproduced with permission of Wiley.

It can be seen from the map that whether one deformation mechanism changes to another depends on temperature and strain rate; the boundaries between the various regions are determined from the constitutive equations for the different mechanisms. In most mechanisms, the flow stress decreases with increasing temperature and increases with increasing strain rate. When the deformation mechanism changes, the microstructure of the material usually also changes correspondingly. Several physically based constitutive relations will be briefly described in the following sections.

Campbell Equation

By considering the stress required for a dislocation to overcome a barrier, a research group in Oxford has proposed a series of constitutive equations. Taking into account strain hardening together with the strain-rate effect, Campbell *et al.* (1977) proposed that

$$\tau = A\gamma^n \left[1 + m\ln\left(1 + \frac{\dot{\gamma}}{B}\right)\right] \tag{4.47}$$

where $A, B, m,$ and n are four material constants which should be determined by experiments. The strain-hardening effect is reflected in the Campbell equation by the term γ^n, where $n < 1$.

The Z-A Model

Zerilli and Armstrong (1987, 1990a,b, 1992) proposed two microstructurally based constitutive equations that show an excellent match with experimental results. The Z-A equations are, for FCC material:

$$\sigma = \sigma_G + C_2\sqrt{\varepsilon}\exp(-C_3T + C_4T\ln\dot{\varepsilon}) + k\sqrt{d} \tag{4.48}$$

and for BCC material:

$$\sigma = \sigma_G + C_1\exp(-C_3T + C_4T\ln\dot{\varepsilon}) + C_5\varepsilon^n + k\sqrt{d} \tag{4.49}$$

where σ_G represents the athermal component of flow stress, and the last terms on the right-hand side represent the influence of grain size, where d is the grain diameter. The principal difference between these two equations resides in the fact that the plastic strain is uncoupled from the strain rate and temperature for the BCC metals.

A constitutive model to describe the FCC crystalline plasticity at the extreme strain rate was given in Gao and Zhang (2012):

$$\sigma = \sigma_{\text{ath}} + \sigma_{\text{th}} \tag{4.50}$$

where σ_{ath} is the athermal component of flow stress, reflecting mainly the long-range barriers, and σ_{th} is the thermal component of flow stress, reflecting mainly the short-range barriers,

$$\sigma_{\text{ath}} = \sigma_G + B\sqrt{1 - \exp(-k_{a0}\varepsilon)} \tag{4.51}$$

$$\sigma_{\text{th}} = \hat{C}\sqrt{1 - \exp\left(-k_0\tilde{\dot{\varepsilon}}^{-c_1T}\varepsilon\right)}\sqrt{\left[1 + \tanh\left(c_0\log\tilde{\dot{\varepsilon}}\right)\right]\tilde{\dot{\varepsilon}}^{c_1T}}\left\{1 - \left[-c_2\ln(\dot{\varepsilon}/\dot{\varepsilon}_0)\right]^{1/q}\right\}^{1/p} \tag{4.52}$$

where $\tilde{\dot{\varepsilon}} = \dot{\varepsilon}/\dot{\varepsilon}_{s0}$. The constant parameters to be determined are (σ_G, B, k_{a0}) for the athermal part and $\left(\hat{C}, k_0, c_0, c_1, c_2, p, q, \dot{\varepsilon}_{s0}, \dot{\varepsilon}_0\right)$ for the thermal part.

This model distinguishes the mobile dislocations from the total dislocations and incorporates the change of mobile dislocation density to count for the microstructural evolution of the material. The flow stress predictions by the unified model show a very good agreement with experiments within the whole strain rate range from 1×10^{-4} to $6.4 \times 10^5\,\text{s}^{-1}$. The flow stress upturn phenomenon in oxygen-free high thermal conductivity copper was satisfactorily described.

4.5 Experimental Validation of Constitutive Equations

Each proposed constitutive equation usually has a few (typically three to five) constants which have to be determined by pre-designed experiments. Once the parameters are determined, it is useful to validate the constitutive equation by comparing its predictions with more experimental results.

The high-strain-rate mechanical testing techniques illustrated in Chapter 3 are usually employed so as to determine the parameters in the constitutive equations, or to validate the equations under other dynamic loading conditions. One simple way to calibrate constitutive equations is to compare the predicted and observed deformation patterns in a Taylor specimen (i.e., a cylindrical specimen that normally impinges on a rigid anvil). In a Taylor specimen, different sections experienced different strains and strain rates, so the information recorded by the deformation history and shown in the final profile of the specimen is very rich and useful in verifying the proposed constitutive equation.

Problems 4

4.1 Try to modify the Cowper–Symonds constitutive relation (Eq. 4.23) so as to consider the effect of temperature.

4.2 Briefly illustrate the plastic deformation mechanisms in the three regimes of strain rates, as sketched in Figure 4.6.

Part 3

Dynamic Response of Structures to Impact and Pulse Loading

In Part 2, we discussed the dynamic response of materials. The main concern there was the behavior of materials, so that only a few problems were related to structural effects. For example, in the experiments using the split Hopkinson pressure bar technique, the bar's dynamic behavior is influenced by the effect of its transverse inertia. In Part 3, the dynamic response of structures to impact and pulse loading is discussed. From now on we will focus on the structural dynamic behaviors.

5

Inertia Effects and Plastic Hinges

5.1 Relationship between Wave Propagation and Global Structural Response

In the case of a structure under quasi-static loading, as the loading speed is very slow, any portion of the structure can be regarded as being in an equilibrium state that does not vary with time. However, in the case of a structure under dynamic loading, the loading speed is fast and/or the external load varies with time. It will take time for the disturbance caused by the external load to reach different parts of the structure. In general, therefore, the various portions of the structure will be in a non-equilibrium state.

When a structure responds to quasi-static loading, its deformation, strain, and stress are our main concerns in the analysis and simulation. However, in the response of a structure to dynamic loading, the wave propagation and inertia will also have essential influences on top of deformation, strain, and stress. It was noted in Chapter 1 that the speed and characteristics of stress wave propagation in solids are dictated by the material properties, such as modulus and density. As the stress wave travels very quickly in most engineering materials, different parts of an engineering structure of finite size will soon achieve their respective stable stress state. Hereby, the effect of the stress wave propagation quickly sweeps across the structure and then fades away due to multiple reflections at the boundaries of the structure. In this process, however, the deformation and the overall motion of structures related to the particle velocity resulting from the stress waves will continue to develop for a much longer period of time, during which the inertia effect plays a major role.

The widely used engineering structures, such as beams, arches, plates, and shells, possess the following characteristics: the dimension in one direction (usually the thickness direction) is much smaller than the dimensions in other directions (length, radius, etc.), while the direction of this small dimension often serves as the main load-carrying direction of the structure, e.g., when a plate is subjected to a force along its normal direction.

A typical example is shown in Figure 5.1, where a transverse dynamic force is applied to the top of a simply supported beam. First, this transverse dynamic load causes an elastic stress wave (and possibly a plastic stress wave as well if the force is large enough), which propagates through the thickness direction. When the incident compressive wave reaches the free surface at the bottom of the beam and is reflected as a tensile wave,

Introduction to Impact Dynamics, First Edition. T.X. Yu and XinMing Qiu.

Published 2018 by John Wiley & Sons Singapore Pte. Ltd.

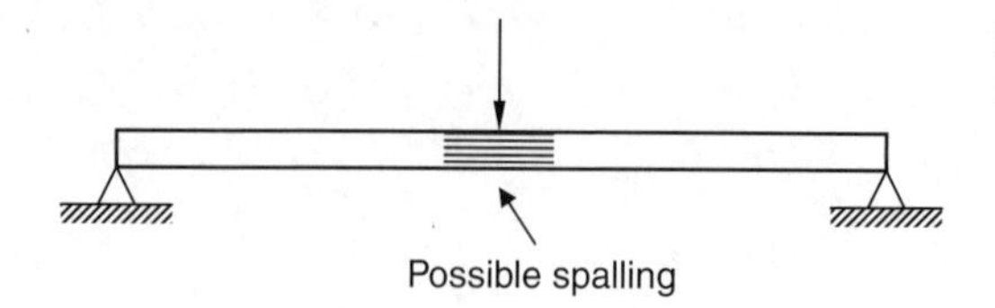

Figure 5.1 A beam subjected to transverse dynamic loading.

it may result in a tensile failure, known as *spalling*, especially if the material has a low tensile strength, such as concrete. Obviously, the wave reflection and even the spalling must occur only in the same order of time as the stress wave propagates through the thickness of the structure, typically a few microseconds. After the through-thickness stress wave reflects on both surfaces several times, the through-thickness normal stress will be dramatically reduced, while the downward particle velocity resulting from the stress wave makes the cross-section move down in its entirety. The subsequent long-term dynamic deformation is called the (global) structural response, which could last several milliseconds or much longer, depending on the span of the beam.

Therefore, for the analysis of a structure under transverse dynamic loading, it is necessary to distinguish its *short-term transient response* and its *long-term structural response*. The transient response is related to the stress wave propagation, reflection, and unloading processes, mainly along the thickness direction of the structure, whereas the structural response refers to the global deformation and motion of the structure as caused by the bending moment and shear force along the length direction. For a structure under impact, the timescale of dynamic deformation process is usually several magnitudes longer than that of the wave propagation, so that the wave propagation process and the structural response could be decoupled and analyzed separately. That is, in the analysis of wave propagation, the structure is assumed to remain in its original configuration, whereas in the analysis of structural response, the early time wave propagation is disregarded and only its global deformation is considered. Obviously such a treatment is extremely beneficial to the analysis of structures under dynamic loading.

The inertia of the structure plays an essential role in its long-term response. With the above understanding in terms of separating global deformation from wave propagation, the external load is supposed to impose a momentum instantaneously to the central line (of a beam or an arch) or the middle surface (of a plate or a shell), so the structural response can be represented by the subsequent motion and deformation of the central line or the middle surface of the structure.

5.2 Inertia Forces in Slender Bars

As shown in Figure 5.2, under the action of a dynamic load, an initially stationary structural member (e.g., a slender bar or a beam) begins to accelerate. It is known from the D'Alembert principle that a distributed inertia force would be produced by the acceleration field. Accordingly, an accelerating structural member can be equivalent to a structural member under "static equilibrium" if the "inertial force" is added to it. This "static equilibrium" state of the structural member can then be analyzed by considering

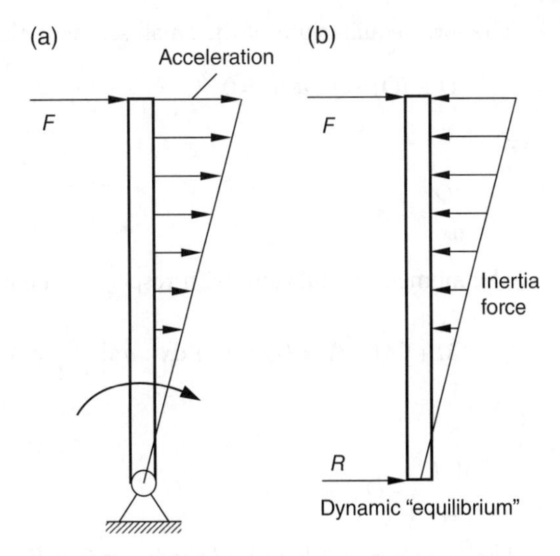

Figure 5.2 Pin-jointed slender link under impact.

the real external loads, the "inertial force" as well as the reaction forces/moments at the structural boundaries. Thus, a dynamic structural problem could be solved by referring to an equivalent static system.

Together with the external loads, the inertia forces also produce bending, shear, and axial forces in the structure generally. In order to assess the response of the structure, it is necessary to calculate the shear force and bending moment under both the external loads and the inertia forces.

5.2.1 Notations and Sign Conventions for Slender Links and Beams

As shown in Figure 5.3, the longitudinal direction of the rod or beam is assigned as the x-axis, while the transverse direction perpendicular to the x-direction is assigned as the y-axis.

The internal force components are defined as follows. At position x in the cross-section, the shear force and bending moment are $Q(x)$ and $M(x)$, respectively. Consider a small beam segment of length dx, under distributed transverse load $q(x)$. Figure 5.3 shows a free-body diagram and the sign conventions of $Q(x)$ and $M(x)$.

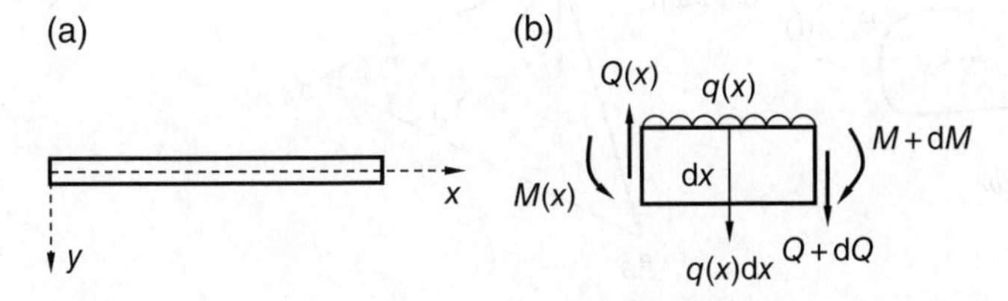

Figure 5.3 The definition of internal force components.

The force equilibrium of the small segment along the transverse direction gives

$$(Q + \mathrm{d}Q) - Q + q\,\mathrm{d}x = 0$$

i.e.,

$$\frac{\mathrm{d}Q}{\mathrm{d}x} = -q \tag{5.1}$$

The moment equilibrium with respect to the left side of the small segment leads to,

$$(M + \mathrm{d}M) - M + (Q + \mathrm{d}Q)\,\mathrm{d}x + q\,\mathrm{d}x\,\frac{\mathrm{d}x}{2} = 0$$

i.e.,

$$\frac{\mathrm{d}M}{\mathrm{d}x} = -Q \tag{5.2}$$

The mass per unit length of the beam is defined as

$$\rho(x) = \rho_0 A(x) \tag{5.3}$$

where ρ_0 is the mass density of material, and $A(x)$ is the cross-sectional area at position x.

5.2.2 Slender Link in General Motion

Figure 5.4 shows a schematic plot of a slender link (rod) AB in general motion. The acceleration components at point A is $\ddot{x}_A$ and $\ddot{y}_A$; the angular velocity and the angular acceleration of the rod about A are ω and $\dot{\omega}$, respectively. Therefore, the acceleration vector at point P, which is a distance x away from position A, could be expressed as

$$\boldsymbol{a}_P = \left(\ddot{x}_A - \omega^2 x\right)\boldsymbol{i} + \left(\ddot{y}_A + \dot{\omega}x\right)\boldsymbol{j} \tag{5.4}$$

where $\boldsymbol{i}$ and $\boldsymbol{j}$ are the unit vectors along the x- and y-directions, respectively. For a small segment adjacent to point P, the inertia forces created by accelerations are plotted in Figure 5.5. In general the inertia forces vary along the x-direction.

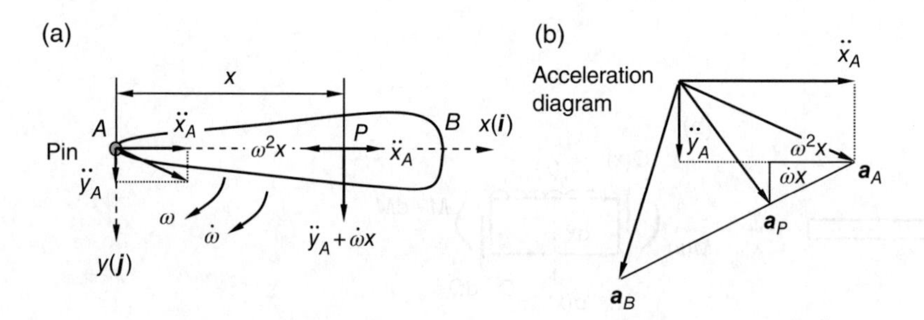

Figure 5.4 The general motion of a beam.

Figure 5.5 Inertia force on a small segment adjacent to point P.

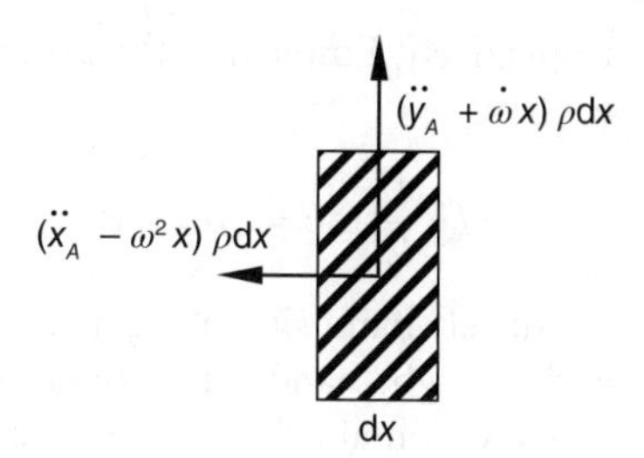

5.2.3 Examples of Inertia Force in Beams

Example 1 Collapse of a Chimney
A chimney AB of length L is made up of a homogeneous material of mass density ρ. Suppose the chimney starts to fall soon after a fracture takes place at the support point A. The question is where another fracture would take place inside the chimney during its collapse.

The collapse of the chimney is a dynamic process. The basic procedure for the dynamic analysis of such a structural member is as follows: (a) determine the acceleration field according to the kinematics; (b) obtain the inertia forces distribution based on the D'Alembert principle; (c) integrate the transversely distributed inertia forces along the axial direction to find the distributions of shear force and bending moment; (d) determine the position where the bending moment renders a maximum; and (e) during the collapse, the fracture would happen at this position when the maximum bending moment reaches the allowed critical value for the material of the chimney.

As shown in Figure 5.6, we should first specify the distribution of acceleration of the chimney according to the kinematics. During the collapse, the external forces applied to the chimney include the supporting (reaction) force at point A and its self-weight. Before the occurrence of the internal fracture, it is reasonable to assume that the chimney AB rotates about A as a rigid body. Here θ is the angle between the axis of the chimney AB and the vertical, and $\dot{\theta}$ and $\ddot{\theta}$ are its angular velocity and angular acceleration,

Figure 5.6 Inertia forces and moments during the collapse of a chimney.

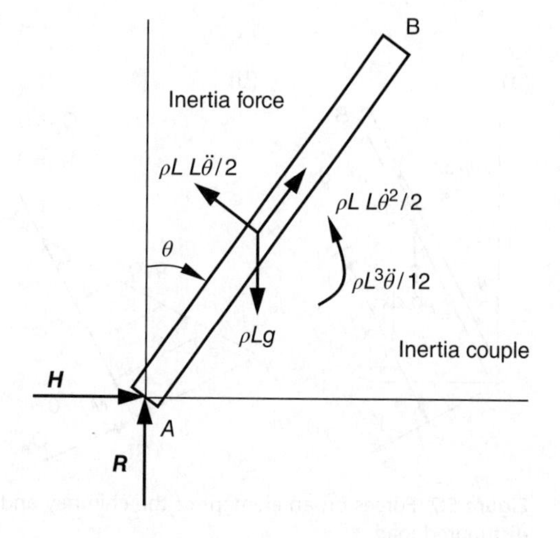

respectively. Employing the principle of moment of momentum with respect to point A, we have

$$\rho L g \frac{L}{2}\sin\theta = \frac{1}{3}\rho L\, L^2\ddot{\theta} \tag{5.5}$$

The left-hand side of Eq. (5.5) is the moment of the gravitational force about point A and the right-hand side is the moment of inertia multiplied by the angular acceleration. Thus, we obtain the angular acceleration of the chimney, which varies with θ, as

$$\ddot{\theta} = \frac{3g}{2L}\sin\theta \tag{5.6}$$

The corresponding angular velocity can be obtained by integrating the above equation and using the initial conditions.

Next, let us look at the inertia forces on a generic element of length dx in the chimney, at a distance x from point A, as shown in Figure 5.7. In addition to the gravity force $\rho g\ dx$, a centrifugal force $x\dot{\theta}^2\rho\ dx$ and a tangential inertia force $x\ddot{\theta}\rho\ dx$ are also applied to the element. Among these forces, only the gravity and the tangential inertia forces contribute to the transversely distributed load, which can be easily found as

$$q(x) = \rho\left(g\sin\theta - x\ddot{\theta}\right) \tag{5.7}$$

The upper end of the chimney, B, is a free end, so the shear force at point B must be equal to zero. By using this boundary condition in the integration of Eq. (5.7), the shear force at point A can be determined as

$$Q_A = \int_0^L q(x)\ dx = \rho\left(gL\sin\theta - L^2\ddot{\theta}/2\right) \tag{5.8}$$

Therefore, the distribution of shear forces along the whole beam is expressed as

$$Q(x) = Q_A - \int_0^x q(x)\ dx = \rho\left[g\sin\theta(L-x) - \left(L^2 - x^2\right)\ddot{\theta}/2\right] \tag{5.9}$$

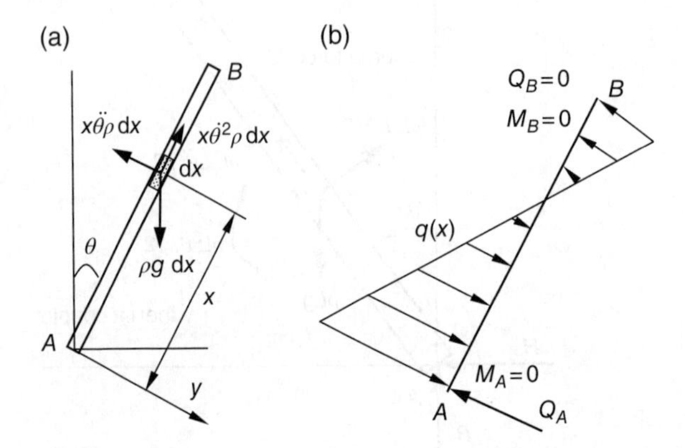

Figure 5.7 Forces on an element of the chimney and the plot of the equivalent transversely distributed load.

By substituting the expression of the angular acceleration, Eq. (5.6), into Eq. (5.9), the shear force can be rewritten as

$$Q(x) = \frac{\rho g}{4L}\sin\theta (L - x)(L - 3x) \tag{5.10}$$

Point A is a pin joint, so the bending moment at point A should be zero. By using this boundary condition, the bending moment can be obtained by integrating $Q(x)$:

$$M(x) = -\int_0^x Q(x)\,\mathrm{d}x = -\frac{\rho g}{4L}\sin\theta\left(L^2x - 2Lx^2 + x^3\right) \tag{5.11}$$

Now, the location of the maximum (minimum) bending moment can be determined from Eq. (5.11). In view of the relationship between shear forces and bending moments, it is easy to see that the extreme of the bending moment must take place at the locations where the shear force is equal to zero, $\frac{\mathrm{d}M}{\mathrm{d}x} = -Q = 0$. Thus, from inspection of Eq. (5.10) it can be seen that $Q(x) = 0$ happens if and only if $x = L$ or $x = L/3$. The case $x = L$ indicates that the bending moment becomes zero at point B, which pertains to a minimum of $M(x)$. Therefore, the maximum bending moment (by its magnitude) takes place at $x = L/3$, where the bending moment is

$$|M|_{\max} = \left|M\left(\frac{L}{3}\right)\right| = \frac{\rho g}{4L}\sin\theta\left(\frac{L^3}{3} - \frac{2L^3}{9} + \frac{L^3}{27}\right) = \frac{1}{27}\rho L^2 g\sin\theta \tag{5.12}$$

The results show that $|M|_{\max}$ increases with increasing θ. As the collapse continues and $|M|_{\max}$ reaches the critical bending moment, which is dictated by the material properties and the cross-section of the chimney, the fracture occurs at the cross-section at the position $x = L/3$.

Example 2 Impact Pendulum

Consider an impact pendulum AB of uniform cross-section and length $2L$, so that its weight is $m = 2L\rho$ (Figure 5.8). It rotates about a pivot point O with angular

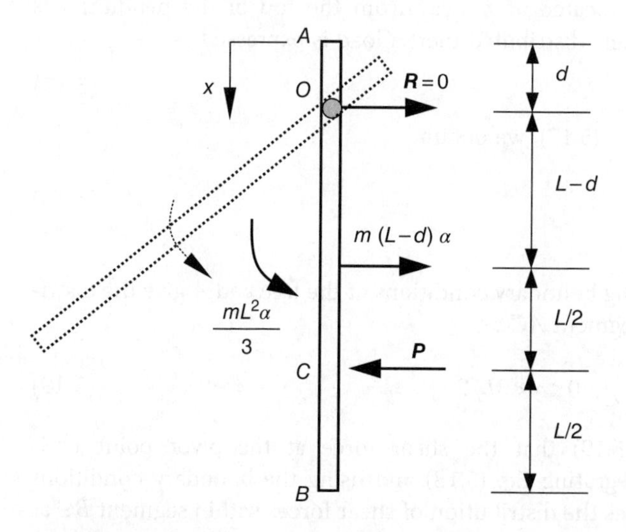

Figure 5.8 Force diagram of a pendulum subjected a concentrated impact force.

acceleration α. When the pendulum impinges on a target at point C, it encounters an impact force P with a direction perpendicular to the pendulum. In the following, we will prove that if there is no reaction force at the pivot point O, the distance between points O and A should be $d = L/3$. We will also calculate the maximum bending moment and its location when the impact happens.

In this problem, the pendulum rotates about a fixed axis O with angular acceleration α, so that the acceleration at the centroid of the pendulum is $(L-d)\alpha$. By noting that the reaction force at point O is required to be zero, the principle of linear momentum along the horizontal direction leads directly to

$$P = m(L-d)\alpha \tag{5.13}$$

On the other hand, the principle of moment of momentum with respect to the centroid gives

$$\frac{1}{2}PL = \frac{1}{3}mL^2\alpha \tag{5.14}$$

Hence, the angular acceleration of the pendulum is found to be

$$\alpha = \frac{3P}{2mL} \tag{5.15}$$

Substituting Eq. (5.15) back into Eq. (5.13) gives $P = m(L-d)\dfrac{3P}{2mL}$, which results in

$$d = \frac{L}{3} \tag{5.16}$$

This proves that Eq. (5.16) is a necessary and sufficient condition for the reaction force at pivot point O being zero.

Next, let us examine the maximum bending moment and its location in the pendulum. Since the pendulum rotates about the pivot point O with angular acceleration α, the tangential acceleration at a point located at x away from the top of the pendulum is $(x-L/3)\alpha$. Hence, the transversely distributed inertia load is expressed as

$$q(x) = -\rho(x-L/3)\alpha \tag{5.17}$$

Substituting Eq. (5.15) into Eq. (5.17), we obtain

$$q(x) = -\frac{P}{4L^2}(3x-L) \tag{5.18}$$

which is plotted in Figure 5.9.

Integrating Eq. (5.18) and using boundary conditions at the free end A give the distribution of shear forces within segment AC as

$$Q(x) = \frac{P}{8}\left[3(x/L)^2 - 2(x/L)\right], \quad 0 \le x < 3L/2 \tag{5.19}$$

It can be seen from Eq. (5.19) that the shear force at the pivot point O is $Q(L/3) = -P/24$. Similarly, integrating Eq. (5.18) and using the boundary conditions at the free end B, $Q(2L) = 0$, gives the distribution of shear forces within segment BC as

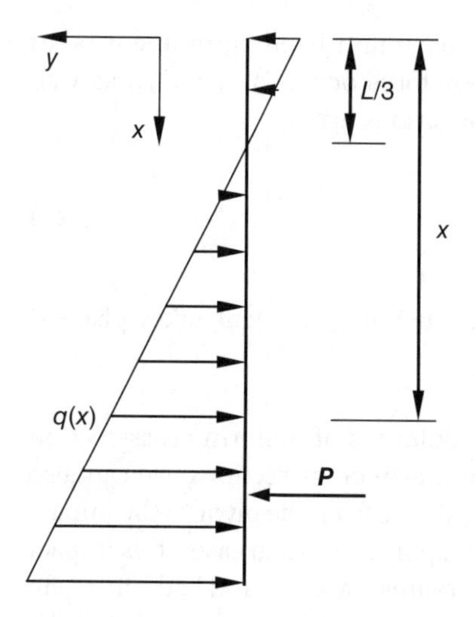

Figure 5.9 Distribution of the transversely distributed load in the impact pendulum.

$$Q(x) = \frac{P}{8}\left[3(x/L)^2 - 2x/L - 8\right], \; 3L/2 < x \leq 2L \tag{5.20}$$

It should be emphasized that the shear force has a jump at point C where a concentrated force P is applied. Figure 5.10 plots the distribution of shear forces along the axial direction of the pendulum with a concentrated force acting at $d = L/3$, which verifies that the jump in shear forces at point C is equal to P.

With the boundary condition of $M(0) = 0$ at point A, integrating the shear force with respect to x leads to the distribution of bending moments along the axial direction of the pendulum:

$$M(x) = -\int_0^x Q(x)\,\mathrm{d}x = \frac{PL}{8}\left[-(x/L)^3 + (x/L)^2\right] \tag{5.21}$$

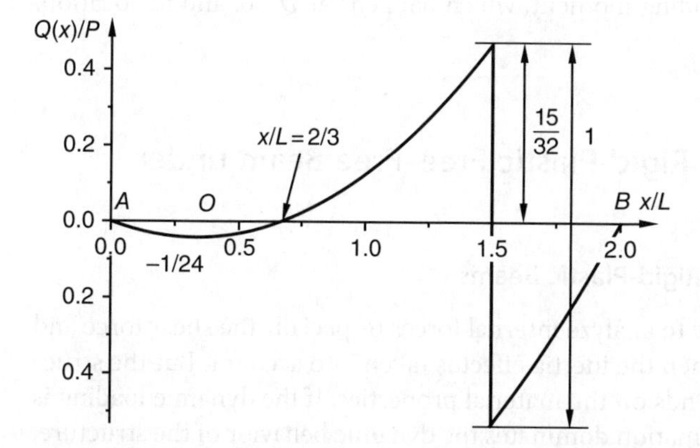

Figure 5.10 Distribution of shear force in the impact pendulum.

In order to determine the position at which the maximum bending moment takes place, it can be seen from Figure 5.10 that zero shear force occurs at $x = 2L/3$ as well as at $x = 3L/2$. The bending moments at these two locations are

$$M\left(\frac{2}{3}L\right) = \frac{PL}{54} \quad \text{and} \quad M\left(\frac{3}{2}L\right) = -\frac{9}{64}PL \tag{5.22}$$

respectively. Therefore, by magnitude, the maximum bending moment takes place at the impact point C, where $|M|_{\max} = \left|M\left(\frac{3L}{2}\right)\right| = \frac{9}{64}PL$.

In this example, it is assumed that the impact pendulum is of uniform cross-section along its axial direction. For a pendulum with non-uniform cross-sections, we can also establish a geometric condition, which indicates the distance from the pivot to the impact position, to ensure no reaction force occurs at the pivot point. For such cases, this impact position is called the *impact center*. For sports such as tennis, baseball, and badminton, in which the sportsman needs a racket to hit balls, if the sportsman holds the racket at the pivot point O and hits the ball at impact center C, the reacted shock to the sportsman is minimal.

A Summary of the Methodology

Considering the two examples in the previous sections, the methodology for the analysis of a slender bar or a beam under dynamic loading can be summarized as follows:

1) Identify the acceleration field using the kinematics.
2) Determine the transversely distributed load $q(x)$ from the acceleration field based on the D'Alembert principle.
3) Integrate the transversely distributed load $q(x)$ and use the boundary conditions to obtain the shear force distribution $Q(x)$.
4) Integrate the shear force $Q(x)$ and use the boundary conditions again to obtain the bending moment distribution $M(x)$.
5) Calculate the maximum bending moment, which happens at $Q = 0$, and its location.

5.3 Plastic Hinges in a Rigid-Plastic Free–Free Beam under Pulse Loading

5.3.1 Dynamic Response of Rigid-Plastic Beams

In Section 5.2 we discussed how to analyze internal forces (especially the shear force and bending moment) in a beam when the inertia effect is taken into account. But the structural response also largely depends on the material properties. If the dynamic loading is so intense that the plastic deformation dominates the dynamic behavior of the structure, then the rigid plastic idealization is usually adopted for the material modeling.

Assumptions

1. The material is rigid-perfectly plastic and rate-independent.

This assumption implies that both the elastic deformation and the strain-hardening effect of the plastic deformation are disregarded. That is, $E = \infty$ and $E_P = 0$. As seen in Chapter 3, the yield stress and the strain-hardening character of materials may both change with the strain rate. However, to simplify the dynamic analysis of structures, the effect of strain rate is approximately considered by taking the yield stress Y as the dynamic yield stress at the average strain rate over the response process.

Figure 5.11(a) plots the relationship between stress and strain for a rigid-plastic material. For a beam composed of such material, the relationship between bending moments and curvature also becomes very simple, as shown in Figure 5.11(b). In fact, when a cross-section of the beam approaches its fully plastic state, the bending moment at this cross-section reaches the *fully plastic bending moment* M_P. For example, the fully plastic bending moment of a beam with a rectangular cross-section is

$$M_P = \frac{1}{4} Ybh^2 \tag{5.23}$$

where b and h are the width and height of cross-sections of the beam, respectively. For a beam with circular tubular cross-section,

$$M_P = 4YR^2h \tag{5.24}$$

where R and h are the radius and thickness of the pipe, respectively.

2. The influences of shear and axial forces on yielding are neglected, so the yielding only depends on the magnitude of bending moments.

As shown in Figure 5.11(b), if $|M| < M_P$, no plastic deformation will occur; if $|M| = M_P$, the cross-section will be in a fully plastic deformation state and the curvature can increase without limit, implying that the cross-section can rotate arbitrarily as a hinge. Hence, the moment–curvature relationship based on the rigid-perfectly plastic idealization of materials, as shown in Figure 5.11(b), naturally brings an important concept, termed the *plastic hinge*, into the plastic analysis of structures.

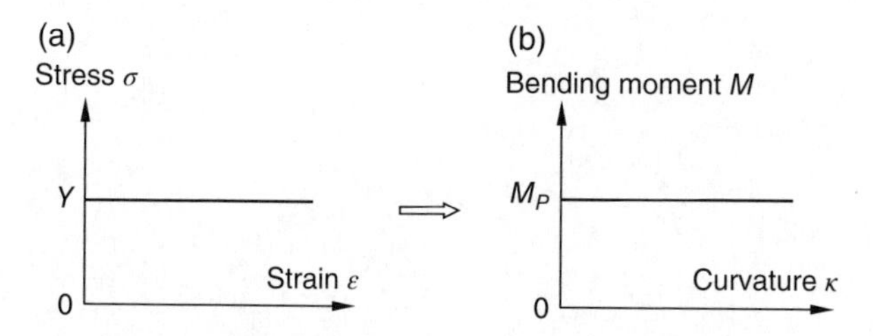

Figure 5.11 (a) Stress–strain relationship of rigid-perfectly plastic materials; (b) bending moment–curvature relationship of rigid-perfectly plastic beams.

3. *The deflection of the beams is sufficiently small compared with their length.*

Therefore, the geometric relationship and the equations of motion can be formulated based on the initial configuration of the beams.

5.3.2 A Free–Free Beam Subjected to a Concentrated Step Force

Consider a beam of length $2L$ with both ends free of any support subjected to a concentrated step force P at its midpoint, as shown in Figure 5.12(a). We are going to analyze its dynamic response immediately after the application of this step force starting at $t = 0$.

According to Newton's second law, under the application of step force P, this free–free beam has a uniform acceleration

$$a_0 = \frac{P}{2L\rho} \tag{5.25}$$

where ρ is the mass per unit length of the beam. Hence, the corresponding inertia force is

$$q = \rho a = \frac{P}{2L} \tag{5.26}$$

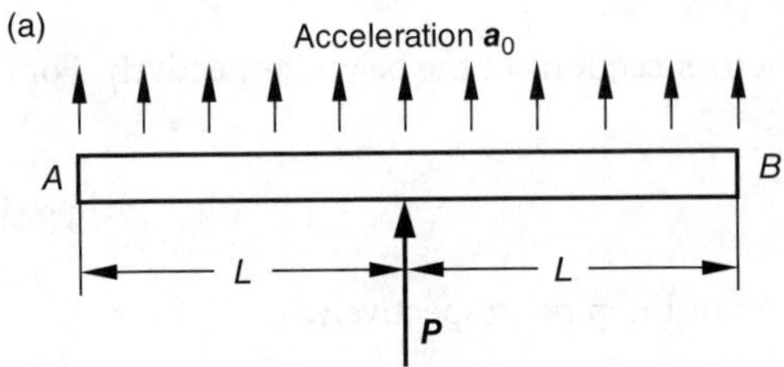

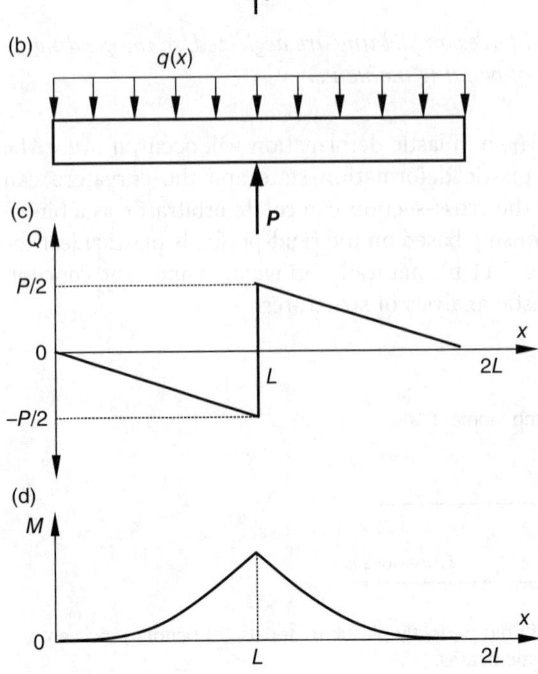

Figure 5.12 A free–free beam subjected to a concentrated step force at its midpoint.

Integrating the uniformly distributed inertia forces yields the distribution of shear force. For the left half of this beam, by noting the boundary conditions at free end A, $Q(0) = 0$, it is easy to derive that

$$Q(x) = -\int_0^x q \,\mathrm{d}x = -\frac{Px}{2L}, \quad 0 \le x \le L \tag{5.27}$$

It is noted that the shear force has a jump at the midpoint of the beam where the concentrated force applies. The magnitude of the jump is equal to the concentrated force P. Similarly, the shear force in the right half of the beam can be expressed as

$$Q(x) = P - \int_0^x q \,\mathrm{d}x = P\left(1 - \frac{x}{2L}\right), \quad L \le x \le 2L \tag{5.28}$$

Figure 5.12(c) shows the distribution of the shear force along the beam. It is seen that the shear force passes its zero axis at the midpoint $x = L$.

Using the boundary conditions at the free ends A and B, $M(0) = M(2L) = 0$, integrating the shear force expressed in Eqs (5.27) and (5.28) leads to the following distribution of bending moments in the beam:

$$M(x) = -\int_0^x Q(x) \,\mathrm{d}x = \begin{cases} \dfrac{P}{4L}x^2, & 0 \le x \le L \\ \dfrac{P}{4L}(2L - x)^2, & L \le x \le 2L \end{cases} \tag{5.29}$$

which is plotted in Figure 5.12(d). It is easy to find that with the increase of x, the bending moment first increases and then decreases. The maximum value occurs at the midpoint of the beam and is equal to

$$M_{\max} = M(L) = PL/4 \tag{5.30}$$

In the following, the dynamic responses of the free–free beams to different magnitudes of step force P are considered.

1) If $P < 4M_P/L$, according to Eq. (5.30), the maximum bending moment is smaller than the fully plastic bending moment of the beam, i.e. $M_{\max} < M_P$. In this case, no part of the beam enters a plastic state, and the whole beam moves as a rigid body without any deformation. The effect of the step force P on the beam is merely to generate a uniform acceleration and force the beam to move along its translational direction as a whole.
2) If $P = 4M_P/L$, the maximum bending moment at the midpoint of the beam reaches the fully plastic bending moment, $M_{\max} = M_P$. Hence, in this case, a plastic hinge forms at $x = L$.
3) If $P > 4M_P/L$, plastic yielding takes place and a plastic hinge forms at the midpoint $x = L$. Simultaneously, each half of the beam rotates about the plastic hinge at the midpoint C, as shown in Figure 5.13. In this case, the kinematics previously proposed in Figure 5.12 has to be switched to the new deformation mechanism shown in Figure 5.13.

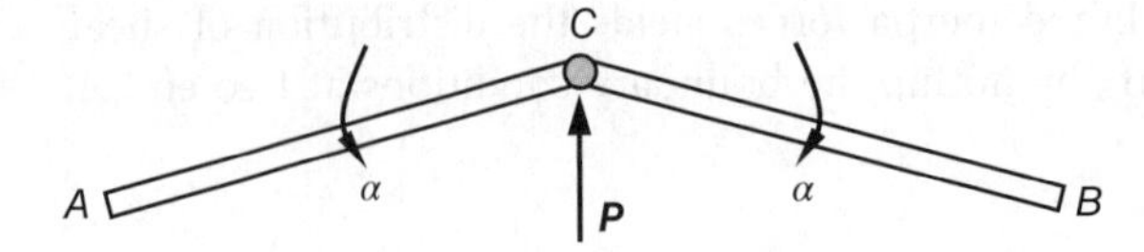

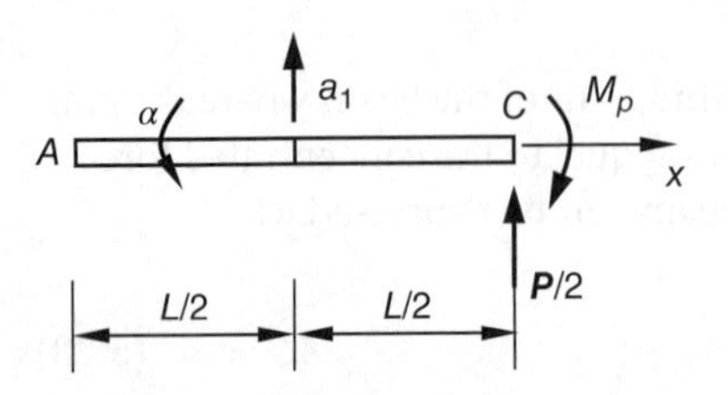

Figure 5.13 Deformation mechanism and acceleration of a free–free beam subjected to a concentrated force ($P > 4M_P/L$) at its midpoint.

In view of the symmetry, only the left half of the beam, segment AC, is selected from the initial configuration as the free body for analysis. The principle of linear momentum in the vertical direction requires that

$$\frac{P}{2} = \rho L a_1 \tag{5.31}$$

where a_1 is the acceleration at the midpoint of segment AC. Using the principle of moment of momentum with respect to the midpoint of AC, we have

$$\frac{P}{2}\frac{L}{2} - M_P = \frac{1}{12}\rho L^3 \alpha \tag{5.32}$$

The left-hand side of Eq. (5.32) is the moment with respect to the midpoint of AC and the right-hand side is the moment of inertia of AC multiplied by the angular acceleration α.

Combination of Eqs (5.31) and (5.32) gives the linear acceleration at the midpoint of segment AC and the angular acceleration as

$$a_1 = \frac{P}{2\rho L} \quad \text{and} \quad \alpha = \frac{3(PL - 4M_P)}{\rho L^3} \tag{5.33}$$

respectively. Thus, the acceleration distribution in segment AC can be obtained from Eq. (5.33) as

$$a(x) = a_1 - \left(\frac{L}{2} - x\right)\alpha = \frac{1}{\rho L^2}\left[(6M_P - PL) + \frac{3x}{L}(PL - 4M_P)\right], 0 < x < L \tag{5.34}$$

which also gives the distribution of the inertia force as the transversely distributed load to the beam:

$$q(x) = \rho a(x) \tag{5.35}$$

Using the boundary conditions at free end A, i.e., $Q(0) = M(0) = 0$, integration of the transversely distributed load once and twice, respectively, yields the distributions of shear force and bending moment:

$$Q(x) = -\int_0^x q(x)\,dx = -\frac{1}{L^2}\left[(6M_P - PL)x + \frac{3x^2}{2L}(PL - 4M_P)\right] \tag{5.36}$$

Figure 5.14 Distribution of bending moment of a free–free beam subjected to a concentrated load ($P > 4M_P/L$) at its midpoint.

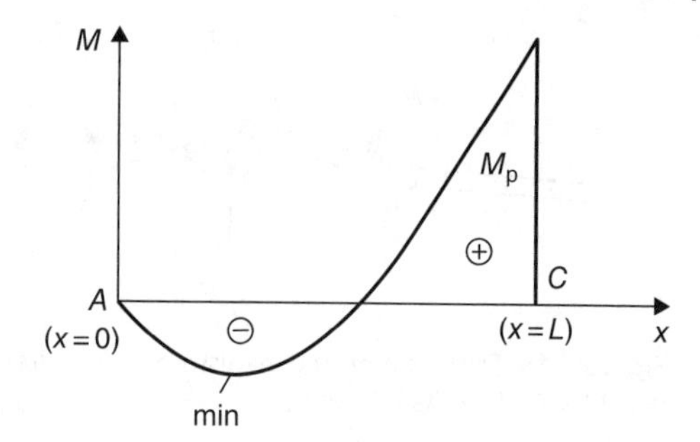

$$M(x) = -\int_0^x Q(x)\,\mathrm{d}x = \frac{1}{2}(6M_P - PL)(x/L)^2 + \frac{1}{2}(PL - 4M_P)(x/L)^3 \tag{5.37}$$

Figure 5.14 plots the bending moment distribution expressed by Eq. (5.37). A check confirms that $M(L) = M_P$ at point C where a plastic hinge has already formed. From the free end A, the bending moment decreases first and increases subsequently, gradually turning the negative bending moment to positive.

If we set $Q(x) = 0$ in Eq. (5.36), it is found that M takes a minimum at $x/L = \dfrac{2(\mu-6)}{3(\mu-4)}$, where $\mu \equiv PL/M_P$ is a non-dimensional load parameter. The previous analysis showed that when $\mu = 4$, a plastic hinge forms at point C. When $\mu > 4$, whether other hinges, in addition to the hinge at C, form at the locations of the minimum bending moments depends on the magnitude of μ. If $4 < \mu < 6$, $Q(x)$ given by Eq. (5.36) will not change its sign within the range of $0 < x < L$, so that the bending moment increases monotonically and the minimum bending moment actually takes place at the free end A, indicating that no interior hinge appears in segment AC. If $\mu > 6$, the cross-section at which the minimum bending moment occurs is located within segment AC. Therefore, when the magnitude of this negative minimum bending moment is equal to the fully plastic bending moment, the second plastic hinge forms inside segment AC, i.e.

$$M\left(\frac{2(\mu-6)}{3(\mu-4)}\right) = -M_P \tag{5.38}$$

Combining Eq. (5.37) with (5.38) gives

$$2(\mu-6)^3 - 27(\mu-4)^2 = 0 \tag{5.39}$$

i.e.

$$2\mu^3 - 63\mu^2 + 432\mu - 864 = 0 \tag{5.40}$$

The valid solution of the above cubic equation is $\mu = 22.9$, which means that when $\mu = 22.9$, the second plastic hinge in the left half of the beam forms at $\dfrac{x}{L} = \dfrac{2(\mu-6)}{3(\mu-4)} = 0.596$.

Hence, if the step force applied at the midpoint is intense and reaches the level $P > 22.9M_P/L$, the dynamic deformation mechanism of the free–free beam will contain three plastic hinges, one central hinge and two side hinges. The direction of rotation of these two side hinges is opposite to that of the central one. The location of these two side

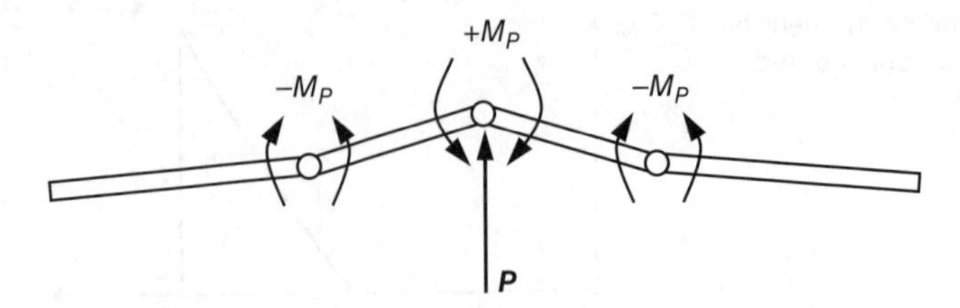

Figure 5.15 Deformation mechanism of a free–free beam subjected to a concentrated load ($P > 22.9M_P/L$) at its midpoint.

hinges depends on the magnitude of the step force applied: the larger the magnitude of P, the closer these two side hinges are to the central point. Figure 5.15 shows a sketch of the dynamic deformation mechanism containing three plastic hinges.

5.3.3 Remarks on a Free–Free Beam Subjected to a Step Force at its Midpoint

1) The dynamic deformation mechanism of a free–free beam subjected a concentrated step force P depends on the magnitude of P:
   i) $PL/M_P < 4$: no plastic deformation and hinge occur in the beam, while the free-free beam makes a translational motion with a uniform linear acceleration as a rigid body.
   ii) $4 \le PL/M_P < 22.9$: one plastic hinge forms at the midpoint of the beam where the step force is applied.
   iii) $PL/M_P \ge 22.9$: three plastic hinges appear, including one hinge at the midpoint and two side hinges rotating in the opposite direction.
2) A free–free beam does not have load-carrying capacity under static loading, but it can sustain dynamic loading, owing to the inertia effect.
3) If force P changes its magnitude with time, the deformation mechanism may vary with time accordingly.

The above analysis on a free–free beam subjected to a step force can be regarded as a simplified version of a pioneering study made by Lee and Symonds in 1952. In that paper, they analyzed the dynamic response of a free–free beam subjected to a triangular-shaped impact force as shown in Figure 5.16. Because the magnitude of the applied force changes

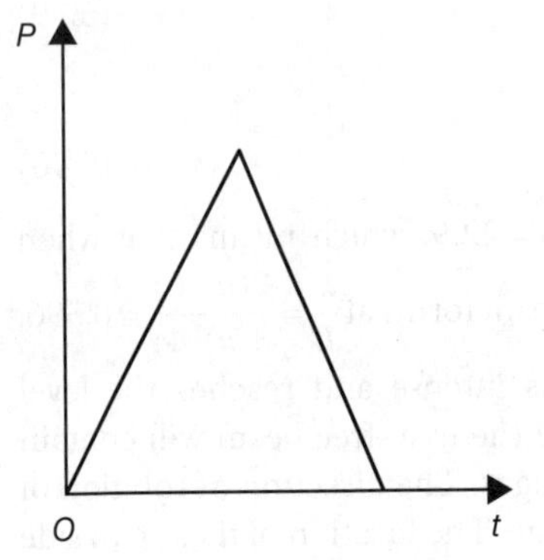

Figure 5.16 A triangular-shaped pulse load.

with time, the dynamic deformation mechanism of the beam varies with time, too, which will make the deformation history very complicated.

Briefly, in the analysis of Lee and Symonds, the dynamic response of a free–free beam subjected to triangular-shaped pulse load at the midpoint contains the following phases:

1) When the magnitude of the impact load is small, the free–free beam translates as a rigid body.
2) With the increase of the impact load, a plastic hinge forms at the midpoint, so along with the rigid-body translation, each half of the free–free beam also rotates about this plastic hinge.
3) When the impact load increases to $P = 22.9M_P/L$, three plastic hinges appear.
4) If the magnitude of P increases further, the positions of the two side hinges will move towards to the midpoint of the beam.
5) When P decreases after its peak magnitude in the triangular shape, the two side hinges move away from the midpoint of the beam towards the respective free ends.
6) When the relative angular velocity at the side hinges reduces to zero, these two side hinges will disappear.
7) When the relative angular velocity at the central hinge reduces to zero, this hinge will disappear, too, but the plastically deformed beam will continue its translational motion as a rigid body.

In the history of structural impact dynamics, Lee and Symonds's paper is regarded as a pioneering work. This is not only because it was almost the first paper on the dynamic analysis of beams under dynamic loading, but also because an important concept, the *travelling plastic hinge* (or the *moving plastic hinge*), was first proposed in the structural response to varying dynamic load. In the following sections, based on the rigid-perfectly plastic idealization, the concepts of stationary plastic hinges and traveling plastic hinges will be extensively employed to construct dynamic deformation mechanisms of structures under constant and varying dynamic loads, respectively.

5.4 A Free Ring Subjected to a Radial Load

This section will discuss the dynamic response of a free ring subjected to a radial dynamic load. The related content can be referred to the work of Owens and Symonds (1955) and Hashmi *et al.* (1972).

As shown in Figure 5.17, to form a mechanism, a close ring should contain four hinges. Due to the symmetry with respect to the vertical diameter, the shear force $Q = 0$ at points A and B, so the bending moment M renders extreme values at these two points. Hence, two plastic hinges form at points A and B. Two other hinges appear at points C and C', which are symmetric with respect to vertical diameter AB.

Applying the principle of linear momentum in the vertical direction leads to

$$P = 2\pi R\rho a \tag{5.41}$$

where a is the acceleration at the center of mass, which can be expressed as

$$a = \frac{P}{2\pi R\rho} \tag{5.42}$$

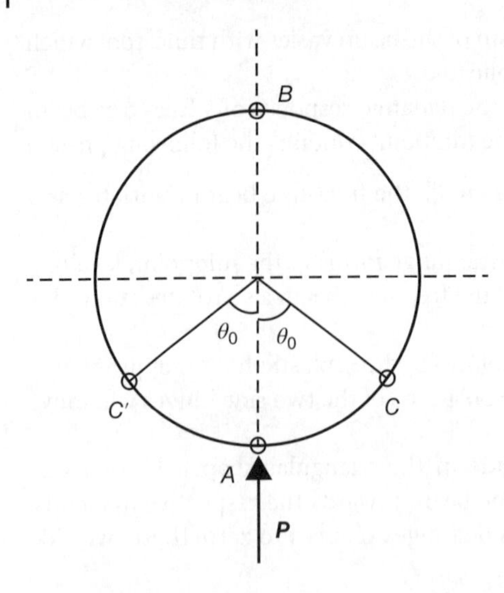

Figure 5.17 Deformation mechanism of a ring subjected to a radial dynamic load.

Figure 5.18 depicts the forces applied to the right half of the circular ring when it is subjected to a radial impact. Because the forces in the circumferential direction should be taken into account, too, the problems related to rings are more complex than those of beams. At points A and B, there are circumferential forces N from the left half and a fully plastic bending moment M_P. For an element $Rd\theta$ on the ring, its inertia force is

$$\rho R\,\mathrm{d}\theta\, a = \frac{P}{2\pi}\,\mathrm{d}\theta \tag{5.43}$$

Based on the force analysis diagram shown in Figure 5.18, the equilibrium of moments about point A yields

$$N = \frac{1}{2R}\int_0^{\pi} \frac{P}{2\pi} R\sin\theta\,\mathrm{d}\theta = \frac{P}{2\pi} \tag{5.44}$$

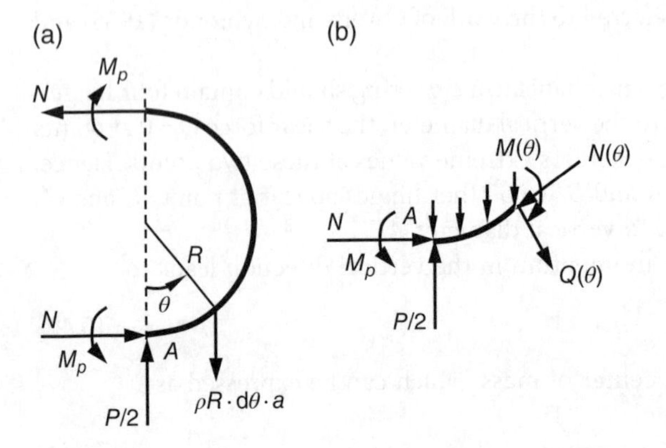

Figure 5.18 Force analysis of a circular ring subjected to a radial impact.

Referring to Figure 5.18(b), the bending moment at the cross-section, which is an angle θ away from point A, is

$$M(\theta) = M_P + NR(1 - \cos\theta) - \frac{P}{2}R\sin\theta + \int_0^{\theta} \frac{P}{2\pi}R\left(\sin\theta - \sin\tilde{\theta}\right) \mathrm{d}\tilde{\theta} \tag{5.45}$$

The last term on the right-hand side of Eq. (5.45) represents the moment of inertia forces. Using Eq. (5.44) and integrating, Eq. (5.45) can be simplified as

$$M(\theta) = M_P - \frac{PR}{2}\sin\theta + \frac{PR}{2\pi}\theta\sin\theta \tag{5.46}$$

If the plastic hinge C is located at angle θ_0, the following two conditions have to be satisfied:

$$M(\theta_0) = -M_P \tag{5.47}$$

$$\left.\frac{\mathrm{d}M}{\mathrm{d}\theta}\right|_{\theta=\theta_0} = 0 \tag{5.48}$$

Taking a derivative of the bending moment expressed by Eq. (5.46) yields

$$\frac{\mathrm{d}M}{\mathrm{d}\theta} = \frac{PR}{2\pi}[\sin\theta - (\pi - \theta)\cos\theta] \tag{5.49}$$

By combining Eq. (5.48) with Eq. (5.49), the location, θ_0, of the plastic hinge should satisfy

$$\tan\theta_0 + \theta_0 = \pi \tag{5.50}$$

which results in $\theta_0 = 1.113$ rad or 63.77°.

Now, we can calculate the dynamic load associated with the mechanism whereby the plastic hinge is formed at θ_0. Combining Eq. (5.47) with Eq. (5.46) gives

$$2M_P = \frac{PR}{2\pi}\sin\theta_0\,(\pi - \theta_0) \tag{5.51}$$

Rearrangement of Eq. (5.51) gives

$$\frac{PR}{M_P} = \frac{4\pi}{\sin\theta_0\tan\theta_0} = 6.90 \tag{5.52}$$

In other words, only when the magnitude of the step force exceeds $P = 6.90M_P/R$ will a dynamic deformation mechanism occur, with four plastic hinges being formed at points A, B, C and C'.

If the circular ring is subjected to opposite radial forces P and P' as shown in Figure 5.19, the positions of plastic hinges at θ_0 can be determined as follows. If $P \geq P'$,

$$\tan\theta_0 + \theta_0 = \frac{\pi P}{P - P'} \tag{5.53}$$

For the special case of $P = P'$, the plastic hinges form on both sides of the vertical diameter at $\theta_0 = \pi/2$, and the corresponding minimum step force is $P = 4M_P/R$.

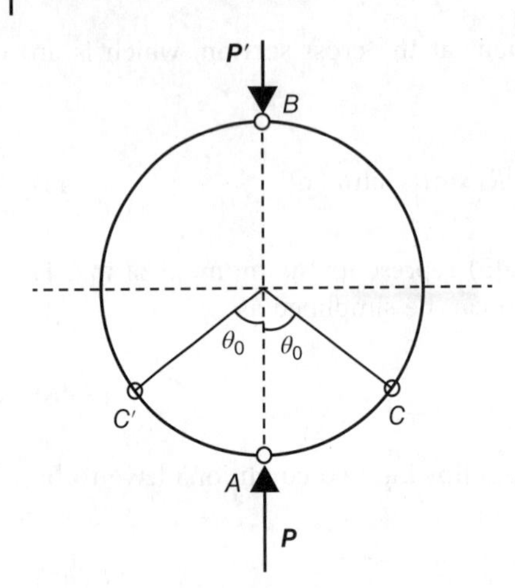

Figure 5.19 Deformation mechanism of a circular ring subjected to opposite radial forces.

5.4.1 Comparison between a Supported Ring and a Free Ring

If a circular ring is supported at point B, then any force $P \geq 4M_P/R$ applied at point A can cause the ring's dynamic plastic deformation, but if the ring is free of supports, its dynamic deformation occurs only if P is greater than $6.90M_P/R$. In the latter case, the additional input energy will be transferred into the kinetic energy of the translational motion of the ring.

This analysis is also applicable to the collision between two moving rings or the collision of a moving ring to a wall.

Questions 5

1 In the first example in Section 5.2, after the fracture of the chimney at $x = L/3$, can the chimney be fractured further, yes or no? Why?

2 For the case of a free ring subjected to a radial force P, would the four plastic hinges A, B, C and C' form simultaneously, or would plastic hinges A and B would form before others?

Problems 5

5.1 In the example of a free–free beam subjected to a concentrated step force at its midpoint, a portion of external work is transferred into the kinetic energy of the beam whilst the remainder is dissipated by the plastic hinge. What is the maximum percentage of the plastic dissipation compared with the total external work done by the force?

5.2 A uniform, free–free beam of length L and mass per unit length ρ is subjected to a constant force F suddenly applied at its left end as shown in the figure. Show that the acceleration, a, at the mass center of the beam and angular acceleration, α, of the beam are given by $a = F/\rho L$ and $\alpha = 6F/(\rho L^2)$. Calculate the distance from the left end of the beam to the position at which the bending moment reaches a maximum. If the beam is made of a rigid, perfectly plastic material with the fully plastic bending moment M_P at the cross-section, calculate the magnitude of F under which a plastic hinge will form within the beam.

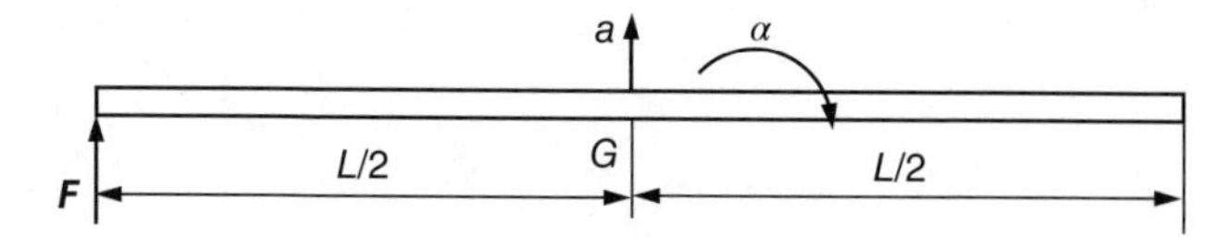

6

Dynamic Response of Cantilevers

As a commonly used structural component in engineering, here we use a cantilever as an example of how to analyze the dynamic response of structures to various loading conditions.

6.1 Response to Step Loading

In Figure 6.1(a), a straight cantilever AB of uniform cross-section is subjected to a dynamic transverse force $F(t)$ at its tip. Suppose the beam is of length L and the fully plastic bending moment of the cross-section is M_P. Here the case of step-loading is considered first.

From the static analysis, the static plastic collapse load, i.e., the limit load, of a cantilever is $F_c = M_P/L$, under which a plastic hinge forms at the root of the cantilever, B, i.e., $M_B = M_P$.

When the step force is lower than the limit load, $F < F_c$, the cantilever remains undeformed as the material is supposed to be rigid-plastic. When $F(t)$ slightly exceeds F_c, it is reasonable to assume that there will be a plastic hinge formed at the root and the cantilever will rotate about this hinge. Let ρ denote the mass per unit length and v be the velocity at the tip; then the rigid body motion and the velocity distribution diagram are shown in Figure 6.2.

The translational equation of motion for the entire cantilever gives

$$\frac{1}{2}\rho L\frac{\mathrm{d}v}{\mathrm{d}t} = F + Q_B \tag{6.1}$$

In Eq. (6.1), Q_B is the reaction force at the clamped end B (i.e. the root) of the cantilever, with the positive direction being downwards.

Taking moments about the root B results in a second equation of motion

$$\frac{1}{3}\rho L^2\frac{\mathrm{d}v}{\mathrm{d}t} = FL - M_P \tag{6.2}$$

Introduction to Impact Dynamics, First Edition. T.X. Yu and XinMing Qiu.

Published 2018 by John Wiley & Sons Singapore Pte. Ltd.

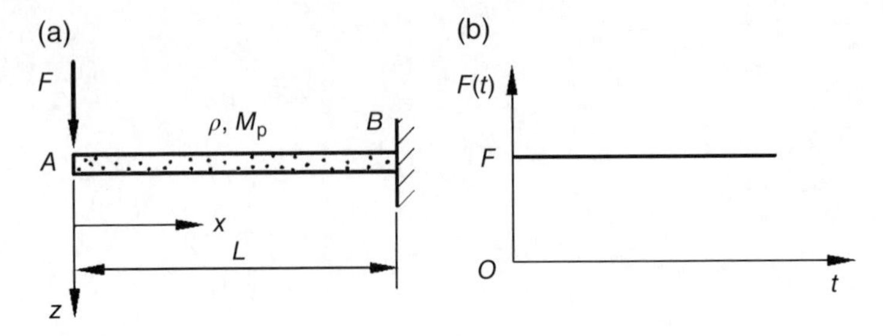

Figure 6.1 A cantilever beam under step loading.

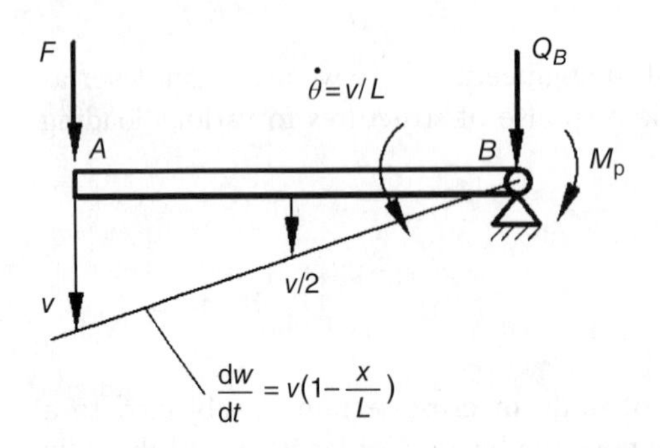

Figure 6.2 The rigid body motion and the velocity distribution diagram of a cantilever under a moderate step force at its tip ($F_c < F < 3F_c$).

This gives a constant acceleration at the tip A,

$$\frac{\mathrm{d}v}{\mathrm{d}t} = \frac{3(FL - M_P)}{\rho L^2} = \frac{3(F - F_c)}{\rho L} \tag{6.3}$$

Substituting Eq. (6.3) into Eq. (6.1) gives the shear force at root B

$$Q_B = \frac{1}{2}(F - 3F_c) \tag{6.4}$$

As shown in Figure 6.2, similar to the velocity distribution, the acceleration is linearly distributed along the cantilever, i.e.

$$\frac{\mathrm{d}^2 w}{\mathrm{d}t^2} = \left(1 - \frac{x}{L}\right)\frac{\mathrm{d}v}{\mathrm{d}t} \tag{6.5}$$

The distribution of acceleration is proportional to the distributed inertia forces along the beam. Hence, from the relation between the distributed inertia force and the shear force,

$$\frac{\mathrm{d}Q}{\mathrm{d}x} = -q = \rho\frac{\mathrm{d}^2 w}{\mathrm{d}t^2} \tag{6.6}$$

Considering the boundary condition, $Q(x) = -F$ at $x = 0$, integration of Eq. (6.6) leads to the distribution of shear force:

$$Q(x) = -F + \int_0^x \rho \frac{\mathrm{d}^2 w}{\mathrm{d}t^2} \cdot \mathrm{d}x = -F + 3(F - F_c)\left(\frac{x}{L} - \frac{x^2}{2L^2}\right) \tag{6.7}$$

By integrating $\frac{\mathrm{d}M}{\mathrm{d}x} = -Q$, the bending moment distribution is found to be

$$M(x) = -\int_0^x Q(x)\ \mathrm{d}x = Fx - \frac{3}{2}(F - F_c)\left(\frac{x^2}{L} - \frac{x^3}{3L^2}\right) \tag{6.8}$$

It is clear from the relationship between bending moment and shear force, $\frac{\mathrm{d}M}{\mathrm{d}x} = -Q$, that M is a maximum at the point where $Q = 0$. The shear force and bending moment of a cantilever under a moderate step force at its tip are depicted in Figure 6.3. When $F < 3F_c$, Eq. (6.7) gives $Q_B < 0$. Therefore, as the distance to tip A increases, the shear force monotonically increases but is always negative, while the bending moment increases monotonically from zero, i.e., $M(x) \leq M_B = M_P$, with the maximum bending moment occurring at the root of the cantilever.

However, when $F > 3F_c$, Eq. (6.7) gives $Q_B > 0$. Thus, an interior point $\bar{x}$ will exist where $Q(\bar{x}) = 0$ and $M(\bar{x}) = M_{max} > M_B$. Therefore, the present deformation mechanism with a plastic hinge at the root is only valid for the moderate dynamic load case, i.e., $F_c < F < 3F_c$.

From the above analysis, for an intense dynamic force, $F > 3F_c$, it is implied that a plastic hinge with $M = M_{max} = M_P$ could form between tip A and root B. Hence, let us examine an alternative deformation mechanism with a plastic hinge at an interior point H $(x = \lambda)$ (see Figure 6.4). Thus, the angular acceleration of segment AH rotating about the hinge at H is $(\mathrm{d}v/\mathrm{d}t)/\lambda$, whilst the segment HB remains stationary.

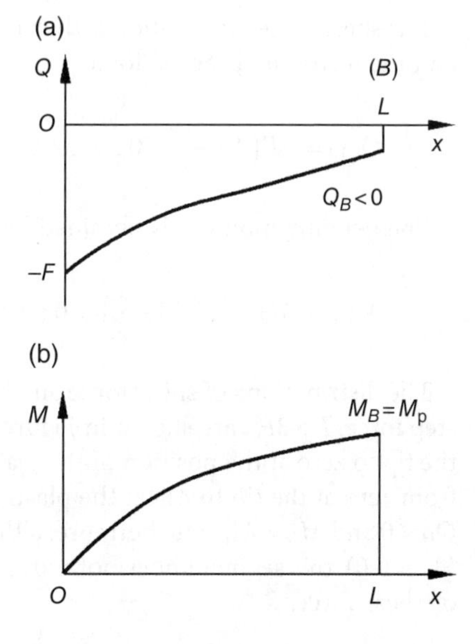

Figure 6.3 The shear force and bending moment of a cantilever under a moderate step force F $(F_c < F < 3F_c)$ applied at its tip.

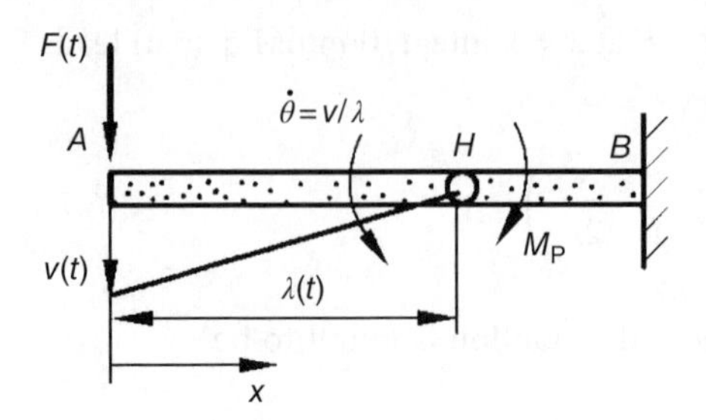

Figure 6.4 The rigid body motion and the velocity distribution diagram of a cantilever under an intense step force ($F > 3F_c$) at its tip.

At the position of plastic hinge H, $M(\lambda) = M_{max} = M_P$ and $Q(\lambda) = 0$; hence, referring to Eqs (6.1) and (6.2), the equations of motions for segment AH are

$$\frac{1}{2}\rho\lambda\frac{dv}{dt} = F \tag{6.9}$$

$$\frac{1}{3}\rho\lambda^2\frac{dv}{dt} = F\lambda - M_P \tag{6.10}$$

in which $v(t)$ and $\lambda(t)$ are two unknowns. In fact, the acceleration of the free end A can be easily found from Eq. (6.9):

$$\frac{dv}{dt} = \frac{2F}{\rho\lambda} \tag{6.11}$$

Also, the transverse acceleration of the cross-section located at x is

$$\frac{d^2w}{dt^2} = \left(1 - \frac{x}{\lambda}\right)\frac{dv}{dt} = \frac{2F}{\rho\lambda}\left(1 - \frac{x}{\lambda}\right) \tag{6.12}$$

The shear force distribution is then obtained from the relation between the shear force and the distributed inertia force:

$$Q(x) = -F\left(1 - \frac{x}{\lambda}\right)^2, \quad 0 \le x \le \lambda \tag{6.13}$$

The bending moment is obtained by integrating the above shear force:

$$M(x) = M_P - \frac{1}{3}F\lambda\left(1 - \frac{x}{\lambda}\right)^3, \quad 0 \le x \le \lambda \tag{6.14}$$

The distributions of shear force and bending moment of a cantilever under an intense step force, $F > 3F_c$, are shown in Figure 6.5. Clearly, the shear force decreases from F at the tip to zero at the position of the plastic hinge, H, and the bending moment increases from zero at the tip to M_P at the plastic hinge H. At the position of the plastic hinge H, $Q_H = 0$ and $M_H = M_P$. Furthermore, although $M(x) = M_P$ is held throughout segment HB ($\lambda \le x \le L$), this segment does not move or deform as it is not subjected to any transverse or shear force.

Figure 6.5 The shear force and bending moment of a cantilever under an intense step force ($F > 3F_c$).

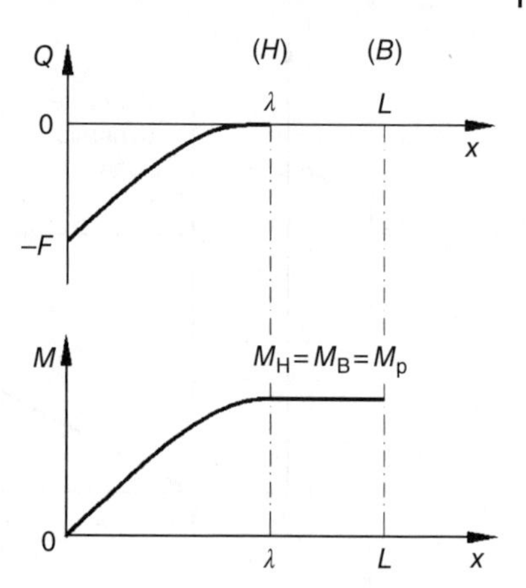

The relationship between plastic hinge position and the applied force can be easily established. Substituting Eq. (6.11) into Eq. (6.10) results in $F\lambda/3 = M_P$, i.e.

$$\lambda = 3M_P/F \tag{6.15}$$

which clearly indicates the dependence of the position of the plastic hinge on the intensity of the applied force. The more intense the force F, the closer the plastic hinge is to the tip A where the force is applied.

The combination of Eqs (6.15) and (6.11) leads to a relationship between the acceleration at the tip and the magnitude of the applied force:

$$\frac{\mathrm{d}v}{\mathrm{d}t} = \frac{2F^2}{3\rho M_P} \tag{6.16}$$

In this range of force, the acceleration at the tip is proportional to the square of the applied force F.

Remarks

The dynamic deformation mechanism changes with the magnitude of the step-force F; when F is much higher than F_c, i.e., $F > 3F_c$, the dynamic deformation mechanism is different from the static collapse mechanism.

- The more intense the dynamic load, the closer the deformed region is to the loading point.
- The acceleration of the tip is expressed as

$$\frac{\mathrm{d}v}{\mathrm{d}t} = \begin{cases} 0, & 0 \le F \le F_c \\ 3(F - F_c)/(\rho L), & F_c \le F \le 3F_c \\ 2F^2/(3\rho L M_P), & F > 3F_c \end{cases} \tag{6.17}$$

which displays a non-linear relation between $\mathrm{d}v/\mathrm{d}t$ and F, as plotted in Figure 6.6.

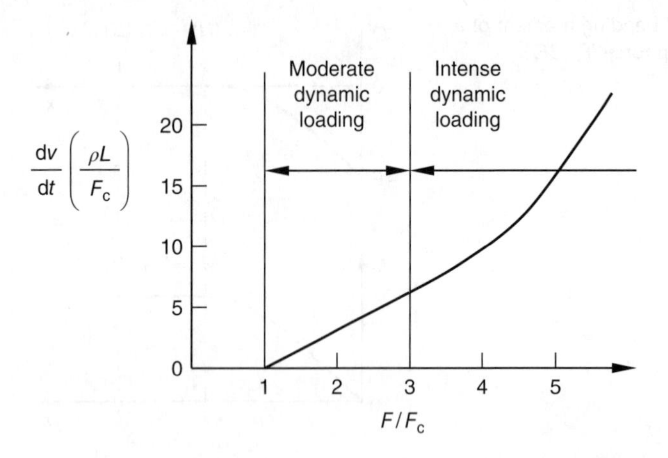

Figure 6.6 The acceleration at the tip of a cantilever varies with the magnitude of the applied force.

6.2 Response to Pulse Loading

In this section, we discuss the dynamic response of a cantilever to pulse loading. Assuming the deformation of the cantilever remains small in comparison with its length, the initial configuration is used when formulating equilibrium and geometrical relationships.

6.2.1 Rectangular Pulse

As shown in Figure 6.7, a rectangular pulse is defined by

$$F(t) = \begin{cases} F_0, & 0 \le t \le t_d \\ 0, & t > t_d \end{cases} \tag{6.18}$$

If this pulse is applied at the tip of a cantilever, and if the force is intensive, i.e., $F_0 > 3F_c = 3M_P/L$, then in *phase I* ($0 \le t \le t_d$), a stationary plastic hinge forms at $\lambda_0 = 3M_P/F_c < L$, as obtained earlier.

After the sudden removal of the force at $t = t_d$, the plastic hinge may move away from $x = \lambda_0$, and the hinge position $\lambda = \lambda(t)$ may vary with time t, i.e., a traveling hinge is formed.

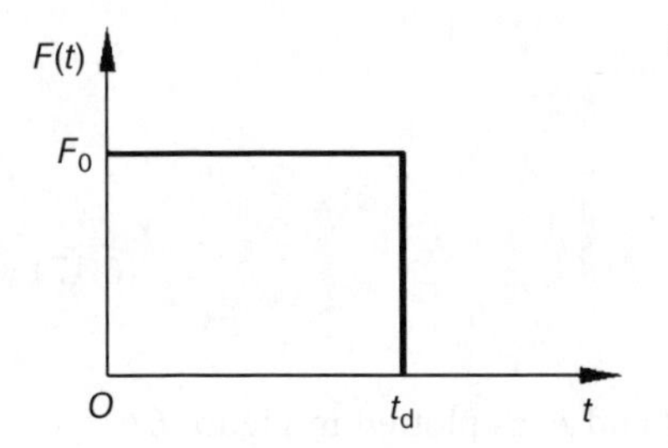

Figure 6.7 Rectangular pulse.

Considering the change in the translational momentum of segment AH from 0 to t, we have

$$\frac{1}{2}\rho\lambda(t)v(t)=\int_0^t F(\bar{t})\,\mathrm{d}\bar{t}=\begin{cases}F_0 t, & 0\le t\le t_\mathrm{d}\\ F_0 t_\mathrm{d}\equiv I, & t>t_\mathrm{d}\end{cases}\tag{6.19}$$

where $I\equiv\int_0^{t_\mathrm{d}} F(\bar{t})\,\mathrm{d}\bar{t}$ is the total impulse imparted during the loading duration. Secondly, by considering the change in the moment of momentum of segment AH about the initial tip position A during the period before the hinge moves to a location λ, we obtain

$$\int_0^\lambda \rho\frac{\mathrm{d}w}{\mathrm{d}t}x\,\mathrm{d}x=M_\mathrm{P}t\tag{6.20}$$

and the velocity distribution is

$$\frac{\mathrm{d}w}{\mathrm{d}t}=v\left(1-\frac{x}{\lambda}\right)\tag{6.21}$$

From Eqs (6.20) and (6.21) it is found that

$$\frac{1}{6}\rho\lambda^2(t)v=M_\mathrm{P}t\tag{6.22}$$

In the first phase of the dynamic response, *phase I*, $(0\le t\le t_\mathrm{d})$, Eqs (6.20) and (6.22) give the same results as those obtained in the case of step-loading, i.e.,

$$\lambda_0=\frac{3M_\mathrm{P}}{F_0},\quad\frac{\mathrm{d}v}{\mathrm{d}t}=\frac{2F_0^2}{3\rho M_\mathrm{P}}\tag{6.23}$$

Hence, at the instant of unloading, $t=t_\mathrm{d}$, the velocity and deflection at the tip are obtained by integrating Eq. (6.23):

$$v_1=\frac{2F_0^2 t_\mathrm{d}}{3\rho M_\mathrm{P}}=\frac{2F_0 I}{3\rho M_\mathrm{P}}\tag{6.24}$$

$$\Delta_1=\frac{1}{2}v_1 t_\mathrm{d}=\frac{F_0^2 t_\mathrm{d}^2}{3\rho M_\mathrm{P}}=\frac{I^2}{3\rho M_\mathrm{P}}\tag{6.25}$$

and at this instant, the rotation angle at the stationary hinge H_0 is

$$\theta_1=\frac{\Delta_1}{\lambda_0}=\frac{I^2F_0}{9\rho M_\mathrm{P}^2}\tag{6.26}$$

After the removal of the force, i.e. during $t\ge t_\mathrm{d}$, the equation of motion, Eq. (6.19), gives

$$\frac{1}{2}\rho\lambda(t)v(t)=I\tag{6.27}$$

By comparing Eq. (6.27) with Eq. (6.22), the position of the traveling plastic hinge is found to be

$$\lambda(t)=\frac{3M_\mathrm{P}}{I}t\tag{6.28}$$

which implies that the hinge moves from λ_0 towards the root ($x = L$) at a constant speed

$$\frac{d\lambda}{dt} = \frac{3M_P}{I} \tag{6.29}$$

This traveling hinge arrives at the root B at

$$t_2 = \frac{IL}{3M_P} \tag{6.30}$$

Thereby, the deformation during $t_d \le t \le t_2$ is called the second phase response, *phase II*; in this phase, the velocity and displacement at the tip are calculated as

$$v(t) = \frac{2I}{\rho\lambda} = \frac{2I^2}{3\rho M_P t} \tag{6.31}$$

$$\Delta(t) = \Delta_1 + \int_{t_d}^{t} \frac{2I^2}{3\rho M_P t}\, dt = \frac{I^2}{3\rho M_P}\left\{1 + 2\ln\left(\frac{t}{t_d}\right)\right\} \tag{6.32}$$

It can be seen from this relation that the velocity at the tip decreases with time. By substituting the terminal time of phase II, i.e., t_2 given by Eq. (6.30), into Eqs (6.31) and (6.32), the velocity and displacement at the tip are obtained:

$$v_2 = v(t_2) = \frac{2I}{\rho L} \tag{6.33}$$

$$\Delta_2 = \Delta(t_2) = \frac{I^2}{3\rho M_P}\left\{1 + 2\ln\left(\frac{F_0 L}{3M_P}\right)\right\} \tag{6.34}$$

In *phase III* ($t \ge t_2$), i.e., after the traveling hinge reaches the root, the plastic hinge is fixed at the root whilst the cantilever rotates about the root as a rigid body. Taking the change in the moment of momentum about the root for the period after t_2,

$$\frac{1}{3}\rho L^2 (v - v_2) = -M_P (t - t_2) \tag{6.35}$$

From Eq. (6.35), the velocity at the tip is given by

$$v(t) = \frac{3}{\rho L}\left(I - \frac{M_P t}{L}\right) \tag{6.36}$$

Obviously, $v(t)$ is gradually reduced to zero, which marks the termination of the dynamic response. Therefore, the total response time is

$$t_f = \frac{IL}{M_P} = 3t_2 \tag{6.37}$$

The motion of the cantilever ceases at $t = t_f$, and the final tip deflection is

$$\Delta_f = \Delta(t_f) = \Delta_2 + \int_{t_2}^{t_f} \frac{3}{\rho L}\left(I - \frac{M_P t}{L}\right) dt \tag{6.38}$$

which can be rewritten into

$$\Delta_f = \frac{I^2}{\rho M_P}\left\{1 + \frac{2}{3}\ln\left(\frac{F_0}{3F_c}\right)\right\} \tag{6.39}$$

Accordingly, the rotation angle at the root during phase III is

$$\theta_3 = \frac{2I^2}{3\rho L M_P} \tag{6.40}$$

Figure 6.8 shows the variations of the hinge position and the tip velocity during the dynamic response, for the rectangular pulse $F = 12F_c$. In phase I, a stationary plastic hinge forms at position λ_0, the deformation is a rigid body rotation with respect to this stationary plastic hinge, and the tip velocity increases with time; in phase II, the plastic hinge moves towards the root at a constant traveling speed, while the tip velocity decreases with time; and finally in phase III, the plastic hinge remains stationary at the root, and the tip velocity decreases linearly with time, until it is reduced to zero, corresponding to the termination of the entire dynamic response.

To find the deformed shape, we can first calculate the curvature produced in phase II, using

$$\frac{d\theta}{dt} = \kappa(x)\frac{d\lambda}{dt} = \frac{v}{\lambda} \tag{6.41}$$

which gives the changing rate of the rotation angle, i.e., the angular velocity of any cross-section along the cantilever. Here the kinematic relation between $d\theta/dt$ and v is based on the initial configuration of the beam, combined with the definition of curvature $\kappa = d\theta/d\lambda$. Therefore, the curvature $\kappa(x)$ is related to the position of plastic hinge λ by Eq. (6.41). Noting the expression of curvature of a straight beam,

$$\kappa(x) = \frac{d^2w}{dx^2} \tag{6.42}$$

and integrating it twice with respect to x, we find the deflection produced in phase II:

$$w_2(x) = \frac{2I^2}{3\rho L M_P}\left(\ln\frac{L}{x} - \frac{x}{L} - 1\right), \quad \lambda_0 \le x \le L \tag{6.43}$$

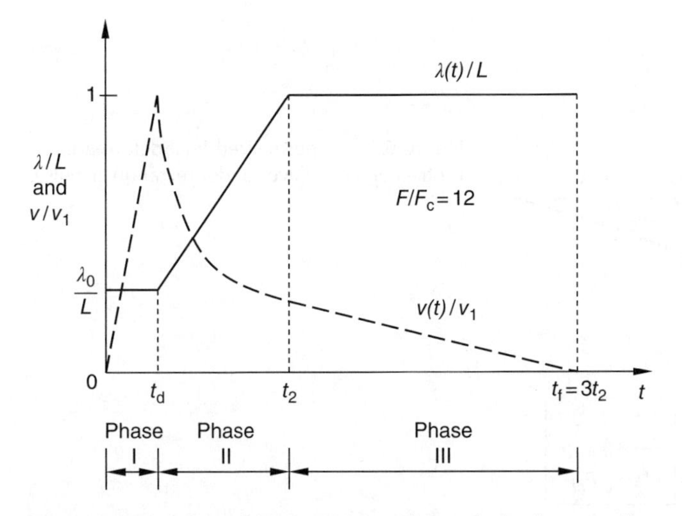

Figure 6.8 The plastic hinge position and tip velocity for a cantilever under a rectangular pulse.

Adding the deflections in phase I and phase III, the final deformation of beam is given by

$$w_f(x) = \begin{cases} \dfrac{I^2}{\rho L M_P}\left\{1 - \dfrac{x}{\lambda_0} + \dfrac{2}{3}\ln\left(\dfrac{L}{\lambda_0}\right)\right\}, & 0 \le x \le \lambda_0 \\ \dfrac{2I^2}{3\rho L M_P}\ln\left(\dfrac{L}{x}\right), & \lambda_0 \le x \le L \end{cases} \tag{6.44}$$

In Figure 6.9, the normalized final deflections of the cantilever are given for $f = 6$ and 12, where $f \equiv F_0/F_c$ is the normalized load intensity. The initial positions of plastic hinge are $\lambda_0/L = 0.5$ and $\lambda_0/L = 0.25$, for $f = 6$ and $f = 12$, respectively. Clearly, the initial position of the plastic hinge depends on the load intensity. In addition, the normalized final profiles near the root are coincident for $f = 6$ and 12, whereas near the tip, the deflection under the intensive load $f = 12$ is larger than that of the moderate load $f = 6$.

Let us now analyze the energy dissipation during the dynamic response. First we need to find the total input energy. For a rectangular pulse, the external work input only takes place in phase I. The displacement in each phase has already been obtained from the above derivation. At time t_d, the deflection of the tip is Δ_1 (see Eq. 6.25), so the total input energy from the rectangular pulse is

$$E_{in} = F_0\Delta_1 = \frac{I^2 F_0}{3\rho M_P} \tag{6.45}$$

Now consider the energy dissipation in each phase. Among the three phases, the deformations in phases I and III are the rotations about the stationary hinge. Hence, from the rotation angles given by Eqs (6.26) and (6.40), i.e., θ_1 and θ_3, the plastic energy dissipations of phases I and III are

$$D_1 = M_P\theta_1 = \frac{I^2 F_0}{9\rho M_P} \tag{6.46}$$

and

$$D_3 = M_P\theta_3 = \frac{2I^2}{3\rho L} \tag{6.47}$$

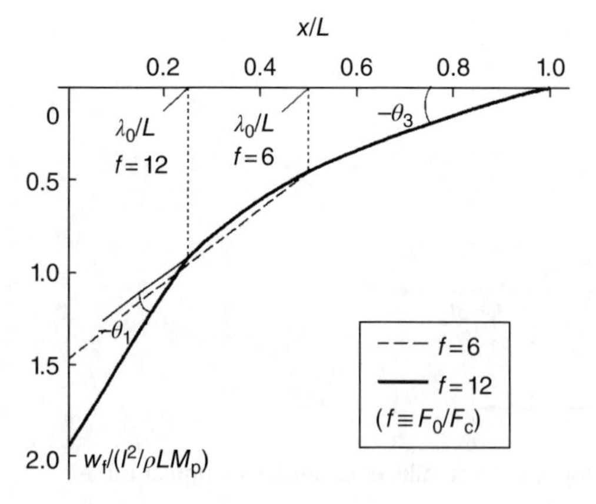

Figure 6.9 The normalized final deformation profiles of a cantilever under rectangular pulse.

respectively. From the energy conservation, the total input energy due to the rectangular pulse must be equal to the total plastic energy dissipations in the three phases, $E_{in} = D_1 + D_2 + D_3$; thus, the partitioning of the dissipations among the three phases is

$$\frac{D_1}{E_{in}} : \frac{D_2}{E_{in}} : \frac{D_3}{E_{in}} = \frac{1}{3} : \left(\frac{2}{3} - \frac{2F_c}{F_0}\right) : \frac{2F_c}{F_0} \tag{6.48}$$

This clearly indicates that for a cantilever subjected to a rectangular pulse at its tip, the plastic deformation in phase I always dissipates one-third of the total input energy, while the energy partitioning of phase II and phase III depends on the load intensity: the larger the value of F_0/F_c, the more energy is dissipated in phase II than in phase III. Phase II dissipates the most significant portion of the total energy for a very intense load, i.e. $F_0/F_c >> 1$; therefore, in this case, the traveling hinge will dissipate more energy than will stationary hinges.

6.2.2 General Pulse

In this section, we discuss the dynamic response of a cantilever under a general pulse loading.

Now consider a *non-negative* force pulse of *arbitrary shape*, $F(t) \geq 0$, which is applied at the tip of a cantilever (Figure 6.10). As any force that is smaller than the static collapse force F_c does not make the rigid-plastic cantilever deform, it is convenient to take $t = 0$ at the instant when $F(t)$ first reaches F_c.

After the force is applied at $t = 0$, the deformation mechanism with a plastic hinge at an interior point H $(x = \lambda)$, shown in Figure 6.4, is valid. For $t > 0$, as the applied force is varying with time, the position of the plastic hinge will also vary with time. To determine how the hinge position $\lambda(t)$ and tip velocity $v(t)$ depend upon $F(t)$, we assume that at instant t a single hinge H appears at $x = \lambda(t)$, as sketched in Figure 6.4.

The equations of motion for segment AH are

$$\frac{\mathrm{d}}{\mathrm{d}t}\left(\frac{1}{2}\rho\lambda v\right) = F(t) \tag{6.49}$$

$$\frac{\mathrm{d}}{\mathrm{d}t}\left(\frac{1}{6}\rho\lambda^2 v\right) = M_P \tag{6.50}$$

Figure 6.10 Schematic of a general pulse.

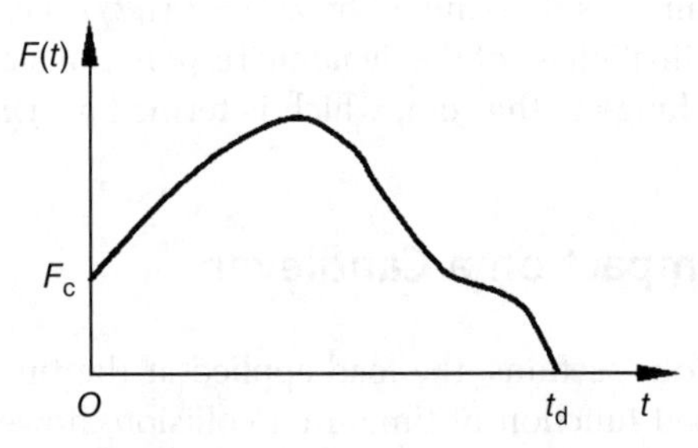

Integrating Eqs (6.49) and (6.50) with respect to time leads to

$$\frac{1}{2}\rho\lambda v = \int_0^t F(t)\,\mathrm{d}t \tag{6.51}$$

$$\frac{1}{6}\rho\lambda^2 v = M_P t \tag{6.52}$$

The position of the plastic hinge and the velocity at the tip are obtained by solving Eqs (6.51) and (6.52), which gives

$$\lambda(t) = \frac{3M_P t}{\int_0^t F(t)\,\mathrm{d}t} \tag{6.53}$$

$$v(t) = \frac{2}{3\rho M_P t}\left\{\int_0^t F(t)\,\mathrm{d}t\right\}^2 \tag{6.54}$$

Differentiating Eq. (6.53) with respect to time gives the speed of the traveling plastic hinge as

$$\frac{\mathrm{d}\lambda}{\mathrm{d}t} = 3M_P\left\{\int_0^t F(t)\,\mathrm{d}t - tF(t)\right\}/\left\{\int_0^t F(t)\,\mathrm{d}t\right\}^2 \tag{6.55}$$

It is evident from Eq. (6.55) that the direction and speed of the hinge's travel depend on the sign of $\mathrm{d}\lambda/\mathrm{d}t$, i.e., the sign of $\int_0^t F(t)\,\mathrm{d}t - tF(t)$. Hereafter

1) If the force is constant, i.e., $F(t) = F_0$, which is the step-loading case discussed in Section 6.1, then $\int_0^t F(t)\,\mathrm{d}t = tF_0$ gives the traveling speed of the hinge $\mathrm{d}\lambda/\mathrm{d}t \equiv 0$, so the hinge is stationary at position $\lambda = 3M_P/F_0$.
2) If $F(t)$ is a monotonically increasing function of time, then $\int_0^t F(t)\,\mathrm{d}t < tF$ and $\mathrm{d}\lambda/\mathrm{d}t < 0$; i.e., the plastic hinge is a traveling hinge moving toward the free end A.
3) By contrast, if $F(t)$ is a monotonically decreasing function of time, then $\int_0^t F(t)\,\mathrm{d}t > tF$ and $\mathrm{d}\lambda/\mathrm{d}t > 0$, i.e., the traveling hinge moves towards the root B.

After the termination of any pulse, $t > t_d$, then $\int_0^t F(t)\,\mathrm{d}t = I$; now the position of the plastic hinge is determined by $\lambda(t) = 3M_P t/I$, i.e., the hinge travels at a constant speed, and the final phase of the dynamic response of a cantilever must be a rigid-body rotation about a hinge at the root, which is termed a *dynamic mode*.

6.3 Impact on a Cantilever

In previous sections, the load applied at the tip of the cantilever was assumed to be a prescribed function of time. In a collision, however, the force pulse is not prescribed; it depends on the compliance of colliding bodies.

Parkes (1955) first analyzed the dynamic response of a cantilever to impact. His analysis assumes that a rigid "particle" of mass m impinges on the tip of a cantilever with initial velocity v_0 and attaches to the tip during the dynamic deformation of the cantilever. The attachment assumption is very strong, which simplifies the problem greatly. Therefore, this problem is often referred to as the *Parkes problem.* In Parkes' original article, the equations were obtained by equilibrium analysis of the representative element. Here, the solution of the Parkes problem will be given by directly using the response of a cantilever under the general pulse loading, which was already obtained in Section 6.2.

Thus, the Parkes problem is idealized as shown in Figure 6.11. A cantilever of total length L, material density per unit length ρ, and fully plastic bending moment M_P is impacted at the tip A by a concentrated mass m with initial transverse velocity v_0.

Let $v(t)$ denote the velocity of the tip mass at any time $t \geq 0$ and $v(0) = v_0$. For the concentrated mass, the equation of motion gives the reaction force by the cantilever:

$$F(t) = -m\frac{dv}{dt} \tag{6.56}$$

In fact, Eq. (6.56) is the interaction shear force between the mass and the cantilever, expressed as the variation of the velocity of the colliding mass, which is also the velocity of the free end of the cantilever. In this way, the analysis of the unknown shear force applied to the cantilever is avoided. Here Eq.(6.56) is taken as a general pulse, from the equations of motion for segment AH as given in Eqs (6.49) and (6.50):

$$\frac{1}{2}\rho\lambda v = \int_0^t F(t)\ dt = -m\int_0^t \frac{dv}{dt}\ dt = m(v_0 - v) \tag{6.57}$$

$$\frac{1}{6}\rho\lambda^2 v = M_P t \tag{6.58}$$

The velocity of the mass, i.e., the velocity of the free end of cantilever, is obtained from Eq. (6.57):

$$v = \frac{v_0}{1 + \rho\lambda/(2m)} = \frac{v_0}{1 + \dfrac{1}{2\gamma}\dfrac{\lambda}{L}} \tag{6.59}$$

where $\gamma \equiv m/\rho L$ denotes the *mass ratio* of the concentrated mass to the mass of the cantilever. By substituting the velocity of the mass (Eq. 6.59) into the expression of the equation of moment of momentum (Eq. 6.58), we obtain the relation between the plastic hinge position and time:

$$t = \frac{\rho v \lambda^2}{6M_P} = \frac{\rho v_0}{6M_P}\lambda^2\left(1 + \frac{1}{2\gamma}\frac{\lambda}{L}\right)^{-1} \tag{6.60}$$

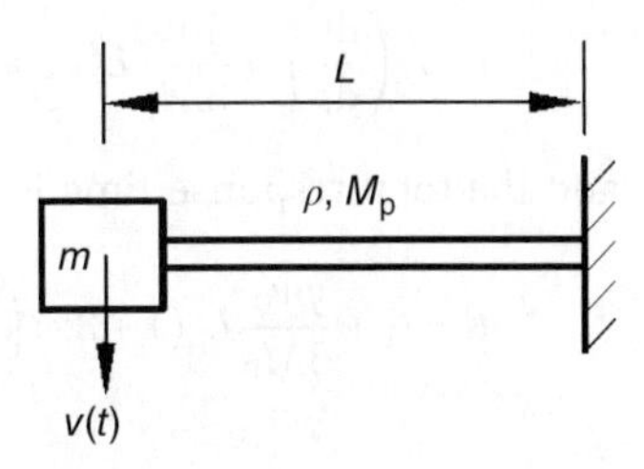

Figure 6.11 A cantilever impacted by a concentrated mass at the free end.

It should be noted that Eq. (6.60) is not an explicit function of plastic hinge position expressed by time, $\lambda(t)$. On the contrary, λ instead of t serves as a process parameter in the analysis of dynamic response, which makes the problem much simpler. The variables' tendency is clearly obtained from Eqs (6.59) and (6.60), i.e., as time increases, the plastic hinge travels from the free end towards the root, while the velocity of the free end decreases.

Differentiating Eq. (6.60) we obtain

$$\frac{\mathrm{d}t}{\mathrm{d}\lambda} = \frac{\rho v_0}{6M_\mathrm{P}}\lambda\left(1+\frac{\lambda}{2\gamma L}\right)^{-2}\left(2+\frac{\lambda}{2\gamma L}\right) \tag{6.61}$$

Therefore, the moving speed of the plastic hinge is the reciprocal of Eq. (6.61):

$$\frac{\mathrm{d}\lambda}{\mathrm{d}t} = \frac{6M_\mathrm{P}}{\rho v_0}\lambda^{-1}\left(2+\frac{\lambda}{2\gamma L}\right)^{-1}\left(1+\frac{\lambda}{2\gamma L}\right)^{2} \tag{6.62}$$

Equations (6.60) and (6.62) indicate that at the initial instant $t = 0$, the plastic hinge is supposed to be at the free end, $\lambda = 0$, with infinite moving speed, $\dot{\lambda} = \infty$; when the hinge travels from the tip towards the root, its speed decreases.

The deformation process of the plastic hinge moving from the free end to the root is denoted as *phase I*, or the *transient phase*, of the structure. This traveling hinge phase ends when the hinge arrives at the root B, i.e. $\lambda = L$, so

$$t_1 = \frac{\rho v_0}{6M_\mathrm{P}}L^2\left(1+\frac{1}{2\gamma}\right)^{-1} \tag{6.63}$$

$$v_1 = v_0\left(1+\frac{1}{2\gamma}\right)^{-1} \tag{6.64}$$

When $t \geq t_1$, the plastic hinge fixes at the root B and the cantilever-mass system rotates about the hinge as a rigid body. This is termed *phase II*, or the *modal phase.*

Taking the change in the moment of momentum about the root, we have

$$m\frac{\mathrm{d}v}{\mathrm{d}t}L + \frac{1}{3}\rho L^3\left(\frac{1}{L}\frac{\mathrm{d}v}{\mathrm{d}t}\right) = -M_\mathrm{P} \tag{6.65}$$

The acceleration of the tip is given by Eq. (6.65):

$$\frac{\mathrm{d}v}{\mathrm{d}t} = -\frac{M_\mathrm{P}/L}{m+\rho L/3} = -\frac{3M_\mathrm{P}}{\rho L^2}\frac{1}{1+3\gamma} \tag{6.66}$$

The right side of Eq. (6.66) is a constant, which implies a uniformly decelerating motion of the cantilever mass system. When the tip velocity is reduced to zero, the dynamic response of the structure is terminated. Therefore, the duration of phase II is

$$v_1\Big/\left(\frac{\mathrm{d}v}{\mathrm{d}t}\right) = \frac{\rho v_0}{3M_\mathrm{P}}L^2(1+3\gamma)\left(1+\frac{1}{2\gamma}\right)^{-1} \tag{6.67}$$

and the total response time is

$$t_\mathrm{f} = t_1 + \frac{\rho v_0}{3M_\mathrm{P}}L^2(1+3\gamma)\left(1+\frac{1}{2\gamma}\right)^{-1} = \frac{mv_0L}{M_\mathrm{P}} \tag{6.68}$$

As $mv_0 = I$ is the total impulse impacted on mass m, t_f predicted by Eq. (6.68) is simplified to

$$t_f = \frac{IL}{M_P} \tag{6.69}$$

which is exactly the same as the total response time in the case of a rectangular pulse (see Eq. 6.37).

The angle of rotation at the root during phase II is

$$\theta_{Bf} = \frac{1}{2}\frac{v_1}{L}\frac{v_1}{-dv/dt} = \frac{\rho L^2 v_0}{6M_P}(1+3\gamma)\left(1+\frac{1}{2\gamma}\right)^{-2} \tag{6.70}$$

The above analysis indicates that all response variables depend on the mass ratio $\gamma \equiv m/\rho L$. The variation of the tip velocity and the hinge locations with time of several mass ratios, $\gamma = 0.2, 1, 5$, are given in Figure 6.12. For a large colliding mass, phase I (transient phase) is very short and negligible, i.e., phase II (modal phase) is important, but for a small colliding mass, the transient response in phase I is important. For example, for a bullet of small mass and with high velocity impacting the free end of a cantilever, the transient response is important, and the profile of the deformed beam will obviously be curved.

To find the distributions of shear force and bending moment in the beam, we start from the velocity distribution in phase I:

$$\dot{w} = v\left(1-\frac{x}{\lambda}\right), \quad 0 \le x \le \lambda \tag{6.71}$$

where the dot denotes differentiation with respect to time, i.e., $\dot{w} = dw/dt$ etc. Now differentiating Eq. (6.71) with respect to time leads to

$$\ddot{w} = \dot{v}\left(1-\frac{x}{\lambda}\right) + \frac{vx}{\lambda^2}\dot{\lambda}, \quad 0 \le x \le \lambda \tag{6.72}$$

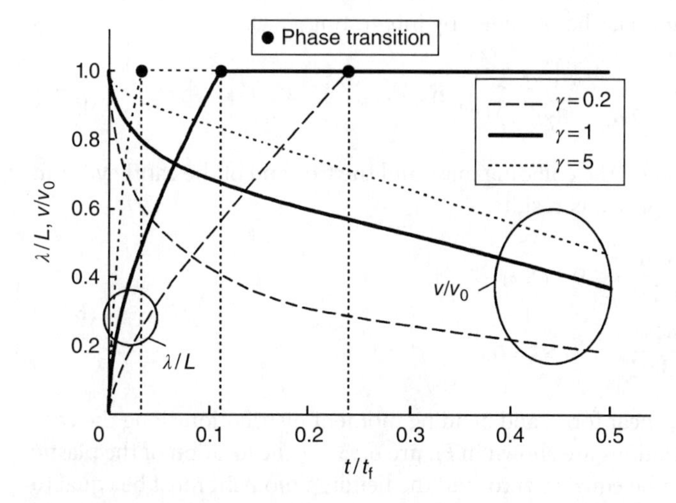

Figure 6.12 The hinge location and tip velocity versus time for a cantilever impacted by a mass at its free end.

Equation (6.72) is the acceleration distribution of the cantilever. The first term on the right-hand side is caused by the acceleration of the concentrated mass, and the second term is caused by the increasing length of the rotating segment *AH*. Using the velocity of the free end given by Eq. (6.59), the acceleration of the concentrated mass $\dot{v}$ is given by

$$\dot{v} = \frac{dv}{d\lambda}\frac{d\lambda}{dt} = -\frac{6M_P}{\rho\lambda(\lambda + 4\gamma L)} \tag{6.73}$$

The moving speed of the plastic hinge $\dot{\lambda}$ is given in Eq. (6.62), and substituting Eqs (6.62) and (6.73) into Eq. (6.72) leads to the acceleration distribution of the cantilever:

$$\ddot{w} = \frac{6M_P}{\rho\lambda(\lambda + 4\gamma L)}\left[-1 + 2\lambda^{-2}(\lambda + \gamma L)x\right] \tag{6.74}$$

From Eq. (6.74), the sign of the acceleration is switched from the free end to the fixed end, i.e., $\ddot{w}|_{x=0} < 0$ and $\ddot{w}|_{x=\lambda} > 0$. As discussed in Chapter 5, the inertia pressure produced by the distribution of acceleration can be obtained. The shear force is equal to the integration of the transversely distributed pressure caused by the inertia force. With the boundary condition of the shear force at the free end,

$$Q(0) = -m\dot{v} = -\frac{6mM_P}{\rho\lambda(\lambda + 4\gamma L)} \tag{6.75}$$

the shear force distribution is given by

$$\begin{aligned} Q(x) &= Q(0) + \int_0^x \rho\ddot{w}\, dx \\ &= -\frac{6\gamma M_P L}{\lambda(\lambda + 4\gamma L)}\left(1 + \frac{x}{\gamma L} - \frac{\lambda + \gamma L}{\gamma L}\frac{x^2}{\lambda^2}\right), 0 \le x \le \lambda \end{aligned} \tag{6.76}$$

The bending moment at the free end *A* is zero. By using this boundary condition, the bending moment distribution can be obtained by integrating $Q(x)$:

$$M(x) = \frac{6\gamma M_P L x}{\lambda(\lambda + 4\gamma L)}\left(1 + \frac{x}{2\gamma L} - \frac{\lambda + \gamma L}{3\gamma L}\frac{x^2}{\lambda^2}\right), \quad 0 \le x \le \lambda \tag{6.77}$$

The interaction force between the colliding mass and the free end of the cantilever can be found by combining the previous results:

$$F(t) = -m\dot{v} = \begin{cases} \dfrac{6\gamma M_P L}{\lambda(\lambda + 4\gamma L)}, & 0 \le t \le t_1 \\ \dfrac{3\gamma M_P}{(1 + 3\gamma)L}, & t_1 \le t \le t_f \end{cases} \tag{6.78}$$

The velocity, acceleration, shear force, and bending moment distribution along the cantilever given by the above relations are shown in Figure 6.13. At the location of the plastic hinge λ, the shear force must be equal to zero, and the bending moment must be equal to M_P. Clearly, from the free end to the root, the bending moment increases monotonically. The segment *HB* has neither plastic deformation nor relative rotation.

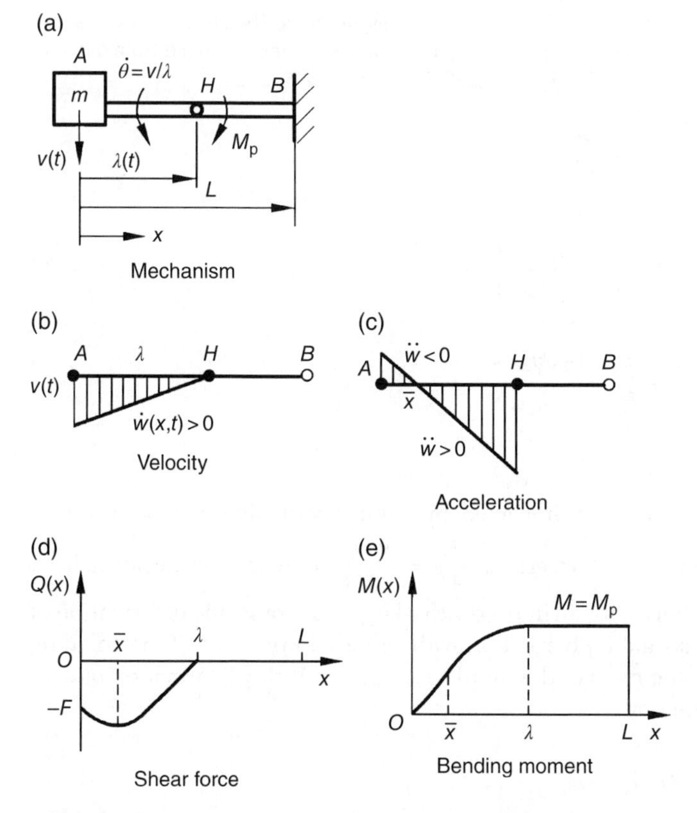

Figure 6.13 The velocity, acceleration, shear force, and bending moment distribution along the cantilever.

To find the deformed shape of the cantilever after impact, we note that the curvature generated by the traveling hinge can be determined from Eq. (6.41), i.e., $\frac{d\theta}{dt} = \kappa\frac{d\lambda}{dt} = \frac{v}{\lambda}$; the distribution of curvature is

$$\kappa(x) = \frac{\rho v_0^2}{6M_P}\left(2 + \frac{x}{2\gamma L}\right)\left(1 + \frac{x}{2\gamma L}\right)^{-3} = w_f''(x) \tag{6.79}$$

where $'$ denotes partial differential with respect to position, i.e., $\partial()/\partial x$.

Using the boundary condition at the root, $w_f'(L) = -\theta_{Bf}$ and $w_f(L) = 0$. Integrating Eq. (6.79) twice, we obtain the distribution of deflection along the cantilever:

$$w_f(x) = \frac{\rho v_0{}^2 L^2 \gamma^2}{3M_P}\left\{\frac{1}{1+2\gamma} - \frac{x}{x+2\gamma L} + 2\ln\left(\frac{1+2\gamma}{x/L+2\gamma}\right)\right\} \tag{6.80}$$

The *final tip deflection* is therefore

$$\Delta_f \equiv w_f(0) = \frac{\rho v_0{}^2 L^2 \gamma^2}{3M_P}\left\{\frac{1}{1+2\gamma} + 2\ln\left(1 + \frac{1}{2\gamma}\right)\right\} \tag{6.81}$$

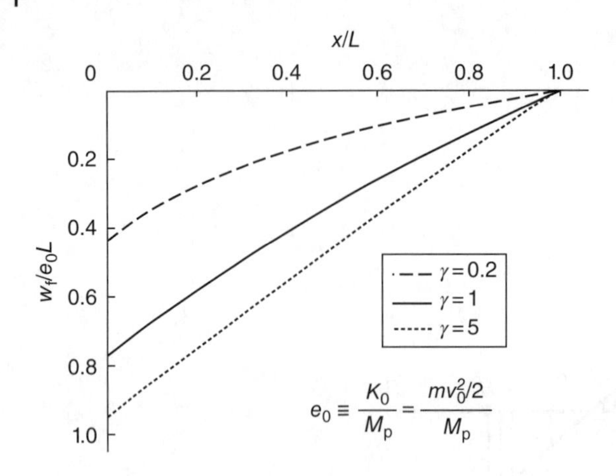

Figure 6.14 The final profiles of a cantilever under impacts from different mass ratios.

The final profiles of a cantilever impacted by a mass with different mass ratios, $\gamma = 0.2, 1, 5$, are given in Figure 6.14, where $e_0 \equiv \dfrac{K_0}{M_P} = \dfrac{m\,v_0^2/2}{M_P}$ is defined as the normalized initial kinetic energy. Obviously, for an impact with a large mass ratio, the deformation of the cantilever is close to a rigid body rotation with respect to the root. If the colliding mass is much larger than the mass of the cantilever, i.e., $\gamma \gg 1$, the deformation of cantilever can be approximated by

$$w_f(x) \approx \frac{\rho {v_0}^2 L^2 \gamma}{2M_P}(1 - x/L) \tag{6.82}$$

The transverse deflection given by this relation is linearly distributed along the beam length, which implies that the cantilever remains almost straight as a result of a rigid body rotation about the root. The maximum deflection at the tip is

$$\Delta_f \approx \frac{\rho {v_0}^2 L^2 \gamma}{2M_P} = \frac{1}{2} m v_0^2 \frac{L}{M_P} = e_0 L \tag{6.83}$$

By contrast, if the colliding mass is very light compared with that of the cantilever itself, i.e., $\gamma >> 1$, then $x/(\gamma L) << 1$ yields an approximate deflection curve

$$w_f(x) \approx \frac{2\rho\, v_0^2 L^2 \gamma^2}{3M_P} \ln(L/x) \tag{6.84}$$

In this case the cantilever is ultimately deformed into a logarithmic curve with a final tip deflection of

$$\Delta_f \approx \frac{2\rho\, v_0^2 L^2 \gamma^2}{3M_P} \ln\left(\frac{1}{2\gamma}\right) \tag{6.85}$$

Finally, the energy conversion during the dynamic response is discussed below. The initial kinetic energy of the impinging mass, $K_0 = mv_0^2/2$, has to be dissipated in phases I and phase II, so

$$K_0 = D_1 + D_2 \tag{6.86}$$

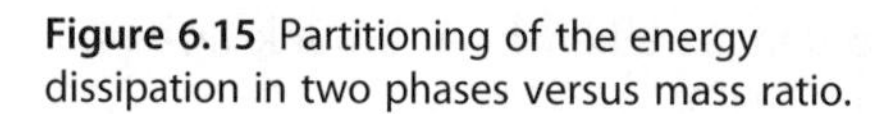

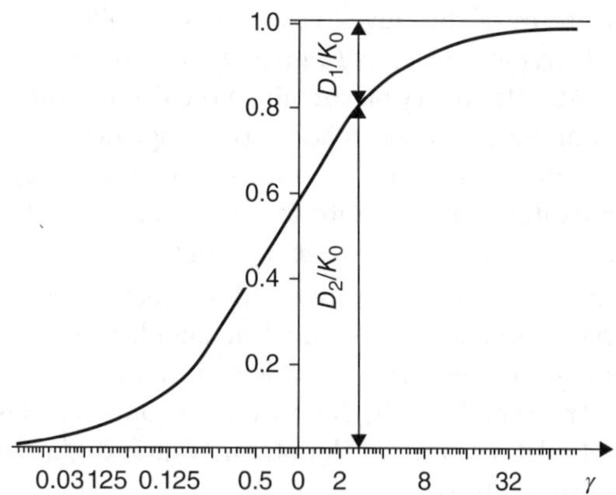

Figure 6.15 Partitioning of the energy dissipation in two phases versus mass ratio.

where D_1 and D_2 denote the energy dissipated in phase I and II, respectively, and we have

$$D_2 = M_P\, \theta_{Bf},\quad D_1 = K_0 - D_2 \tag{6.87}$$

with θ_{Bf} being the final rotation angle at root B (see Eq. 6.70). Thus,

$$\frac{D_2}{K_0} = \frac{1}{6}\rho L v_0^2 (1+3\gamma)\left(1+\frac{1}{2\gamma}\right)^{-2} \Big/ \left(\frac{1}{2}mv_0^2\right) \tag{6.88}$$

Therefore, the normalized plastic energy dissipation in the two phases is

$$\frac{D_2}{K_0} = \frac{4\gamma(1+3\gamma)}{3(1+2\gamma)^2},\ \frac{D_1}{K_0} = 1 - \frac{D_2}{K_0} \tag{6.89}$$

The partition of the total energy dissipation in two phases versus the mass ratio $\gamma \equiv m/\rho L$ is plotted in Figure 6.15. It is known from Eq. (6.89) that the partition of energy dissipation depends only on the mass ratio γ. For a small colliding mass, the energy dissipated in the *transient phase* (phase I) dominates. With the increase of the colliding mass, the energy dissipated in the modal phase (phase II) will increase. If the colliding mass is much larger than the mass of the cantilever itself, i.e., $\gamma >> 1$, almost all the initial kinetic energy will be dissipated in the modal phase.

6.4 General Features of Traveling Hinges

In the last section, we saw that a rigid-perfectly plastic cantilever has all the deformation concentrated at a plastic hinge. The hinge is stationary if the applied force is constant or if the momentum distribution has converged to a mode form; if the force varies in magnitude, however, there is a transient phase of motion wherein the hinge travels along the cantilever. The traveling hinge is an important concept in rigid-plastic structural dynamics, which was first introduced by Lee and Symonds in 1952 (see the discussion in Section 5.3). They analyzed flexural deformation of a rigid-plastic free–free beam subjected to a force pulse at its midpoint. In this section, we will discuss some general features of the traveling hinge. The following content is taken from a book *Dynamic Models for Structural Plasticity* by Strong and Yu (1993).

At a stationary plastic hinge the displacement is continuous but the inclination is discontinuous. Rotation about the hinge develops a kink in the neutral axis of the beam at the hinge location. In contrast, a continuously traveling hinge leaves behind a residual curvature that is finite at every point. While both the displacement and inclination are continuous at a traveling hinge, the inclination $\theta = \mathrm{d}w/\mathrm{d}x$ may have weak discontinuities, i.e., the derivative with respect to time, $\dot{\theta} = \mathrm{d}\theta/\mathrm{d}t$, and that with respect to a spatial coordinate, $\kappa = \mathrm{d}\theta/\mathrm{d}x$, (the angular velocity and the curvature, respectively) may be discontinuous although θ is continuous.

In general, let $\lambda(t)$ denote the location of a plastic hinge. Furthermore, for any function $\Phi(x)$ let Φ^+ and Φ^- denote the values of the function on either side of the plastic hinge position; thus

$$\Phi^+ \equiv \Phi(\lambda+0) \equiv \lim_{\varepsilon\to 0}\Phi(\lambda+\varepsilon),\ \Phi^- \equiv \Phi(\lambda-0) \equiv \lim_{\varepsilon\to 0}\Phi(\lambda-\varepsilon) \tag{6.90}$$

The schematic of a function $\Phi(x)$ is shown in Figure 6.16. At the location of plastic hinge $x = \lambda(t)$, the function values may be different on different sides, i.e., Φ^+ and Φ^-, respectively. Accordingly, let $[\Phi] \equiv \Phi^+ - \Phi^-$ denote the jump or discontinuity in magnitude of $\Phi(x)$ that develops as the hinge passes the point $x = \lambda(t)$. Clearly, $[\Phi] = 0$ implies continuity, while $[\Phi] \neq 0$ implies discontinuity.

At the location of the plastic hinge the transverse deflection $w(\lambda, t)$ is continuous, so

$$[w] = w^+ - w^- = 0 \tag{6.91}$$

Otherwise the beam is broken.

Furthermore as required by Euler beam assumption, the shear deformation is negligible, i.e., $\mathrm{d}w/\mathrm{d}x \approx 0$. The transverse velocity $\mathrm{d}w/\mathrm{d}t$ must also be continuous, i.e.,

$$[\dot{w}] = \dot{w}^+ - \dot{w}^- = 0 \tag{6.92}$$

Equations (6.91) and (6.92) apply to both stationary and traveling hinges.

At a hinge, Eq. (6.91) is always valid, i.e., $[w] = 0$, so its derivative with respect to time must vanish:

$$\frac{\mathrm{d}}{\mathrm{d}t}[w(\lambda,t)] = 0 \tag{6.93}$$

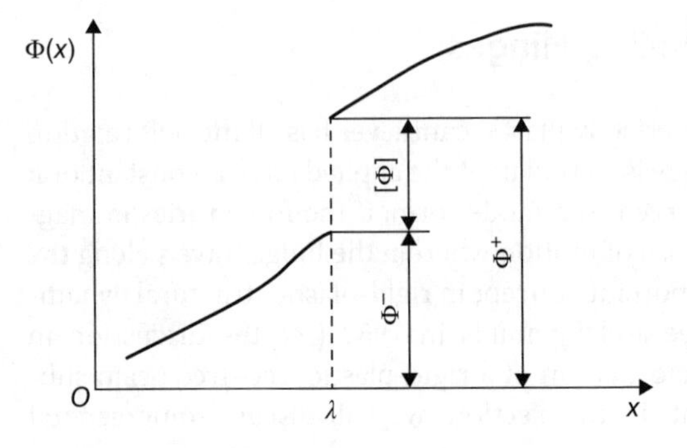

Figure 6.16 Schematic of a function Φ(*x*).

where $\lambda = \lambda(t)$. Hence, the above relation gives

$$0 = \frac{\mathrm{d}}{\mathrm{d}t}[w] = \frac{\mathrm{d}}{\mathrm{d}t}\{w^+(\lambda,t) - w^-(\lambda,t)\} = \left(\frac{\partial w^+}{\partial t} - \frac{\partial w^-}{\partial t}\right) + \left(\frac{\partial w^+}{\partial \lambda}\frac{\partial \lambda}{\partial t} - \frac{\partial w^-}{\partial \lambda}\frac{\partial \lambda}{\partial t}\right) \tag{6.94}$$

i.e.,

$$0 = \left(\frac{\partial w^+}{\partial t} - \frac{\partial w^-}{\partial t}\right) + \dot{\lambda}\left(\frac{\partial w^+}{\partial \lambda} - \frac{\partial w^-}{\partial \lambda}\right)$$

which can be simplified as

$$[\dot{w}] + \dot{\lambda}[w'] = 0 \tag{6.95}$$

where the transverse velocity, $\dot{w} = \partial w/\partial t$, and the inclination, $w' = \partial w/\partial x$, are derivatives for a material particle when the hinge passes it. Moreover, $\dot{\lambda}$ is the traveling speed of the plastic hinge. As Eq. (6.95) indicates that the jump of velocity and the jump of inclination are related, at the traveling hinge, Eq. (6.92) gives $[\dot{w}] = 0$, so Eq. (6.95) yields

$$\dot{\lambda}[w'] = 0 \tag{6.96}$$

Equation (6.96) indicates that:

1) for a stationary hinge, $\dot{\lambda} = 0$, so $[w'] \neq 0$, i.e., the slope is discontinuous at the hinge;
2) for a traveling hinge, $\dot{\lambda} \neq 0$, so $[w'] = 0$, i.e., the slope remains continuous at the hinge.

Similarly, differentiation of Eq. (6.92), $[\dot{w}] = 0$ yields

$$[\ddot{w}] + \dot{\lambda}[\dot{w}'] = 0 \tag{6.97}$$

Equation (6.97) further indicates that:

1) for a stationary hinge, $\dot{\lambda} = 0$, hence $[\ddot{w}] = 0$ must hold, i.e., the acceleration remains continuous;
2) for a traveling hinge, $\dot{\lambda} \neq 0$, and both acceleration $\ddot{w}$ and the angular velocity $\dot{w}' = \dot{\theta}$ can be discontinuous.

In the case of traveling hinge, $\dot{\lambda} \neq 0$, starting from $[w'] = 0$ and carrying out a similar differentiation with respect to space x, we obtain

$$[\dot{w}'] + \dot{\lambda}[w''] = 0 \tag{6.98}$$

which indicates that the angular velocity $\dot{\theta} = \dot{w}'$ is related to the curvature $\kappa = w''$. Eliminating $[\dot{w}']$ from Eqs (6.97) and (6.98) leads to

$$[\ddot{w}] = \dot{\lambda}^2[w''] \tag{6.99}$$

The above relation indicates that the acceleration $\ddot{w}$ is related to the curvature κ.

Note that Eqs (6.98) and (6.99) are only valid for traveling hinges, $\dot{\lambda} \neq 0$. Expressions (6.96)–(6.99) provide the features of discontinuities at plastic hinges, which are summarized in Table 6.1.

Table 6.1 Discontinuities at plastic hinges

	Stationary Hinge $\dot{\lambda}=0$	**Traveling Hinge** $\dot{\lambda}\neq 0$
Deflection w	C	C
Slope or inclination $w' = \theta$	D	C
Curvature $w'' = \kappa$	D*	D
Velocity $\dot{w}$	C	C
Angular velocity $\dot{w}' = \dot{\theta}$	D	D
Acceleration $\ddot{w}$	C	D

Note: C, continuous; D, discontinuous.
$^*\kappa(\lambda+0)=\kappa(\lambda-0), \kappa(\lambda)=\infty$

Suppose a plastic hinge forms at a cross-section of a rigid-perfectly plastic beam, where the yield criterion $|M(\lambda)|=M_P$ is satisfied. As the energy dissipation at the plastic hinge $D=M_P|\dot{\theta}|>0$ is positive, the associated flow rule requires $\dot{\theta}\neq 0$. Indeed, Table 6.1 indicates that *a discontinuity in angular velocity is a common and essential feature of all plastic hinges*, no matter whether they are stationary or traveling. If the hinge is stationary, a finite θ results; if it is traveling, $\dot{\theta}\neq 0$ will result in a curvature $\kappa = w''$.

The expressions concerning discontinuities at traveling hinges can be useful in analysis. For instance, the configuration that was shown in Figure 6.4 contains a static and an undeformed segment HB; hence Eq. (6.98), $[\dot{w}']+\dot{\lambda}[w'']=0$ leads to

$$\dot{w}'(\lambda-0)+\dot{\lambda}w''(\lambda-0)=0 \tag{6.100}$$

For segment AH, which has plastic deformation, the location near the plastic hinge H is $x=\lambda-0$. The corresponding angular velocity $\dot{w}'(\lambda-0)$ and curvature $w''(\lambda-0)$ are

$$\dot{w}'(\lambda-0)=-v/\lambda, w''(\lambda-0)=\kappa(\lambda) \tag{6.101}$$

By substituting Eq. (6.101) into Eq. (6.100), the final shape of the deformed cantilever can be obtained.

Problems 6

6.1 A cantilever beam has length L, mass per unit length ρ and fully plastic bending moment M_P of the cross-section. The cantilever is subjected to a triangular pulse, as shown in Figure 5.16, at its free end. Please determine the plastic hinge location and the motion of the traveling hinge. The peak value of the triangular pulse is F, and the duration of the pulse is T.

6.2 Consider two cantilevers made of rigid-perfectly plastic material, as shown in the figure. They are of mass per unit length ρ and fully plastic bending moment M_P of the cross-section. In case (a), the beam of length L, with a concentrated mass

$m = \rho L/4$ at its tip, is subjected to a rectangular pulse of amplitude $F = 12M_P/L$ and time duration T at its tip. In case (b), the beam of length $5L/4$ is subjected to the same rectangular pulse at a distance L from the root.

Compare the following variables of these two cantilevers: (i) the initial location of the plastic hinge; (ii) the time at which the traveling hinge reaches the root; and (iii) the final rotation angle at the root.

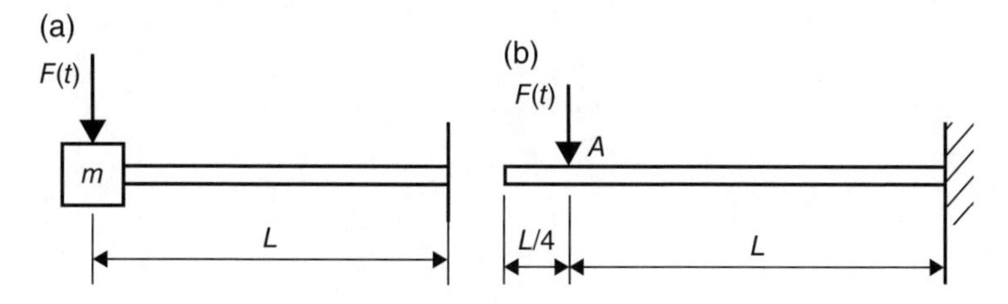

7

Effects of Tensile and Shear Forces

7.1 Simply Supported Beams with no Axial Constraint at Supports

In Chapter 6, the dynamic responses of a uniform cantilever under several kinds of pulse (step loading, rectangular pulse, general pulse) and rigid-mass impact were analyzed. The main focus was the rigid-plastic deformation mechanism, deformed shape and energy partitioning. In the case of a cantilever, one of its ends is free, and so the deflection along the transverse direction will not produce any extension along the longitudinal (axial) direction, and it will cause no tensile force. However, for a beam that is clamped at both ends, extension of the central axis and the tensile force will be generated by the deflection along the transverse direction. In this chapter, we look at the effects of the tensile forces on the dynamic response of beams. The majority of the content of Sections 7.1 and 7.2 is from the papers by Symonds and Mentel (1958) and Symonds and Jones (1972).

In Chapter 6, several kinds of dynamic pulse, such as a rectangular pulse and a general pulse, were discussed. If the pulse duration is extremely short while the pulse amplitude is extremely high, the limit case is a pulse being a Delta function of time. In this case, the effect of loading is equivalent to an initial velocity distribution being applied to the structure (e.g., a beam). This special case of dynamic pulse loading is termed *impulsive loading*.

Whether a transverse deflection induces axial force in the beam or not depends upon the axial restraint at the supports. Figure 7.1 shows a simply supported beam of length $2L$, mass per unit length ρ and fully plastic bending moment M_P. As the free extension along the longitudinal (axial) direction is not constrained, no tensile force will be produced when the beam deflects in the transverse direction. An impulsive loading brings a uniformly distributed initial velocity v_0 to the entire span of the beam at the initial instant. The dynamic response of this simply supported beam without axial constraint at supports will be discussed further below.

7.1.1 Phase I

The deformation mechanism in phase I of a simply supported beam without axial constraint is depicted in Figure 7.2. The supporting ends A and A' are natural hinges, i.e., rotation is free at these supports, but no transverse displacement or transverse velocity is allowed. Under the uniform impulsive loading at time $t = 0$, except for the two ends, the

Introduction to Impact Dynamics, First Edition. T.X. Yu and XinMing Qiu.

Published 2018 by John Wiley & Sons Singapore Pte. Ltd.

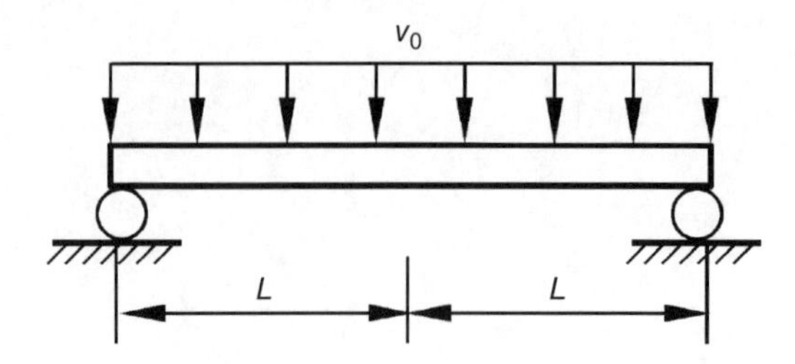

Figure 7.1 A simply supported beam with no axial constraint under uniform impulsive loading.

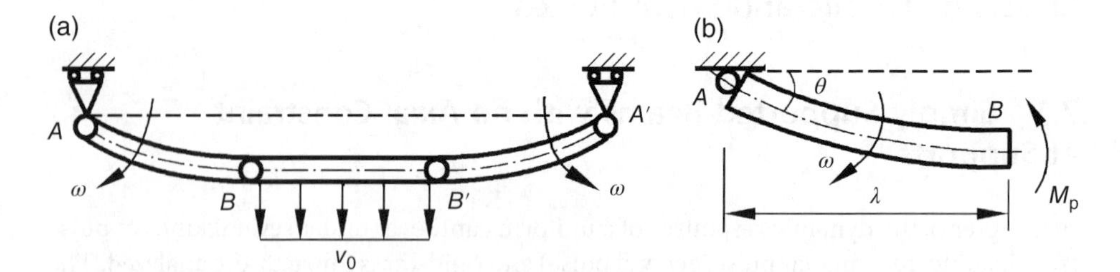

Figure 7.2 Response phase I of a simply supported beams with no axial constraint.

entire beam gains the same initial velocity v_0 along the traverse direction. At $t > 0$, two traveling plastic hinges B and B' move from the two ends towards the middle of the beam, with current distance to the end being $AB = A'B' = \lambda$. The middle segment BB' continues to move at transverse velocity v_0, while the two side segments AB and $A'B'$ rotate about the supports A and A', respectively, with angular velocity ω. As, initially, the whole beam has velocity v_0, it is known that $\lambda = 0$ at $t = 0$.

The equation of moment of momentum for segment AB with respect to point A is

$$\frac{1}{3}\rho\lambda^3\frac{\mathrm{d}\omega}{\mathrm{d}t} = -M_P \tag{7.1}$$

where $\mathrm{d}\omega/\mathrm{d}t$ is the angular acceleration. Two points should be noted here:

1) The equation of moment of momentum given in Eq. (7.1) is in the form of the instantaneous angular acceleration α multiplied by the moment of inertia J, i.e., $M = J\alpha$. As the middle segment BB' possesses non-zero velocity v_0, the derivation should not be based on $\frac{\mathrm{d}}{\mathrm{d}t}\left(\frac{1}{3}\rho\lambda^3\omega\right) = -M_P$.
2) The equation of motion in the transverse direction does not help because the reaction force at A is unknown.

Here we have two unknowns, i.e., plastic hinge position λ and angular velocity ω, but continuity of the transverse velocity at hinge B requires

$$\omega\lambda = v_0 \tag{7.2}$$

This kinematic relation leads to

$$\omega = \frac{v_0}{\lambda} \tag{7.3}$$

The differentiation of the angular velocity with respect to time is

$$\frac{d\omega}{dt} = -\frac{v_0}{\lambda^2}\frac{d\lambda}{dt} \tag{7.4}$$

Substituting Eq. (7.4) into Eq. (7.1) results in

$$\frac{1}{3}\rho\lambda v_0 \frac{d\lambda}{dt} = M_P \tag{7.5}$$

Equation (7.5) can be integrated with respect to time and gives

$$\lambda^2 = 6M_P t/\rho v_0 + C_1 \tag{7.6}$$

The unknown constant C_1 can be determined using the initial condition. As $\lambda = 0$ at $t = 0$, we have $C_1 = 0$. Thus, the traveling hinge position λ is given by

$$\lambda^2 = \frac{6M_P t}{\rho v_0} \tag{7.7}$$

Employing a non-dimensional plastic hinge position defined by $\bar{\lambda} = \lambda/L$, Eq. (7.7) can be rewritten as

$$\bar{\lambda}^2 = \frac{6M_P t}{\rho v_0 L^2} \tag{7.8}$$

which relates the traveling hinge position with time. Then, from Eq. (7.3) the angular velocity is a function of time:

$$\omega^2 = \frac{\rho v_0^3}{6M_P t} \tag{7.9}$$

It should be noted that:

1) From Eq. (7.9), ω is independent of beam length L, whereas from Eq. (7.7), the plastic hinge position is also independent of the beam length L.
2) By calculating the shear force $Q(x)$ and the bending moment $M(x)$ from angular acceleration $d\omega/dt$, it can be verified that the bending moment $|M| \le M_P$ in segment AB, which agrees with the assumed deformation mechanism shown in Figure 7.2.
3) For the normalized plastic hinge position, $\bar{\lambda} = \lambda/L$, the differentiation of Eq. (7.8) gives

$$\dot{\bar{\lambda}} = \frac{3M_P}{\rho v_0 L^2 \bar{\lambda}} = \frac{1}{L}\sqrt{\frac{3M_P}{2\rho v_0 t}} \propto \frac{1}{\sqrt{t}} \tag{7.10}$$

which indicates that the moving speed of the traveling hinge B decreases with time.

Phase I ends when $\lambda = L$, i.e., $\bar{\lambda} = 1$. The response time of phase I is obtained from Eq. (7.8):

$$t_1 = \frac{\rho v_0 L^2}{6M_P} \tag{7.11}$$

The deflection of the midpoint of the beam at the end of phase I is,

$$\Delta_1 = v_0 t_1 = \frac{\rho v_0^2 L^2}{6M_P} \tag{7.12}$$

Define a non-dimensional parameter *energy ratio* as,

$$e_0 \equiv \frac{K_0}{M_P} = \frac{\frac{1}{2}\rho\, 2L\, v_0^2}{M_P} = \frac{\rho v_0^2 L}{M_P} \tag{7.13}$$

A comparison of Eq. (7.12) and Eq. (7.13) shows a linear relation between the deflection at the end of phase I and the energy ratio:

$$\Delta_1/L = e_0/6 \tag{7.14}$$

7.1.2 Phase II

The deformation mechanism in phase II of a simply supported beam without axial constraint is sketched in Figure 7.3. At the transition from phase I to phase II, two traveling plastic hinges *B* and *B′* meet each other at the midpoint C of the beam, forming a new plastic hinge at C.

During phase II, the velocity of plastic hinge C will vary with time. In fact, from the equation of moment of momentum about point A

$$\frac{1}{3}\rho L^3 \frac{d\omega}{dt} = -M_P \tag{7.15}$$

whilst a kinematic relation is valid:

$$\omega L = v \tag{7.16}$$

Obviously, in the dynamic response of phase II, two unknowns need to be solved, i.e., angular velocity ω and the transverse velocity of stationary plastic hinge C, v. The integration of Eq. (7.15) with respect to time gives

$$\omega = -\frac{3M_P}{\rho L^3} t + C_2 \tag{7.17}$$

Using the initial condition, $\omega_1 = v_0/L$ at $t = t_1$, parameter C_2 is determined as $C_2 = \frac{3v_0}{2L}$. Therefore, Eq. (7.17) can be rewritten as

$$\omega = -\frac{3M_P}{\rho L^3} t + \frac{3v_0}{2L} \tag{7.18}$$

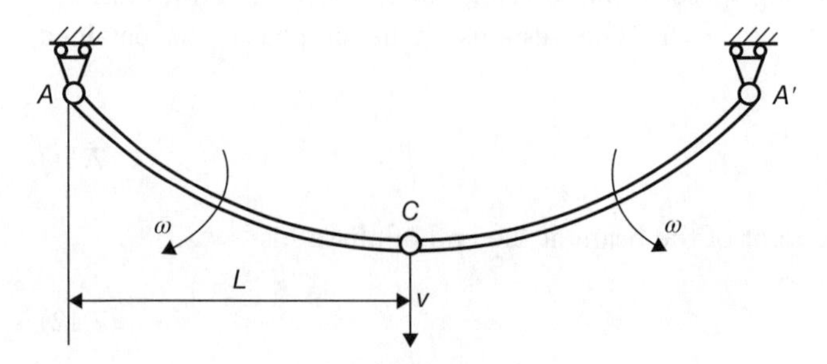

Figure 7.3 Response phase II of a simply supported beam with no axial constraint.

The linear relation of Eq. (7.18) between the angular velocity and time indicates that ω will uniformly decrease with time. Phase II ends when $\omega = 0$; then the termination time of the dynamic response t_f is found from Eq. (7.18):

$$t_f = \frac{\rho v_0 L^2}{2M_P} = 3t_1 \tag{7.19}$$

The final deflection at the middle span C is

$$\Delta_f = \Delta_1 + \int_{t_1}^{t_f} v(t)\, dt = \Delta_1 + \int_{t_1}^{t_f} \omega L\, dt = 2v_0 t_1 = 2\Delta_1 \tag{7.20}$$

Comparing Eq. (7.20) and Eq. (7.14) leads to

$$\Delta_f / L = e_0 / 3 \tag{7.21}$$

which indicates that the final deflection is also proportional to $e_0 = K_0 / M_p$.

Then the rotation angle at supports can be calculated. During phase I

$$\theta_1 = \int_0^{t_1} \omega\, dt = \frac{\rho v_0^2 L}{3M_P} \tag{7.22}$$

and during phase II

$$\theta_2 = \int_{t_1}^{t_f} \omega\, dt = \frac{\rho v_0^2 L}{6M_P} = \frac{\theta_1}{2} \tag{7.23}$$

Therefore, the total rotation angle at the support during the entire dynamic response is

$$\theta_1 + \theta_2 = \frac{\rho v_0^2 L}{2M_P} = \frac{K_0}{2M_P} = \frac{e_0}{2} \tag{7.24}$$

Considering the energy conservation during the entire deformation process, the input energy should be equal to the sum of the energy dissipated at the traveling plastic hinge B and B' during phase I, and that at the stationary plastic hinge C during phase II. Hence

$$K_0 = 2\theta_1 M_P + 2\theta_2 M_P \tag{7.25}$$

The consistency of Eq. (7.24) and Eq. (7.25) clearly confirms the energy balance in the entire dynamic response process.

Table 7.1 summarizes the dynamic response of a simply supported beam with no axial constraint to impulsive loading. Compared with phase I, the time duration of phase II is twice as long; the midspan deflection is the same, and the rotation angle is only a half. The deformation in phase II is slower than that in phase I.

Table 7.1 Summary of the dynamic response of a simply supported beam with no axial constraint

Phase	Time duration	Deflection at midspan	Rotation angle at supports
I	$t_1 - 0$	$\Delta_1 = e_0 L/6$	θ_1
II	$t_f - t_1 = 2t_1$	$\Delta_f - \Delta_1 = \Delta_1$	$\theta_1/2$
total	$t_f = 3t_1$	$\Delta_f = 2\Delta_1 = e_0 L/3$	$\theta_f = 3\theta_1/2$

7.2 Simply Supported Beams with Axial Constraint at Supports

7.2.1 Bending Moment and Tensile Force in a Rigid-Plastic Beam

Consider a uniform straight beam of rectangular section with height h and width b, made of rigid-perfectly plastic material of yield strength Y. If the beam is subjected to bending moment M only, in the fully plastic stress state, the stress distribution on a cross-section is as shown in Figure 7.4. Clearly, the fully plastic bending moment of this beam is

$$M_P = \left(Yb\frac{h}{2}\right)\frac{h}{2} = \frac{1}{4}Ybh^2 \tag{7.26}$$

If the same beam is subjected to tensile force N only, in the fully plastic stress state, the stress distribution on a cross-section is as shown in Figure 7.5. Thus, the fully plastic tensile force is

$$N_P = Ybh \tag{7.27}$$

If this beam is subjected to both bending moment M and tensile force N, the stress distribution on a cross-section of the beam is the superposition of the bending and tension, as shown in Figure 7.6. In this case, the stress distribution on the cross-section in the thickness direction is asymmetric about the central axis of the beam. Suppose the neutral axis, which has zero stress, is located at a distance c from the central axis, then the stress distribution shown in Figure 7.6(a) can be decomposed into two parts, related to the pure

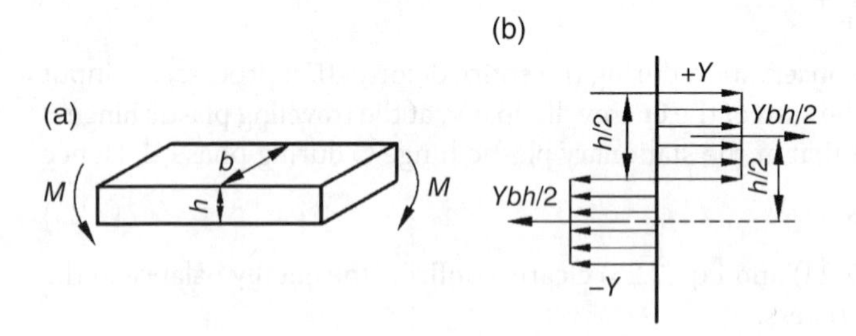

Figure 7.4 The stress distribution on a cross-section of a rectangular beam under bending moment M.

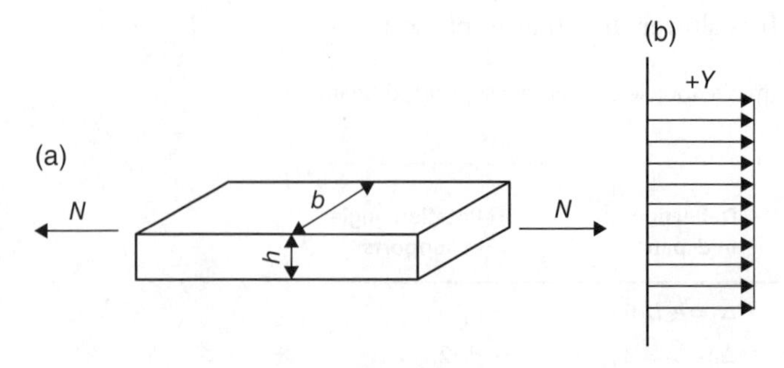

Figure 7.5 The stress distribution on a cross-section of a rectangular beam under tensile force N.

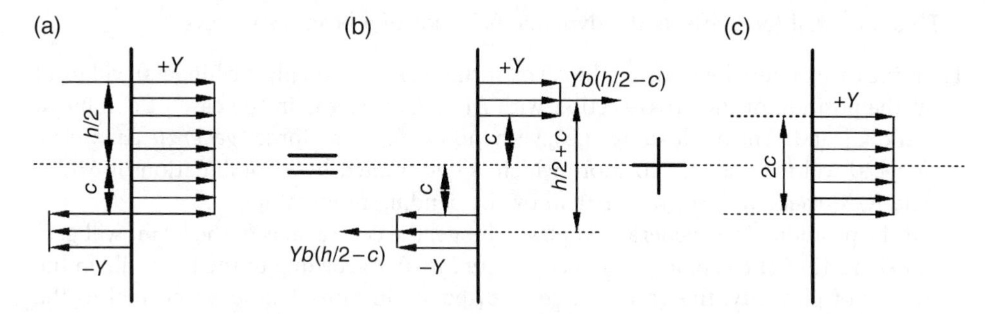

Figure 7.6 The stress distribution on a cross-section of a rectangular beam subjected to both bending moment *M* and tensile force *N*.

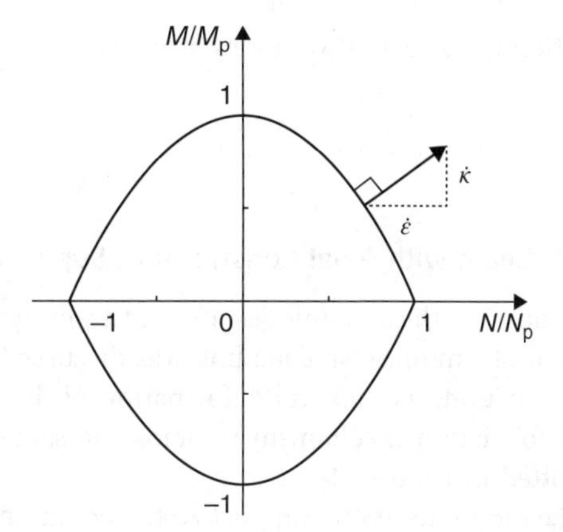

Figure 7.7 The limit locus on the *M*–*N* plane.

bending case and the pure tension case, as shown in Figures 7.6(b) and (c), respectively. According to the stress distributions shown in Figures 7.6(b) and (c), the bending moment M and tensile force N applied at the beam's cross-section are

$$M = Yb\left(\frac{h}{2} - c\right)\left(\frac{h}{2} + c\right) = Yb\left(\frac{h^2}{4} - c^2\right) \tag{7.28}$$

$$N = Yb\ 2c \tag{7.29}$$

By eliminating the eccentricity c from the above two relations, the uniaxial stress yield criterion leads to an interactive limit criterion:

$$\left|\frac{M}{M_P}\right| + \left(\frac{N}{N_P}\right)^2 = 1 \tag{7.30}$$

The limit locus of a beam of rectangular cross-section, made of rigid-perfectly plastic material under bending and tension is plotted in Figure 7.7, on the plane of normalized bending moment and tensile force.

If the value given by the left side of Eq. (7.30) is smaller than 1, the resultant stress state is inside the limit locus, whereas if the value given by the left side of Eq. (7.30) is equal to 1, then the resultant stress is exactly on the limit locus, which will produce further plastic deformation.

Thus the axial force affects the dynamic behavior of a beam as follows:

1) In the case of pure bending loading, i.e., with no axial force, plastic hinge(s) will form at the position of the cross-section with $M = M_P$, whereas in the case of combined bending and tension loading, i.e., with non-zero axial force, *generalized plastic hinge(s)* will form at the position, which is determined by a combination of M/M_P and N/N_P (see Eq. 7.30) rather than by the bending moment alone.
2) At the position of the generalized plastic hinge, the central axis of the beam will gain an elongation rate $\dot{\varepsilon}$, along with a curvature rate $\dot{\kappa}$. According to the flow rule in the theory of plasticity, the vector of generalized strain rates should be normal to the limit curve:

$$\frac{\dot{\kappa}}{\dot{\varepsilon}} = -\frac{\mathrm{d}N}{\mathrm{d}M} = \frac{N_P^2}{2NM_P} \tag{7.31}$$

Equation (7.31) can also be rewritten as

$$\frac{N_P\dot{\varepsilon}}{M_P\dot{\kappa}} = 2\frac{N}{N_P} \tag{7.32}$$

7.2.2 Beam with Axial Constraint at Support

In Section 7.1, the dynamic response of a simply supported beam with no axial constraint under uniform impulsive loading was discussed. If the supports cannot move along the axial direction, the flexural deformation of the beam will induce axial force. Assuming the deformation mechanism remains the same as that shown in Figure 7.2, it can be re-plotted in Figure 7.8.

Take moments about support A, the equation of moment of momentum for segment AB gives

$$\frac{1}{3}\rho\lambda^3\frac{\mathrm{d}\omega}{\mathrm{d}t} = -M_B - N_B\Delta \tag{7.33}$$

From the deformation mechanism shown in Figure 7.8, at the beginning the deformation, the deflection at the midpoint of the beam is equal to the transverse displacement of the traveling plastic hinge B, i.e., $\Delta = v_0 t$. The continuous condition at point B requires

$$\omega\lambda = v_0 \tag{7.34}$$

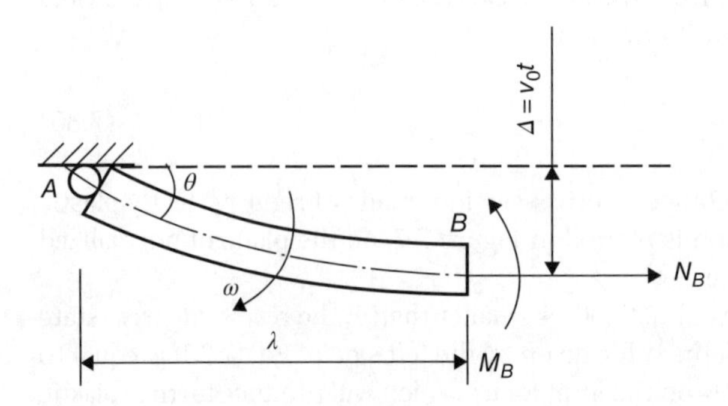

Figure 7.8 Response of a simply supported beam with axial constraint.

At the traveling plastic hinge B, the bending moment and tensile force should satisfy the condition required by the limit locus, Eq. (7.30), i.e.

$$\left|\frac{M_B}{M_P}\right| + \left(\frac{N_B}{N_P}\right)^2 = 1 \tag{7.35}$$

Up to now, we have three equations, i.e., Eqs (7.33), (7.34) and (7.35), but there are four unknowns $(\lambda, \omega, M_B, N_B)$ that need to be solved. Hence, an additional equation has to be found from deformation compatibility.

As shown in Figure 7.9, if the rotation angle of segment AB is denoted by θ and the deflection at the midpoint of the beam is denoted by Δ, at a given instant t, they are related by

$$\theta = \Delta/a \tag{7.36}$$

After a small time increment δt, i.e., at the instant $t + \delta t$, the rotation angle gains an increment $\delta\theta$, and the position change of the traveling plastic hinge, $\delta\lambda(1 + \delta\varepsilon)$, results from two components: the first comes from the motion of the traveling hinge itself, $\delta\lambda$; and the second component is caused by the axial elongation of the beam $\delta\lambda\ \delta\varepsilon$. Based on the geometric relation shown in Figure 7.9, the position change of the traveling plastic hinge is related to the rotation angle θ by

$$\delta\lambda(1 + \delta\varepsilon) = \delta\lambda + a\ \delta\theta\ \theta \tag{7.37}$$

The combination of Eqs (7.36) and Eq. (7.37) leads to

$$\delta\lambda\ \delta\varepsilon = \Delta\ \delta\theta \tag{7.38}$$

Curvature is defined as the angle change along the unit arc length, i.e.

$$\delta\kappa = \delta\theta/\delta\lambda \tag{7.39}$$

Hence, using Eqs (7.38) and Eq. (7.39), the elongation strain rate and the curvature rate are related to the midpoint deflection of the beam as

$$\Delta = \frac{\delta\varepsilon}{\delta\kappa} = \frac{\dot{\varepsilon}}{\dot{\kappa}} = \frac{2N_B M_P}{N_P^2} \tag{7.40}$$

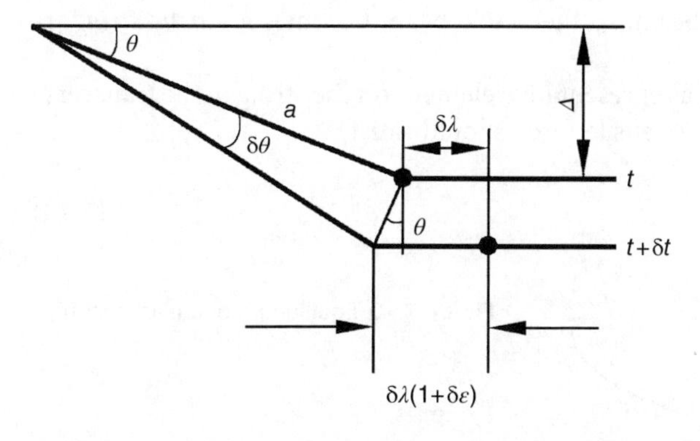

Figure 7.9 Schematic on the compatibility relation of a beam with axial constraint.

Then, for a beam of rectangular cross-section, substituting the expressions of M_P and N_P given in Eqs (7.26) and (7.27) into Eq. (7.40) leads to

$$\frac{\Delta}{h} = \frac{N_B}{2N_P} \tag{7.41}$$

i.e.,

$$\frac{N_B}{N_P} = 2\frac{\Delta}{h} = \frac{2v_0 t}{h} \tag{7.42}$$

which indicates that the tensile force at the plastic hinge is linearly increasing with time.

Therefore, the dynamic response of phase I is solved by the combination of Eqs (7.33), (7.34), (7.35) and (7.42). The deformation history will be traced on the limit locus relating bending and tension, i.e., Figure 7.7. At the initial instant, the traveling plastic hinge only pertains to bending $M = M_P$, with no tensile force. That is, the generalized stress state is on the top point of the interactive limit locus on the M–N plane.

As the beam's deflection increases, the plastic hinge moves towards the middle of the beam, resulting in an increasing tensile force and a decrease in bending moment. Thus, the generalized stress state will depart from the vertical axis on the M–N plane and move down along the limit locus in the first quadrant. As seen from Eq. (7.42), when the midpoint deflection of the beam is close to half of the beam thickness, i.e., $\Delta \to h/2$, the tensile force at the traveling plastic hinge B is close to the fully plastic tensile force, i.e., $N_B \to N_P$; thus, the corresponding bending moment is very small and the general stress state is approaching the right corner of the interactive limit locus. It is evident, therefore, that even though the midpoint deflection of the beam is still less than the beam thickness, the axial force effect has already become dominant.

In relation to the above discussion, before an analysis of phase II, we need to distinguish whether $\Delta = h/2$ or $\lambda = L$ will be satisfied first. To do this, in addition to the energy ratio $e_0 = K_0/M_P = \rho L v_0^2/M_P$, another non-dimensional parameter is introduced as $\Gamma = e_0 L/h = \rho L^2 v_0^2/M_P h$:

1) If $\Gamma = e_0 L/h \leq 4$, which refers to a moderate energy input case, then $\lambda = L$ appears first, i.e., two traveling hinges meet at the midpoint, and phase II starts with $N < N_P$.
2) If $\Gamma = e_0 L/h > 4$, which refers to a higher energy input case, then $\Delta = h/2$ appears first, i.e., $N_B = N_P$ and $M_B = 0$ occur before the two traveling hinges meet at the midpoint.

In the second of these cases, what would happen after N_B reaches N_P? This implies that $N = N_P$ and $M = 0$ in the whole beam, so the whole beam becomes a *plastic string*, as shown in Figure 7.10.

Considering the equilibrium of a representative element on the string in the transverse direction, the inertia force and the tensile force should satisfy

$$N_P \frac{\partial^2 w}{\partial x^2} = \rho \frac{\partial^2 w}{\partial t^2} \tag{7.43}$$

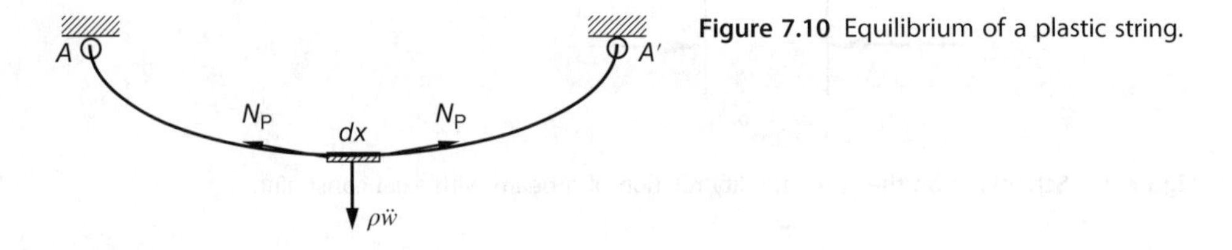

Figure 7.10 Equilibrium of a plastic string.

which can be rewritten into the following form:

$$\frac{\partial^2 w}{\partial t^2} = c^2 \frac{\partial^2 w}{\partial x^2}, \quad c^2 = \frac{N_P}{\rho} \tag{7.44}$$

Clearly, Eq. (7.44) represents a one-dimensional wave propagation, which was discussed in Chapter 1. The general solution of Eq. (7.44) is

$$w(x,t) = f_1(x - ct) + f_2(x + ct) \tag{7.45}$$

where f_1, f_2 can be determined by the "initial" condition of w and $\mathrm{d}w/\mathrm{d}t$ at time $t^* = h/2v_0$ when $\Delta = v_0 t = h/2$.

Regarding the final deflection, Symonds and Mentel (1985) and Symonds and Jones (1972) proved that the final midpoint deflection of the beam, Δ_f, satisfies

$$\frac{1}{2}\left(\sqrt{\Gamma} - 1\right) < \frac{\Delta_f}{h} < \frac{1}{2}\left(\sqrt{\Gamma} - 0.5\right) \tag{7.46}$$

For example, the final deflection will be in the range $1.0 < \Delta_f/h < 1.25$ in the case of $\Gamma = e_0 L/h = 9$, and in the range $2.0 < \Delta_f/h < 2.25$ in the case of $\Gamma = e_0 L/h = 25$.

The expression given by Eq. (7.30) represents the exact limit locus under the combination of bending moment and tensile force. As it is a non-linear function of M and N, this limit locus is complex in application. A simplified analysis can be obtained by employing an independent locus

$$\left|\frac{M}{M_P}\right| = 1, \quad \left|\frac{N}{N_P}\right| = 1 \tag{7.47}$$

Equation (7.47) provides an upper bound to the exact limit locus Eq. (7.30). As observed from Figure 7.11, the limit locus described by Eq. (7.47) is a circumscribing square of the exact limit locus. The sides of this square refer to either bending moment being equal to $\pm\ M_P$, or tensile force being equal to $\pm\ N_P$. If the applied strain rate indicates that the stress point is at one of the four corners of the limit square, both equations

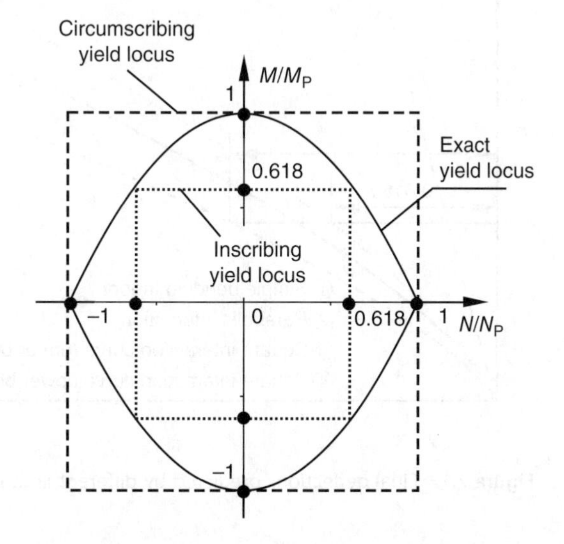

Figure 7.11 Several limit loci of a rectangular beam made of rigid-perfectly plastic material.

in Eq. (7.47) are simultaneously valid. Similar to the upper bound given by Eq. (7.47), a lower bound can be constructed by taking a factor of 0.618 on Eq. (7.47), i.e.

$$\left|\frac{M}{M_{\mathrm{P}}}\right| = 0.618, \quad \left|\frac{N}{N_{\mathrm{P}}}\right| = 0.618 \tag{7.48}$$

As seen from Figure 7.11, the yield locus corresponding to Eq. (7.48) is an inscribing square of the exact limit locus. As the difference between Eq. (7.47) and Eq. (7.48) is only a constant factor 0.618, the upper and lower bounds of the exact solution could be reached by using those approximate limit squares.

The normalized final midpoint deflections versus the input energy, predicted by different limit loci, are depicted in Figure 7.12. For a simply supported beam without axial constraint, there is no tensile force, and the corresponding curve of deflection, i.e., curve ①, is based on simple bending theory, which gives a linear relation between the input energy and the final deflection. For a simply supported beam with axial constraint, the deflection curve ② is based on the exact limit locus, which considers both bending and tension. By comparing curves ① and ②, it is obvious that the tensile force induced by the axial constraint has a significant effect on the final deflection, especially in cases of large energy input.

The curves of final deflection predicted by two approximate limit loci are also plotted in Figure 7.12. For the same input energy, the deflection given by the circumscribing yield locus is smaller (see curve ③), while the deflection given by the inscribing yield locus is larger (see curve ④). By contrast, to produce the same deflection, the input energy required by the circumscribing limit locus is larger (a), and the input energy required by the inscribing limit locus is smaller ($0.618a$). Hereafter, these two approximate limit loci correspond to the upper (curve ③) and lower (curve ④) bounds of the load-carrying capacity. The exact load-carrying capacity (curve ②) is in between of the upper and lower bounds. From this point of view, the approximate limit loci, which are shown as independent relations between bending moment and tension force, are very convenient and practical in engineering applications.

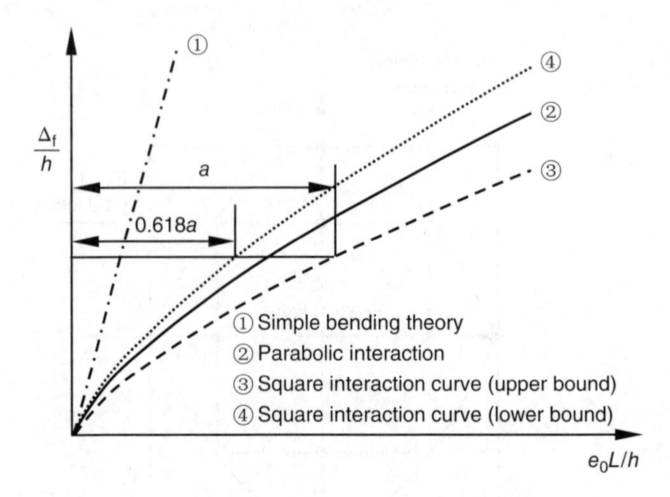

Figure 7.12 Final deflections predicted by different limit loci.

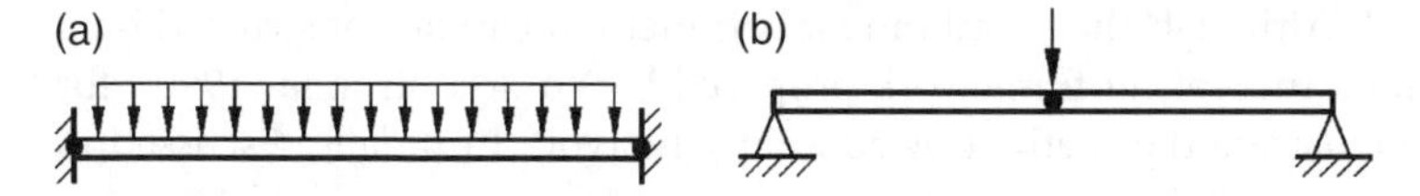

Figure 7.13 The initial position of the traveling hinge.

7.2.3 Remarks

1) A ***traveling hinge*** always departs from the cross-section where the initial momentum distribution has a discontinuity. As shown in Figure 7.13, for a fully clamped beam under uniform impulsive loading, the traveling plastic hinges start from the two clamped supports and move towards the middle of the beam, while for a simply supported beam with axial constraint subjected to a concentrated dynamic force, two traveling plastic hinges start from the loading point and move outwards.
2) A traveling plastic hinge exists in the transient phase (phase I) of the dynamic response, and the mode phase (phase II) of deformation starts when the motion of the traveling hinge(s) ceases. Then the momentum distribution always approaches a modal form, which is usually similar to that associated with the static plastic collapse mechanism. Therefore, an approximate method can be adopted in analyzing the dynamic response of beams. By only considering the deformation mechanism consisting of stationary plastic hinge(s), this method is called the *mode technique* and this is discussed in Chapter 8.
3) If the axial motion is constrained in the beam, the deflection ***induces axial force***, which has significant effect ***in reducing the final deflection*** compared with the solution given by the simple bending theory, i.e., it strengthens the structure. Therefore, in the design of some energy absorbers, it is preferable to generate axial force in beams or membrane force in plates. The effect of axial (or membrane) force can enhance the load-carrying capacity of the structure and the energy absorption efficiency of the energy absorbers.

7.3 Membrane Factor Method in Analyzing the Axial Force Effect

7.3.1 Plastic Energy Dissipation and the Membrane Factor

As discussed in Section 7.2, in general the axial force effect should be considered in the analysis of the dynamic response of beams. The deformation mechanism is constructed by plastic hinges and rigid segments, and the response history consists of two phases: phase I is characterized by traveling hinges, and the succeeding phase II contains stationary plastic hinge(s) only. The so-called plastic hinge here means a generalized one, namely, the cross-section of the beam where plastic flow occurs under the interaction between the bending moment M and the tensile force N. According to the analysis in Section 7.2, in order to consider the interaction between bending moment and axial force, the evolution of the deformation mechanism has to be followed stage-by-stage, which is quite complicated. In this section, a unified treatment of the effects of bending moment and axial force will be presented, which is called the *membrane factor method*

(Chen and Yu, 1993). Although the membrane factor method can be constructed on a wider basis by using a theoretical framework proposed by Yu and Stronge (1990), for simplicity, here we illustrate the method by adopting the typical example discussed in Section 7.2.

To facilitate the derivation, the following non-dimensional parameters are introduced:

$$\overline{M}\equiv M/M_{\mathrm{P}},\overline{N}\equiv N/N_{\mathrm{P}},\bar{\Delta}\equiv\Delta/h \tag{7.49}$$

Taking into account the symmetry of the problem, only a half of the beam is analyzed. The energy dissipation rate due to both the bending moment and tensile force in half of the beam refers to only one plastic hinge:

$$J_{\mathrm{mn}}=M\dot{\kappa}+N\dot{\varepsilon}=\bar{M}M_{\mathrm{P}}\dot{\kappa}+\bar{N}N_{\mathrm{P}}\dot{\varepsilon}=M_{\mathrm{P}}\dot{\kappa}\left(\bar{M}+\bar{N}\frac{N_{\mathrm{P}}\dot{\varepsilon}}{M_{\mathrm{P}}\dot{\kappa}}\right) \tag{7.50}$$

Recalling the analysis in Section 7.3, while using the exact interactive limit locus of the rectangular beam (Eq. 7.30) and the associated normal flow rule (Eq. 7.31 or Eq. 7.32), we have

$$J_{\mathrm{mn}}=M_{\mathrm{P}}\dot{\kappa}\left(\bar{M}+2\bar{N}^{2}\right)=J_{\mathrm{m}}\left(\bar{M}+2\bar{N}^{2}\right)=J_{\mathrm{m}}\left(1+\bar{N}^{2}\right) \tag{7.51}$$

where $J_{\mathrm{m}}=M_{\mathrm{P}}\dot{\kappa}$ represents the energy dissipation due to bending only.

Using the compatibility relation of beams with axial constraint shown in Figure 7.9, the tensile force N and the midpoint deflection Δ are related by Eq. (7.42). Using the non-dimensional parameters, the relation of Eq. (7.42) can be rewritten as

$$\bar{N}=2\bar{\Delta} \tag{7.52}$$

Thus, the combination of Eqs (7.51) and (7.52) leads to

$$J_{\mathrm{mn}}=J_{\mathrm{m}}\left(1+4\bar{\Delta}^{2}\right) \tag{7.53}$$

It should be noted that $N=2\bar{\Delta}$ is only correct for small deflection $\bar{\Delta}\leq 1/2$. If the non-dimensional midpoint deflection $\bar{\Delta}\geq 1/2$, the stress state is at the right corner of the limit locus plotted in Figure 7.7, where $\bar{M}=0$ and $\bar{N}=1$. In this case, a re-derivation gives

$$J_{\mathrm{mn}}=J_{\mathrm{m}}(4\bar{\Delta}) \tag{7.54}$$

Without loss of generality, a *membrane factor* is defined as the ratio of the energy dissipation rate due to the combination of bending and tension, to the energy dissipation rate due to bending alone, i.e.

$$f_{\mathrm{n}}=\frac{J_{\mathrm{mn}}}{J_{\mathrm{m}}} \tag{7.55}$$

With regard to the energy dissipation, taking account of the contribution of the tensile (membrane) force induced by the large deflection is equivalent to regarding the beam as one having a plastic bending moment that varies with the deflection. Hence, the membrane factor, which represents the effect of the membrane force and depends on the deflection of the beam, could be used to modify the energy dissipation by enlarging the fully plastic bending moment from M_{P} to $f_{\mathrm{n}}M_{\mathrm{P}}$.

From the above analysis, the membrane factor for a simply supported beam with axial constraint is

$$f_n = \frac{J_{mn}}{J_m} = \begin{cases} 1 + 4\bar{\Delta}^2, & \bar{\Delta} \le 1/2 \\ 4\bar{\Delta}, & \bar{\Delta} \ge 1/2 \end{cases} \tag{7.56}$$

For a beam fully clamped on both ends, the membrane factor is given by Yu and Stronge (1990), and also Chen and Yu (1993), as

$$f_n = \frac{J_{mn}}{J_m} = \begin{cases} 1 + \bar{\Delta}^2, & \bar{\Delta} \le 1 \\ 2\bar{\Delta}, & \bar{\Delta} \ge 1 \end{cases} \tag{7.57}$$

An important property of this membrane factor is that it is independent of the beam's length, and only depends on the non-dimensional deflection of the beam $\bar{\Delta} \equiv \Delta/h$. Besides, the membrane factor is independent of the type and intensity of the external load, but only depends on the constraint of the beam. That is, when the beam support is specified, the membrane factor as a function of non-dimensional deflection is given. These properties will greatly simplify the solution of the final deflection of beams under various types of dynamic loading.

In Chapter 9, membrane factors will also be introduced for plates with different types of support, and the method will help us to analyze the dynamic response of plates with large deformation.

7.3.2 Solution using the Membrane Factor Method

To demonstrate how to apply the membrane factor method, let us consider a fully clamped beam of length L, which is impinged by a concentrated mass m with initial velocity v_0 at the midpoint of the beam. Suppose the impact velocity is not very high, and hence the deformation mechanism is the same as that in the static case, as shown in Figure 7.14. If we do not care about the detailed history of dynamic deformation, the final deflection of the beam can be obtained directly by using the membrane factor method, as shown below.

From the limit analysis under static loading, clearly the deformation mechanism contains four plastic hinges. Thereby, the energy dissipation by bending that corresponds to the midpoint deflection Δ is

$$D_m = \int_0^{\Delta} J_m \, d\Delta = \int_0^{\Delta} \frac{8M_P}{L} \, d\Delta \tag{7.58}$$

where the relation $M_P \, d\theta = M_P \, d\Delta/(L/2)$ has been used, which is the energy dissipated by a plastic hinge with a rotation angle increment $d\theta$, and $d\Delta$ is the corresponding incremental deflection.

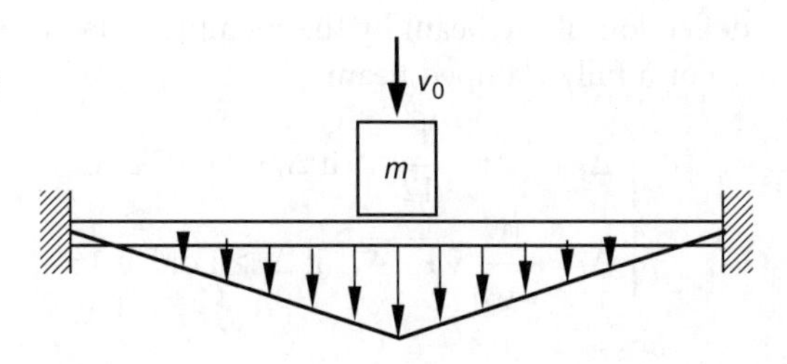

Figure 7.14 The velocity distribution of a fully clamped beam subjected to collision of a mass at the midpoint.

In fact, tension and bending will both dissipate plastic energy. Therefore, with the help of the concept of the membrane factor, the energy dissipation corresponding to the deflection Δ is expressed as

$$D_{mn} = \int_0^{\Delta} J_{mn}\, d\Delta = \int_0^{\Delta} f_n \frac{8M_P}{L}\, d\Delta$$

Or in a normalized non-dimensional form:

$$D_{mn} = \int_0^{\Delta} J_{mn}\, d\Delta = \frac{8M_P}{L/h} \int_0^{\bar{\Delta}} f_n\, d\bar{\Delta} \tag{7.59}$$

Using the membrane factor for a fully clamped beam given by Eq. (7.57), the energy dissipation is found to be

$$D_{mn} = \begin{cases} \dfrac{8M_P}{L/h} \displaystyle\int_0^{\bar{\Delta}} \left(1 + \bar{\Delta}^2\right) d\bar{\Delta} = \dfrac{8M_P}{L/h}\left(\bar{\Delta} + \dfrac{\bar{\Delta}^3}{3}\right), & \bar{\Delta} \le 1 \\ \dfrac{8M_P}{L/h}\left(\dfrac{4}{3} + \displaystyle\int_1^{\bar{\Delta}} 2\bar{\Delta}\, d\bar{\Delta}\right) = \dfrac{8M_P}{L/h}\cdot\left(\dfrac{1}{3} + \bar{\Delta}^2\right), & \bar{\Delta} \ge 1 \end{cases} \tag{7.60}$$

According to the conservation of energy, at the end of the dynamic response, the plastic energy dissipated should be equal to the total input energy, i.e., the initial kinetic energy of the impinging mass is

$$D_{mn}(\bar{\Delta}_f) = K_0 = \frac{1}{2} m v_0^2 \tag{7.61}$$

Employing the parameter introduced in Section 7.2, $\Gamma = e_0 L/h = \dfrac{K_0}{M_P}\dfrac{L}{h}$, the non-dimensional midpoint deflection of the beam, $\bar{\Delta}_f$, is determined by

$$\begin{cases} \bar{\Delta}_f + \dfrac{1}{3}\bar{\Delta}_f^3 = \dfrac{\Gamma}{8} & \text{if } \bar{\Delta}_f \le 1 \text{ or } \Gamma \le 32/3 \\ \bar{\Delta}_f = \dfrac{1}{2\sqrt{6}}\sqrt{3\Gamma - 8} & \text{if } \bar{\Delta}_f \ge 1 \text{ or } \Gamma \ge 32/3 \end{cases} \tag{7.62}$$

It should be emphasized that although the above analysis gives a theoretical expression of the final deflection of the beam, it is based on the assumption that the dynamic deformation mechanism remains the same as that in the static case. Therefore, it is an approximate solution that is only valid for impacts with low initial velocity.

For beams subjected to uniformly impulsive loading, i.e., the whole beam having a uniform initial velocity, by carefully considering the kinematic relation of phase I (with two traveling hinges moving from supports towards the midpoint) and phase II (with a stationary hinge at the midpoint), Chen and Yu (1993) obtained the expression of the final deflection of the beam by the membrane factor method as follows.

For a fully clamped beam

$$\begin{cases} \bar{\Delta}_f + \dfrac{1}{3}\bar{\Delta}_f^3 = \dfrac{\Gamma}{12} & \text{if } \bar{\Delta}_f \le 1 \text{ or } \Gamma \le 16 \\ \bar{\Delta}_f = \dfrac{1}{2\sqrt{3}}\sqrt{\Gamma - 4} & \text{if } \bar{\Delta}_f \ge 1 \text{ or } \Gamma \ge 16 \end{cases} \tag{7.63}$$

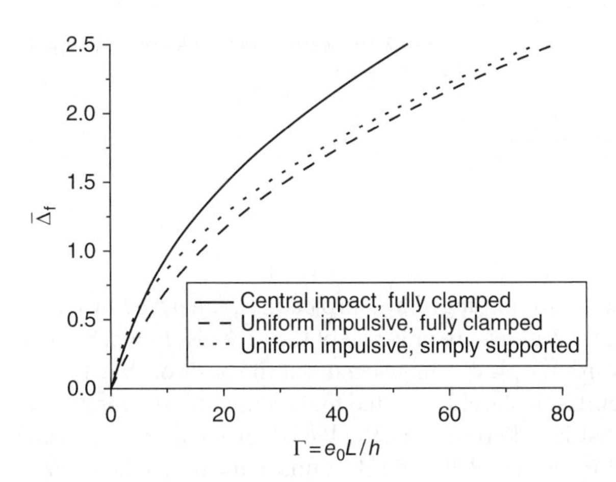

Figure 7.15 Non-dimensional final deflection $\bar{\Delta}_f$ varies with non-dimensional input energy $\Gamma = e_0L/h$.

For a simply supported beam

$$\begin{cases} \bar{\Delta}_f + \dfrac{4}{3}\bar{\Delta}_f^3 = \dfrac{\Gamma}{6} & \text{if } \bar{\Delta}_f \le 1/2 \text{ or } \Gamma \le 4 \\ \bar{\Delta}_f = \dfrac{1}{2\sqrt{3}}\sqrt{\Gamma - 1} & \text{if } \bar{\Delta}_f \ge 1/2 \text{ or } \Gamma \ge 4 \end{cases} \tag{7.64}$$

When the result predicted by the membrane factor method is plotted in Figure 7.12, the curve obtained is between the two curves obtained by the approximate limit loci, ③ and ④. It is close to but slightly lower than the exact solution, i.e., curve ②. Thus, the final deflection of a simply supported beam $\bar{\Delta}_f$ given by the membrane factor method shows a good agreement with the complicated analysis shown in Section 7.2.

The final deflections of the beams, $\bar{\Delta}_f$, predicted by the membrane factor method varying with the normalized energy ratio $\Gamma = e_0L/h$ are plotted in Figure 7.15 for three different loading conditions. From the figure, the effects of the constraint condition and the loading condition of the beam on the final deflection are clearly demonstrated.

In summary, the membrane factor method is able to simplify the analysis of dynamic response of beams. Moreover, the final deflection of beams under different constraints and loading conditions can be expressed analytically.

7.4 Effect of Shear Deformation

In previous examples, shear deformation and the influence of shear force on yield are all neglected. To illustrate the effects of shear deformation, let us now examine an infinite beam under the transverse impact of a rigid projectile.

As shown in Figure 7.16, an infinite long beam with density per unit length ρ and fully plastic bending moment of the cross-section M_P is impinged by a rigid projectile of width $2b$ and mass m_0 with initial velocity v_0.

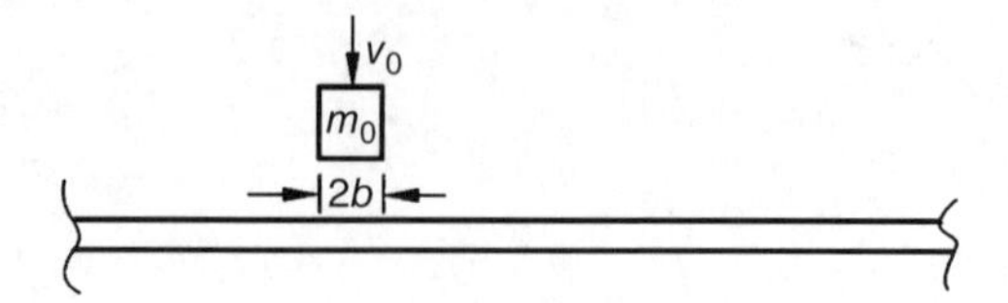

Figure 7.16 Beam of infinite length impinged by a projectile.

7.4.1 Bending-Only Theory

The response of a beam of infinite length impinged by a projectile, as analyzed by bending-only theory, is sketched in Figure 7.17, where only half of the beam is shown. Similar to the Parkes problem, after the impact, a plastic hinge occurs at the position H, which is at a distance λ from the impact point A. It should be noted that another plastic hinge also appears at the impact point A. That is, different from the Parkes problem, now a beam that is half as long has two plastic hinges, A and H. After the impact, the projectile of mass m_0 is attached to the beam's surface and moves with the beam segment immediately underneath it, so that the total mass of the projectile plus the beam segment underneath it is $m = m_0 + 2\rho b$, which will move at a velocity of $\tilde{v}(t)$.

For segment AH, the equation of motion is

$$\frac{\mathrm{d}}{\mathrm{d}t}\left(\frac{1}{2}\rho\lambda v\right) = Q_A \tag{7.65}$$

where v denotes the transverse velocity of point A, and Q_A is the shear force at point A, which is related to the change in velocity of the moving block that includes the projectile and the beam segment underneath it:

$$Q_A = -\frac{1}{2}m\frac{\mathrm{d}\tilde{v}}{\mathrm{d}t} \tag{7.66}$$

If it is assumed that there is no sliding between the moving block and the beam tip A, then $\tilde{v} = v$.

For segment AH, the equation of moment of momentum about H gives

$$\frac{\mathrm{d}}{\mathrm{d}t}\left(\frac{1}{6}\rho\lambda^2 v\right) = 2M_P \tag{7.67}$$

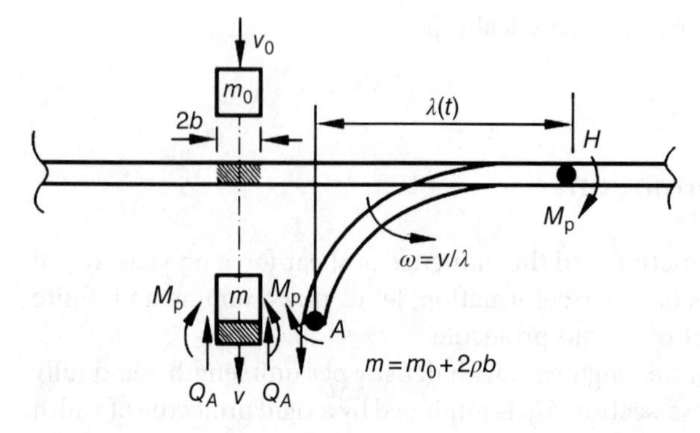

Figure 7.17 Response of a beam of infinite length impinged by a projectile, analyzed by bending-only theory.

By eliminating unknown shear force Q_A from the Eqs (7.65) and (7.66), and using the relation $\widetilde{v} = v$, integration leads to

$$(m + \rho\lambda)v = mv_0 \tag{7.68}$$

The above relation can also be expressed in the following two forms, representing the tip velocity and the positon of plastic hinge, respectively:

$$v = \frac{v_0}{1 + \rho\lambda/m} \tag{7.69}$$

$$\lambda = \frac{m(v_0 - v)}{\rho v} \tag{7.70}$$

Integrating Eq. (7.67) results in

$$\rho\lambda^2 v = 12M_P t \tag{7.71}$$

Substituting the tip velocity Eq. (7.69) into Eq. (7.71), we find

$$\frac{12M_P t}{mv_0} = \frac{\rho\lambda^2}{m + \rho\lambda} \tag{7.72}$$

Then substituting the plastic hinge location Eq. (7.70) into the above relation leads to

$$\frac{12M_P t}{mv_0} = \frac{\rho m^2 (v_0 - v)^2}{\rho^2 v^2 \, mv_0/v} \tag{7.73}$$

i.e.,

$$m^2 (v_0 - v)^2 = 12M_P t \, \rho v \tag{7.74}$$

which can be rewritten in a non-dimensional form:

$$\frac{v_0}{v} + \frac{v}{v_0} - 2 = \frac{12M_P \rho}{m^2 v_0} t \tag{7.75}$$

The differentiation of Eq. (7.75) with respect to time gives

$$-\frac{v_0}{v^2}\dot{v} + \frac{1}{v_0}\dot{v} = \frac{12M_P \rho}{m^2 v_0} \tag{7.76}$$

i.e.,

$$\dot{v} = \frac{12M_P \rho}{m^2 \left[1 - (v_0/v)^2\right]} < 0 \tag{7.77}$$

Based on Eqs (7.77) and (7.66), the shear force is related to the tip velocity by

$$Q_A = -\frac{1}{2} m\dot{v} = \frac{6M_P \rho}{m\left[(v_0/v)^2 - 1\right]} > 0 \tag{7.78}$$

It is known from Eq. (7.78) that at the instant of impact, $t = 0$ and $v = v_0$, hence the shear force Q_A at this instant is infinite.

The above analysis neglects the effect of shear force on the yield condition at hinge A, and so assumes that the material is of infinite shear strength. In fact, no material could have the infinite shear strength, so this model that neglects the shear strength on yielding must be flawed.

7.4.2 Bending-Shear Theory

Now assume a finite shear strength Q_P but neglect the interaction between M and Q at the limit status: then $M_A = M_P$ and $Q_A = Q_P$ simultaneously held at hinge A. The deformation mechanism of a beam of infinite length impinged by a projectile analyzed by bending-shear theory is sketched in Figure 7.18. After the beam is impinged with initial velocity v_0, the projectile of mass m_0 will move with a beam segment immediately underneath it, so that a rigid block of mass $m = m_0 + 2\rho b$ will move downwards with velocity $\tilde{v}(t)$, while the velocity of the beam end A is $v(t)$. In this case, the dynamic response of the beam consists of two phases: a *sliding phase* and a *traveling hinge phase*.

In the sliding phase, it is assumed that the shear force at position A is always equal to the shear strength, $Q_A = Q_P$. For example, the shear strength of a beam of rectangular cross-section is estimated as

$$Q_P = \tau_s\, bh \simeq \frac{1}{2} Ybh \tag{7.79}$$

From the integration of the equation of motion (7.65) and (7.66)

$$\rho\lambda v = 2Q_P t \tag{7.80}$$

$$2Q_P t = m v_0 - m\tilde{v} \tag{7.81}$$

And from the integration of the equation of momentum moment (7.67),

$$\rho\lambda^2 v = 12 M_P t \tag{7.82}$$

Comparing Eqs (7.80) and Eq. (7.82) leads to the position of plastic hinge:

$$\lambda_s = 6M_P / Q_P \tag{7.83}$$

As both the fully plastic bending moment M_P and the shear strength Q_P are constants, Eq. (7.83) indicates that the plastic hinge H is a stationary hinge in the sliding phase. Therefore, at this stage the dynamic response of the beam is similar to the case of step-loading of a cantilever beam which was discussed in Chapter 6.

The transverse velocity of the beam tip A could be found from Eq. (7.80):

$$v = \frac{2Q_P t}{\rho\lambda_s} = \frac{Q_P^2 t}{3\rho M_P} \tag{7.84}$$

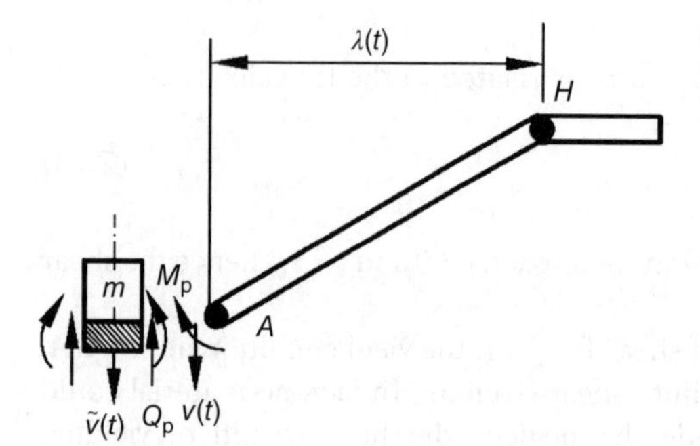

Figure 7.18 Response of a beam of infinite length impinged by a projectile, analyzed by bending-shear theory.

The residual velocity of the projectile is different from the beam tip velocity, and from Eq. (7.81)

$$\tilde{v} = v_0 - \frac{2Q_P t}{m} \tag{7.85}$$

Case A

As shown in Figure 7.19, the sliding phase ceases when the beam tip velocity is equal to the residual velocity of projectile, $v(t_1) = \tilde{v}(t_1)$, which implies that the relative motion at interface A vanishes at instant $t = t_1$. From Eqs (7.80)–(7.83), we obtain

$$m[v_0 - \tilde{v}(t_1)] = \rho\lambda_s v(t_1) = \frac{6M_P\rho}{Q_P}\tilde{v}(t_1) = 2Q_P t_1 \tag{7.86}$$

The above relations give the transverse velocities on both sides of interface A at the end of the sliding phase:

$$v(t_1) = \tilde{v}(t_1) = \frac{v_0}{(1 + 6M_P\rho/mQ_P)} \tag{7.87}$$

Also, the time duration of the sliding phase is determined as

$$t_1 = \frac{mv_0}{2Q_P}\left(1 + \frac{mQ_P}{6\rho M_P}\right)^{-1} \tag{7.88}$$

After $t = t_1$, the rigid block of mass m will move with beam tip A with the same velocity, without further sliding. This marks the beginning of a traveling plastic hinge phase, in which hinge H travels away from its original position towards the far end, and the analysis is similar to that of the Parkes problem.

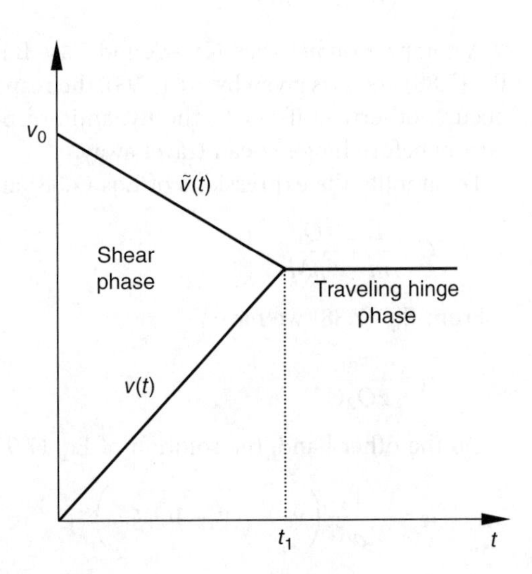

Figure 7.19 History of the transverse velocity of beam tip $v(t)$ and that of the projectile $\tilde{v}(t)$.

Case B

Before the rigid block and the beam tip attain the same transverse velocity, $v(t_1) = \widetilde{v}(t_1)$, a full penetration may be achieved, i.e., the relative displacement at A has reached the beam thickness h.

The above analysis already shows that $m(v_0 - \widetilde{v}) = \rho\lambda_s v = \dfrac{6M_P\rho}{Q_P}v$, hence the velocity of the rigid block could be expressed as

$$\widetilde{v}(t) = v_0 - \frac{6M_P\rho}{mQ_P}v(t) \tag{7.89}$$

Thus, the relative velocity between the two sides of interface A is

$$\widetilde{v} - v = v_0 - \left(1 + \frac{6M_P\rho}{mQ_P}\right)v(t) \tag{7.90}$$

By using Eq. (7.84), the right-hand side of Eq. (7.90) is expressed by the initial velocity v_0 and time t, and so

$$\widetilde{v} - v = v_0 - 2Q_P\left(\frac{Q_P}{6\rho M_P} + \frac{1}{m}\right)t \tag{7.91}$$

By integrating the above relation with respect to time, the relative displacement at A is found to be

$$s = \int_0^t (\widetilde{v} - v)\,\mathrm{d}t = v_0 t - Q_P\left(\frac{Q_P}{6\rho M_P} + \frac{1}{m}\right)t^2 \tag{7.92}$$

Perforation occurs when the relative displacement reaches the beam thickness, i.e., $s = h$. Suppose this happens at time $t = t_2$, then

$$Q_P\left(\frac{Q_P}{6\rho M_P} + \frac{1}{m}\right)t_2^2 - v_0 t_2 + h = 0 \tag{7.93}$$

A comparison between Case A and Case B indicates that if $t_1 < t_2$, where t_1 is given by Eq. (7.88) and t_2 is given by Eq. (7.93), the response will follow Case A and no perforation occurs; otherwise if $t_2 < t_1$, the dynamic response will follow Case B, and perforation occurs before hinge H can travel away.

To simplify the expressions of Eqs (7.88) and (7.93), let

$$C = \frac{1}{m} + \frac{Q_P}{6\rho M_P} \tag{7.94}$$

From Eq. (7.88) we have

$$t_1 = \frac{v_0}{2Q_P C} \tag{7.95}$$

On the other hand, the solution of Eq. (7.93) is

$$t_2 = \frac{1}{2Q_P C}\left(v_0 \pm \sqrt{v_0^2 - 4Q_P Ch}\right) \tag{7.96}$$

Equation (7.96) gives two possible solutions. Unless $v_0^2 < 4Q_PCh$, two positive real solutions will result from Eq. (7.96), and among them perforation should occur at an earlier time. Hence, the negative root is the valid one:

$$t_2 = \frac{1}{2Q_PC}\left(v_0 - \sqrt{v_0^2 - 4Q_PCh}\right) \tag{7.97}$$

Discussion

1) Under a high-velocity impact, $v_0^2 > 4Q_PCh$, t_2 is real and $t_2 < t_1$. This leads to Case B, i.e., a complete severance will occur.
2) If the impact velocity is moderate, $v_0^2 < 4Q_PCh$, no real value of t_2 can be found. This implies a Case A response with no severance.

Therefore, the *ballistic limit* is given by

$$v_0^2 = 4Q_PCh = 4Q_Ph\left(\frac{1}{m} + \frac{Q_P}{6\rho M_P}\right) \tag{7.98}$$

In military applications, the perforation of projectiles is usually involved. Equation (7.98) is very useful in determining the ballistic limit for beams and plates.

In summary, sliding happens when a load or impulse is intensive and the beam is stubby (short and thick). The influence of shear deformation and shear force have been considered in this section. It should be noted that this analysis still has some limitations. For example, the shear strength Q_P is estimated on the assumption of uniformly distributed shear stress over the whole cross-section (see Eq. 7.79). The shear strength Q_P is also assumed to be constant during the entire sliding phase. In fact, in the sliding phase, Q_P will decrease with the reduced connective area on the interface due to sliding. Further studies should consider the interaction between M and Q, as well as the variation of Q_P in the sliding phase.

7.5 Failure Modes and Criteria of Beams under Intense Dynamic Loadings

In this section, the failure modes and criteria of a plastic beam subjected to intense dynamic loads, such as impulsive loading (initial velocity) and impact by projectile, will be reviewed (see also Yu & Chen, 1998). In addition, the three basic failure modes identified by Menkes and Opat (1973) for impulsively loaded beams, and more complicated failure behaviors, as observed in complex structural members and/or under complex dynamic loading conditions, are reviewed.

7.5.1 Three Basic Failure Modes Observed in Experiments

Different failure modes may develop in a structure under various dynamic loading conditions. In 1973, Menkes and Opat conducted a series of experiments for fully clamped beams. As shown in Figure 7.20, they carried out an experimental investigation of fully clamped beams subjected to a uniformly distributed initial velocity over the entire beam

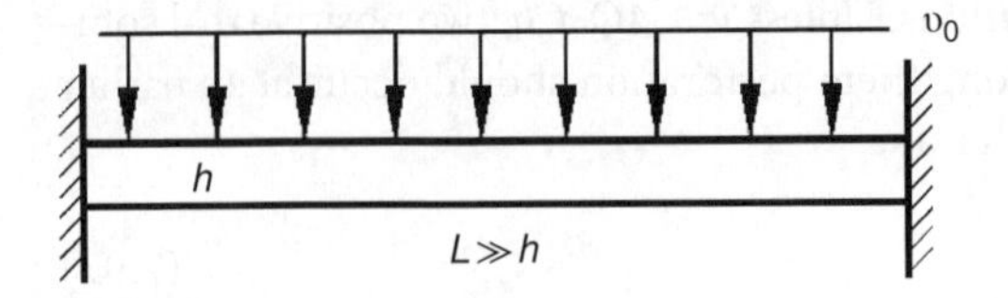

Figure 7.20 A clamped beam subjected to a uniformly distributed velocity v_0.

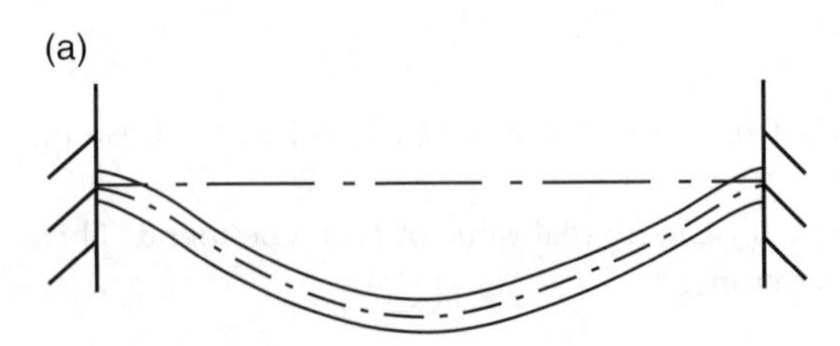

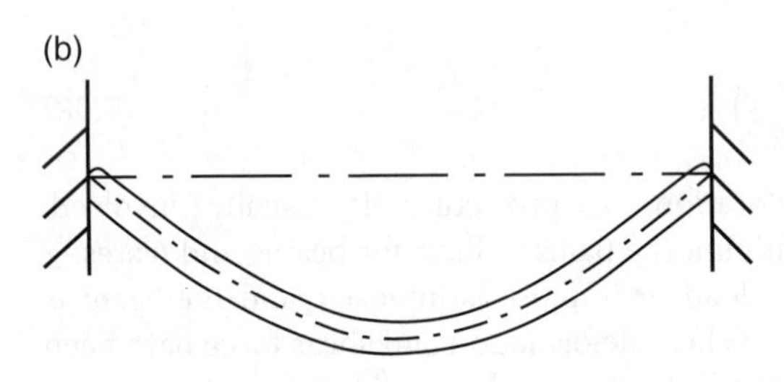

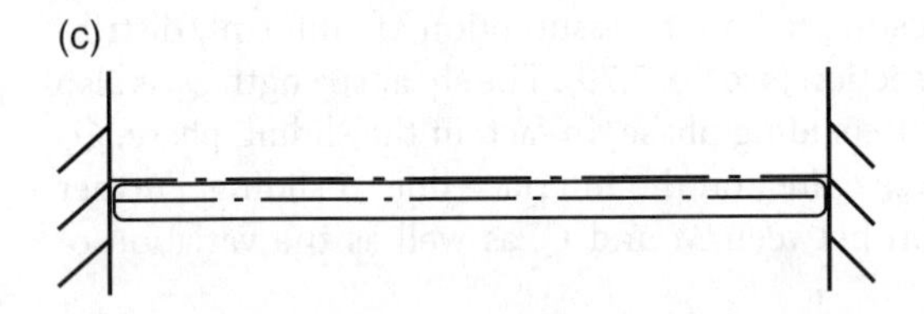

Figure 7.21 Failure modes: (a) mode I; (b) mode II; (c) mode III.

span by detonating sheet explosives over the beam. A change in the intensity of the uniform initial velocity of the beam is obtained by changing the mass of the explosive.

The tested beams were observed to respond in a ductile manner and to acquire a permanently deformed profile when subjected to a velocity less than a certain value (see Figure 7.21a). However, when the impulsive velocity exceeded this critical value, the beams failed due to tensile tearing of the beam material at the supports (Figure 7.21b). As the impulsive velocity was further increased, the plastic deformation of the beams became more localized near the supports until another critical velocity was reached that was associated with shear failure at the supports (Figure 7.21c). With the increase in the impulsive loading, the final deflection at the midpoint of the beam first increases then decreases.

Based on these experiments, Menkes and Opat identified three basic failure modes for impulsively loaded fully clamped beams:

- *mode I* – large inelastic deformation;
- *mode II* – tearing (tensile failure) in outer fibers, at or over the supports;
- *mode III* – transverse shear failure at the supports, with no significant flexural deformation.

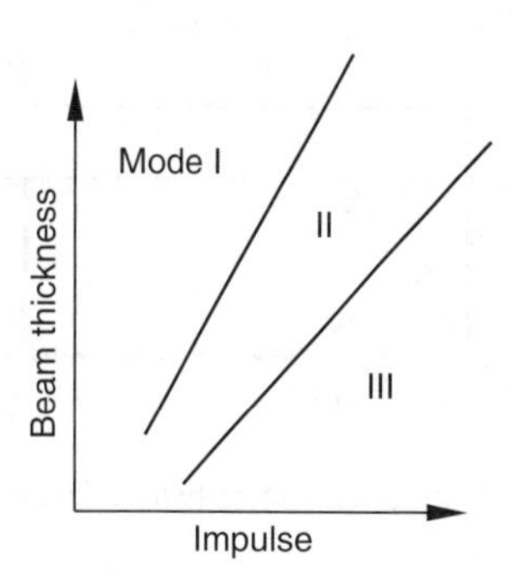

Figure 7.22 A schematic map of failure mode occurrence.

It should be noted that in modes II and III, failure occurred essentially over the cross-sections at the two supports rather than over the whole beam. Figure 7.22 shows a schematic map of the failure mode occurrence changing with the loading intensity. For thin beams under moderate impulsive loading, mode I failure occurs; for thin beams under intensive impulsive loading, mode II failure occurs; and for thick beams under intensive impulsive loading, mode III failure occurs. On the failure map, the boundary between regimes of modes I and II is clear, while the boundary between the regimes of modes II and III is usually not very clear. In experiments carried out by other researchers, it was observed that under sufficiently large impact energy, the flat steel beam was broken or cracked at the impact point with a shear mode, which is termed mode IV. Similarly, three basic failure modes were also observed in impulsively loaded clamped circular and square plates.

7.5.2 The Elementary Failure Criteria

Compared with the experimental investigations, theoretical works have been published concerning structural dynamic failure. Several structural dynamic failure criteria have been proposed and used in the relevant theoretical analysis.

Regarding the basic failure mode I, i.e., large inelastic deformation observed by Menkes and Opat (1973), over the last few decades, several efficient means have been developed to improve the existing small deformation analysis, and the effect of axial force induced by large deformation has been considered (Sections 7.2 and 7.3). Considering the variation of the deformation mechanisms due to a large deformation effect, the final deflection of the beams can be predicted more accurately by the corresponding equations of motion and geometric relations based on an updated configuration (Stronge & Yu, 1993).

In 1976, Jones used elementary failure criteria (critical tensile strain criterion and critical accumulative shear sliding criterion) to estimate the threshold velocities corresponding to failure modes II and III for the dynamically loaded beams tested by Menkes and Opta (1973), as shown in Figure 7.23. The elementary failure criterion related to tensile failure mode II can be expressed as

$$\varepsilon_{\max} = \varepsilon_c \tag{7.99}$$

where $\varepsilon_{\max}$ is the maximum strain within the structure and ε_c is the critical tensile strain of the material, while the elementary failure criterion related to shear failure mode III is

$$\Delta^s_{\max} = h \tag{7.100}$$

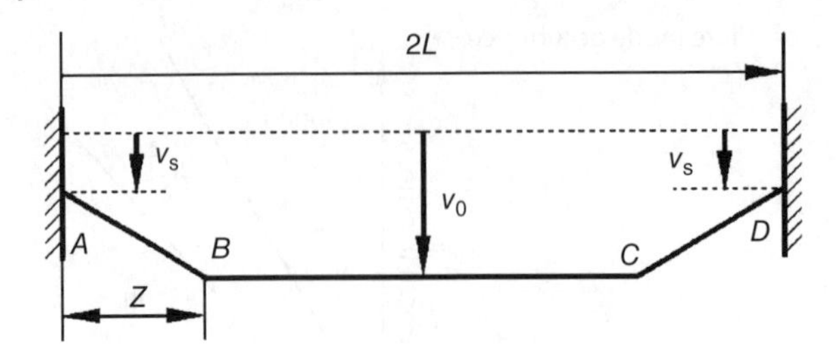

Figure 7.23 Velocity field for an impulsively loaded, fully clamped beam with shear sliding at supports.

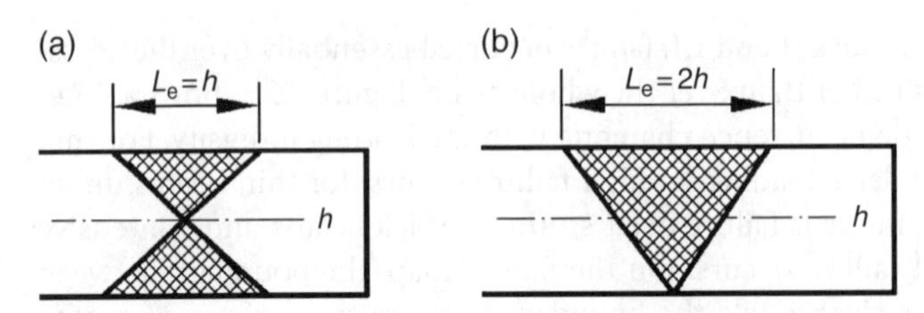

Figure 7.24 The effective length of the plastic hinge (a) bending only (b) with the influence of tensile force.

where Δ^s_{max} is the maximum shear sliding at the cross-section, and h is the thickness of the beam.

If a rigid-perfectly plastic simplification is adopted, deformation is ideally localized at the plastic hinge(s) only, so that the information about the strain distribution in a structure cannot be obtained directly. To calculate the maximum strain within the structure involved in the failure criterion Eq. (7.99), an effective length of the plastic hinge, L_e, has to be defined. At beginning of deformation, the tensile force $N = 0$ and

$$L_e = h \tag{7.101}$$

is taken (see Figure 7.24a). With the influence of membrane force considered, Nonaka (1967) and other authors estimated

$$L_e = (2 - 5)h \tag{7.102}$$

That is, the effective length of the plastic hinge is about two to five times the beam thickness (see Figure 7.24b).

Why is an effective length of the plastic hinge needed? This can be explained by means of the slip-line theory of plasticity. The simplest slip-line field under pure bending is given in Figure 7.24(a), where $L_e = h$, whereas from the slip-line field under the combination of bending and tension shown in Figure 7.24(b), the influencing area of the plastic hinge is clearly larger than the former. Therefore, defining the effective length of the plastic hinge is, in fact, specifying how large the plastic zone should reasonably be in calculating ε_{max}. Generally, the effective length of a plastic hinge will be smaller under pure bending or pure tension, but will be larger under the combination bending and tension.

When applying the elementary failure criterion for mode III, the maximum slipping distance, Δ^{s}_{max}, at a cross-section of the beam (usually at the support or at the edge of the impinging mass) should be calculated according to the method discussed in Section 7.4 (e.g., Eq. 7.92) and then compared with the beam thickness.

Although the theoretical predictions give fairly good agreement with the corresponding experimental observations, these simple theoretical formulations are placed on a far less firm foundation; for example, the interaction among bending moment M, tensile force N and shear force Q is entirely neglected. We discuss the modified failure criterion further in Section 7.5.4.

7.5.3 Energy Density Criterion

Shen and Jones (1992) proposed an *energy density criterion*, which states that rupture occurs in a rigid-plastic structure when the plastic energy dissipation per unit volume, ϑ, reaches a critical value

$$\vartheta = \vartheta_c \tag{7.103}$$

when ϑ contains the plastic energy dissipation related to all the stress components.

If a plastic hinge in a rigid-plastic beam has an effective length L_e, a width b and a thickness h, then the effective volume of the plastic hinge is bhL_e. Thus, the maximum plastic energy that the hinge can dissipate is $\Omega_c = \vartheta_c bhL_e$. If the total amount of energy dissipation at a plastic hinge during the structural response is Ω, the energy density criterion is expressed as

$$\Omega < \Omega_c = \vartheta_c bhL_e \tag{7.104}$$

Assume the plastic energy dissipated by the same plastic hinge through shearing deformation is Ω^s, and the ratio is $\beta = \Omega^s/\Omega$. In order to distinguish various failure modes, Eq. (7.104) can be further written as follows:

$$\Omega < \Omega_c, \beta < \beta_c \quad \text{mode I} \tag{7.105}$$

$$\Omega = \Omega_c, \beta < \beta_c \quad \text{mode II} \tag{7.106}$$

$$\Omega < \Omega_c, \beta > \beta_c \quad \text{mode III} \tag{7.107}$$

Based on the experimental measurement reported by Menkes and Opat (1973), Shen and Jones (1992) estimated $\beta_c = 0.45$ to distinguish between mode II and mode III.

The difficulty in applying the energy density criteria is in how to determine the effective volume of the plastic region around a plastic hinge. As the effective length of a plastic hinge, L_e, found in different problems is usually different, there is great uncertainty in the failure criteria shown in Eq. (7.104). Also, in the calculation by finite element or other numerical methods, it is found that the energy density per volume is very sensitive to the element size and mesh shape. Due to the large strain gradient in the region around a plastic hinge, it is hard to obtain the maximum plastic energy as required by the application of the failure criteria. Consequently, this situation greatly limits the application of the energy density failure criteria.

7.5.4 A Further Study of Plastic Shear Failures

A deeper understanding of the three basic failure modes was given by Yu and Chen (2000). In their study, the plastic shear failure (mode III) of an impulsively loaded, fully clamped beam was re-examined (see Figure 7.20), with the focus on two effects: (i) the interaction between the shear force and the bending moment; and (ii) the weakening of the sliding sections during the failing process.

It was observed that, unlike tensile tearing failure (mode II), which followed the large defection of the beam (mode I), transverse shear failure (mode III) is characterized by insignificant flexural deformation at most cross-sections. Therefore, shear failure occurs at the supports in the early stage of the beam's dynamic response and generally exhibits a type of localized behavior; thus the mode III threshold does not depend on the length of the beam. With the length of the beam being excluded from the theoretical description, if a rigid-perfectly plastic material model is adopted, the material behavior is merely determined by the yield strength Y. The width of the beam b, which is much smaller than the beam length L, is adopted as $b = 1$ in the analysis. Therefore, the maximum shear sliding at the supports, Δ^{s}_{max}, can be expressed as a function of four parameters, i.e., impact velocity, yield stress, material density, and the beam thickness:

$$\Delta^{s}_{max} = \Delta^{s}_{max}(v_0, Y, \rho, h) \tag{7.108}$$

From the π -theorem of dimensional analysis, Δ^{s}_{max} given by Eq. (7.108) can be normalized by the beam thickness h, and then the normalized maximum shear sliding distance should only depend on the non-dimensional combination of the three other parameters:

$$\frac{\Delta^{s}_{max}}{h} = f\left(\frac{v_0}{\sqrt{Y/\rho}}\right) = f\left(\sqrt{\psi}\right) \tag{7.109}$$

Here the non-dimensional parameter $\psi \equiv \rho v_0^2/Y$ is termed the *damage number* or *Johnson's number*, which is commonly used to indicate the intensity of impact. Hence, we may adopt a shear failure criterion in form of

$$\Delta^{s}_{max} = kh \tag{7.110}$$

with k being a constant determined by experimental measurement or theoretical consideration. Thus, the threshold impulsive velocity for transverse shear failure (mode III) can be expressed as

$$v_{c3} = f^{-1}(k)\sqrt{Y/\rho} \tag{7.111}$$

Here $f^{-1}()$ is the inverse function of f which appears in Eq. (7.109). It transpires that for a rectangular cross-section beam with $L/h \gg 1$, the threshold impulsive velocity for a mode III failure given by Eq. (7.111) is independent of the geometry of the beam and only depends on the material property.

From the observations in punching or impact experiments, before the shear sliding at the supports reaches the beam thickness h, the beam section will be penetrated by a crack along the thickness direction. Generally the shear sliding of 30% of the beam thickness, i.e., $\Delta^{s}_{max} = 0.3h$, will lead to through-thickness cracking and shear failure of the whole beam section.

With the aid of the velocity field shown in Figure 7.23, where sliding occurs at the supports of the beam with a finite shear strength Q_P, the threshold velocity for the onset of a mode III failure based on the elementary failure criterion, i.e., $\Delta^s_{max} = h$, which is equivalent to $k = 1$ in Eq. (7.110), is given by

$$v_{c3} = \sqrt{\frac{2\,Q_P^2}{3\rho b M_P}} \tag{7.112}$$

If the von Mises yield criterion is adopted, then for a beam with a rectangular cross-section, $M_P = Ybh^2/4$ and $Q_P = Ybh/\sqrt{3}$, so that

$$v_{c3} = \frac{2\sqrt{2}}{3}\sqrt{\frac{Y}{\rho}} = 0.943\sqrt{\frac{Y}{\rho}} \tag{7.113}$$

Equation (7.113) gives the threshold velocity for a mode III failure by the *elementary failure criteria*, as shown in Section 7.5.2. That is, by adopting the factor as 0.943 in Eq. (7.111), the shear failure criterion is reduced to the elementary failure criteria ($k = 1$).

The elementary failure criterion for mode III can be modified by considering:

1) that the maximum shear sliding is smaller than the beam thickness, i.e., a factor of $k < 1$, e.g., $k = 0.3$ can be taken to predict shear failure;
2) that the bending moment and shear force will both change as a result of the thickness reduction due to sliding, i.e., the effective thickness of the sliding section is reduced from h to $h - \Delta^s$;
3) the interaction between the bending moment and the shear force.

If the above weakening effects are incorporated, then the bending and shear strengths of the sliding sections will be expressed as $M_P' = \zeta^2 M_P$ and $Q_P' = \zeta Q_P$, with weakening factor $\zeta = 1 - (\Delta^s/h)$. Then, the yield condition Eq. (7.30) is modified as

$$\left(\frac{\bar{M}}{\zeta^2}\right)^2 + \left(\frac{\bar{Q}}{\zeta}\right)^2 = 1 \tag{7.114}$$

where $\bar{M} = M/M_P$ and $\bar{Q} = Q/Q_P$.

Taking into account the above modification, the critical value of the damaging number $\psi \equiv \rho v_0^2/Y$ is derived as (see Yu & Chen, 2000)

$$\psi_c = 0.396 \tag{7.115}$$

and the corresponding critical impulsive velocity is identified as

$$v_{c3} = 0.629\sqrt{Y/\rho} \tag{7.116}$$

Thus, the critical impulsive velocity given by the *modified shear failure* is only about two-thirds of that given by the elementary failure criteria (see Eq. 7.113).

A schematic of shear sliding over a finite width of the beam is given in Figure 7.25(a), and the yield locus for the sliding sections taking into account the interaction between bending and shear is depicted in Figure 7.25(b). For the critical impulsive velocity given by Eq. (7.116), the shear energy ratio within the total plastic energy dissipation is found as

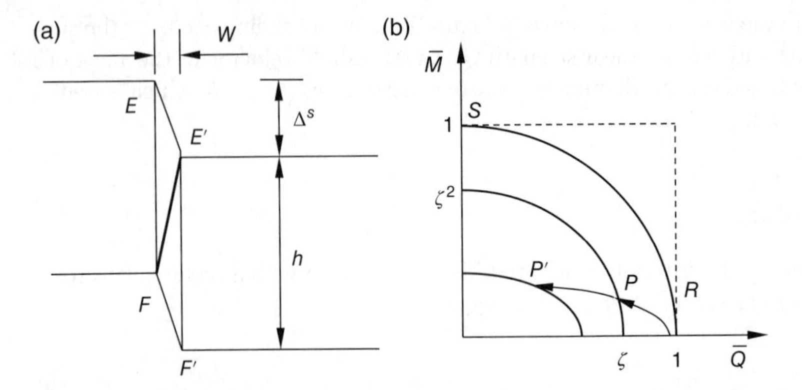

Figure 7.25 (a) Schematic of shear sliding over a finite width W; (b) yield locus for the sliding section.

$$\beta = \Omega^s/\Omega = 0.47 \tag{7.117}$$

Compared with the empirical criterion for the onset of shear failure at the support, $\beta_c = 0.45$, as mentioned in Section 7.5.3, the analysis in this section offered a rational theoretical foundation for better understanding of the ratio β in the shear failure process of impulsively loaded beams.

Questions 7

1 The tip of a curved cantilever beam is subjected to a step-loaded impulse perpendicular to the beam's bent plane. Explain how you might analyze its dynamic response.

Problems 7

7.1 A simply supported beam has length $2L$, mass per unit length ρ, and fully plastic bending moment M_P. Suppose the beam center is impinged by a concentrated mass m with initial velocity v_0. Derive the governing equations of its dynamic response for the following two cases: (i) with no axial constraint at supports; and (ii) with axial constraint at supports.

7.2 An infinite beam has mass per unit length ρ, thickness h, fully plastic bending moment M_P, and shear strength Q_P. The beam is impinged by a rigid mass m with the initial velocity being equal to the ballistic limit. Find the energy dissipated by bending and shear.

8

Mode Technique, Bound Theorems, and Applicability of the Rigid-Perfectly Plastic Model

Many examples in this textbook have illustrated that as a typical feature of impulsively loaded rigid-plastic structures, the dynamic response can be clearly divided into two phases: a transient phase characterized by the presence of traveling plastic hinge(s), and a subsequent modal phase, in which the hinge positions remain stationary.

During the transient phase, the deformation pattern steadily evolves from the initial velocity distribution imposed by the external dynamic loading to a model configuration. In this phase, the deformation mechanism is varying, such as the position of the plastic hinges, the velocity distribution and so forth. This phase generally presents the greater mathematical challenge, as non-linear equations, which include the hinge position as an unknown and incorporate the discontinuity conditions across the hinge, have to be solved. Except for a few simple cases, the solution has to be obtained by step-by-step numerical procedure.

In contrast, the modal phase is relatively straightforward. As the velocity field retains the same shape, the stresses remain constant, and, if geometrical changes are ignored, the acceleration is constant and the problem can be easily solved. In most cases, the deformation mechanism in the modal phase of the dynamic response is the same as that under the static loading condition. The simplicity forms the basis of an approximate method of solving impulsive loading problems called the *mode technique*. In this approximate technique, the entire response of a structure is restricted to be of modal type.

Based on the understanding of the mode technique, several bound theorems of the structural dynamic response will also be introduced, so as to estimate the upper and lower bounds of the final displacement and also the total response time. Some fundamental bound theorems illustrated in Section 8.5 are simple and very practical for use in engineering applications.

In Section 8.6, the applicability of the widely adopted material simplification, i.e. the rigid-perfectly plastic (RPP) model, will be discussed. The dynamic responses of beams made of the elastic-plastic material and the RPP material are compared.

8.1 Dynamic Modes of Deformation

Here the term "mode" is used in its classical sense, meaning that all the kinematic variables (displacement, velocity, and acceleration) are expressible as separable functions of time and space variables, i.e., in the form $f(t)g(x)$.

Introduction to Impact Dynamics, First Edition. T.X. Yu and XinMing Qiu.

Published 2018 by John Wiley & Sons Singapore Pte. Ltd.

Now suppose that due to impulsive loading, a rigid-plastic structure gains an initial velocity distribution $\dot{u}_i^0(x)$, where u_i denotes the displacement consisting of three components ($i = 1,2,3$), and x is the space variable. The dot ($\dot{}$) denotes d()/dt. In a *complete solution* for the dynamic response of a structure, the velocity field $\dot{u}_i(x,t)$ must satisfy the equations of motion, compatibility, and constitutive relations, in addition to the initial condition $\dot{u}_i(x,0) = \dot{u}_i^0(x)$ and the boundary conditions. In general, $\dot{u}_i(x,t)$ is not a separable function of time and space variables. For instance, in the *transient phase* of Parkes solution, $\dot{u}(x,t) = v(t)\ [1 - x/\lambda(t)]$, so it is not in a form $f(t)g(x)$. While in the *model phase* of the dynamic response, the plastic hinge is stationary, e.g., at the root for Parkes problem, $\lambda = L$, and therefore the velocity field can be expressed by separate functions for time and space variables.

A "modal solution" is a velocity field with separate functions for time and space variables,

$$\dot{u}_i^*(x,t) = \dot{w}_*(t)\phi_i^*(x) \tag{8.1}$$

which satisfies the field equations but not necessarily the initial condition. The function ϕ_i^* is called the *mode* or mode shape; the scalar velocity $\dot{w}_*$ of a characteristic point is specified such that $|\phi_i^*(x)| \leq 1$.

In the mode technique, due to the velocity field with separate functions for time and space variables, as shown in Eq. (8.1), the model velocity field $\dot{u}_i^*$ and the associated *acceleration field* $\ddot{u}_i^*$ have the *same shape*, i.e.

$$\ddot{u}_i^*(x,t) = \ddot{w}_*(t)\phi_i^*(x) \tag{8.2}$$

If no external forces act on the structure, mode shapes can be regarded as *natural modes* of dynamic plastic deformation as they are *a property of the structure* and *independent of the initial velocity field.* Modes also exist for structures subjected to statically admissible time-independent surface tractions. For simple structures, the mode shape that is most suitable will be fairly obvious: usually the static collapse mode, e.g., for a cantilever we can choose

$$\phi^*(x) = 1 - x/L \tag{8.3}$$

with $x = 0$ located at the tip. In more complex problems, there are probably several different deformation mechanisms in competition with each other. In this case, the real dynamic mode of a rigid-plastic structure can be identified by means of *the principle of minimum specific dissipation of power*, i.e., among all kinematically admissible velocity fields with the same kinetic energy, modal solutions will be an extremum of Lee's function (Lee & Martin, 1970; Lee, 1972) expressed by

$$J^* = D_P/\sqrt{K} \tag{8.4}$$

where D_P is the plastic energy dissipation, and K is the kinetic energy related to the kinematically admissible velocity field.

8.2 Properties of Modal Solutions

Impulsive loading on a rigid-plastic structure results in dynamic deformation that continually evolves towards a modal solution.

This property of the modal solution indicates that for an ideal rigid-plastic structure under impact loading, its dynamic response has a trend of evolving towards a modal solution. In other words, if, apart from the imposed initial velocity field, no extra load is applied to the structure during the deformation process, the dynamic response of the rigid-plastic structure will finally evolve to the modal solution.

A proof of this property is based on the "*convergence theorem*" given by Martin (1966). Suppose we have two solutions for a structure, $\dot{u}_i^a$ and $\dot{u}_i^b$; both solutions satisfy all the field equations and the boundary conditions but they are associated with different initial velocity distributions. Then we can introduce a *functional* $\widetilde{\Delta}$ to describe their difference at any time t:

$$\widetilde{\Delta} = \widetilde{\Delta}(t) = \frac{1}{2}\int_V \rho\left(\dot{u}_i^{\mathrm{a}} - \dot{u}_i^{\mathrm{b}}\right)\left(\dot{u}_i^{\mathrm{a}} - \dot{u}_i^{\mathrm{b}}\right)\,\mathrm{d}V \tag{8.5}$$

Martin (1966) proved that for rigid-plastic structures, the difference will not increase with time:

$$\frac{\mathrm{d}\widetilde{\Delta}}{\mathrm{d}t} = \int_V \rho\left(\ddot{u}_i^{\mathrm{a}} - \ddot{u}_i^{\mathrm{b}}\right)\left(\dot{u}_i^{\mathrm{a}} - \dot{u}_i^{\mathrm{b}}\right)\,\mathrm{d}V \le 0 \tag{8.6}$$

In Eq. (8.6), the physical variables in the two brackets are the differences in acceleration and the difference in velocity of the two solutions, respectively. Note that $\rho\ddot{u}_i\dot{u}_i$ is the rate of work of "*inertia force*", so that from the principle of virtual work

$$\int_V \rho\left(\ddot{u}_i^{\mathrm{a}} - \ddot{u}_i^{\mathrm{b}}\right)\left(\dot{u}_i^{\mathrm{a}} - \dot{u}_i^{\mathrm{b}}\right)\,\mathrm{d}V = -\int_V \left(Q_j^{\mathrm{a}} - Q_j^{\mathrm{b}}\right)\left(\dot{q}_j^{\mathrm{a}} - \dot{q}_j^{\mathrm{b}}\right)\,\mathrm{d}V \tag{8.7}$$

where Q_j is the *general stress* (bending moment, axial force, etc.) and $\dot{q}_j$ is the associated *general strain rate* (curvature rate, elongation rate, etc.). If the general stress is a bending moment, then the associated general strain rate is the relevant angular velocity.

When solutions $\dot{u}_i^a$ and $\dot{u}_i^b$ are taken as the complete solution and the modal solution, respectively, we have

$$\frac{\mathrm{d}\widetilde{\Delta}}{\mathrm{d}t} = -\int_V \left(Q_j - Q_j^*\right)\left(\dot{q}_j - \dot{q}_j^*\right)\,\mathrm{d}V \le 0 \tag{8.8}$$

which implies that *the difference between the modal solution and the actual solution does not increase with time*; in other words, both solutions approach each other as time progresses.

For an impulsively loaded structure, the modal solution is a uniformly decelerated motion, i.e. velocity $\dot{w}_*(t)$ is a linearly decreasing function of time.

In the case of impulsive loading, the decreasing rate of the kinetic energy should be equal to the dissipation rate:

$$-\frac{\mathrm{d}K}{\mathrm{d}t} = \dot{D}_{\mathrm{P}} = \frac{\mathrm{d}D_{\mathrm{P}}}{\mathrm{d}t} \tag{8.9}$$

i.e.

$$-\int_V \rho\ddot{u}_i\dot{u}_i\,\mathrm{d}V = \int_V Q_j\dot{q}_j\,\mathrm{d}V \tag{8.10}$$

For the modal solution, taking the velocity field with separate functions for time and space variables as $\dot{u}_i = \dot{w}_*(t)\phi_i^*(x)$, and the associated strain rate as $\dot{q}_j = \dot{w}_*(t)k_j^*(x)$, then from Eq. (8.10) we have

$$-\int_V \rho \ddot{w}_* \dot{w}_* \phi_i^* \phi_i^* \, \mathrm{d}V = \int_V Q_j^* \dot{w}_* k_j^* \, \mathrm{d}V \tag{8.11}$$

which leads to the acceleration

$$\ddot{w}_* = -\frac{\int_V Q_j^* \, k_j^* \cdot \mathrm{d}V}{\int_V \rho \, \phi_i^* \, \phi_i^* \cdot \mathrm{d}V} = -a_* \tag{8.12}$$

For a structure made of RPP material, Q_j^* is independent of $\dot{q}_j^*$ and time t, so that $\ddot{w}_*$ and a_* given by Eq. (8.12) are independent of time. This indicates that

1) the modal motion is a motion with single degree-of-freedom;
2) the deceleration is a constant.

Hence, the velocity can be expressed as

$$\dot{w}_*(t) = \dot{w}_*^0 - a_* t \tag{8.13}$$

where $\dot{w}_*^0$ is the initial modal velocity. It is also evident that the final response time t_{f}^* and the final displacement w_*^{f} are

$$t_{\mathrm{f}}^* = \dot{w}_*^0 / a_* \tag{8.14}$$

$$w_*^{\mathrm{f}} = \frac{1}{2}\dot{w}_*^0 t_{\mathrm{f}}^* = \left(\dot{w}_*^0\right)^2 / (2a_*) \tag{8.15}$$

From this illustration we now see that if the modal shape $\phi_j^*(x)$ is known, and the $k_j^*(x)$ in $\dot{q}_j^* = \dot{w}_*(t)k_j^*(x)$ is found, then the deceleration a_* is given by Eq. (8.12). After that, the modal solution is totally determined by Eqs (8.14) and (8.15), provided the initial modal velocity $\dot{w}_*^0$ is specified.

Should we adopt the actual initial velocity of the structure as the initial velocity of the modal solution? It has been verified that generally this is not a good choice. Since, in the modal solution, the kinetic energy in the transient phase is completely ignored, adopting the actual initial velocity as the initial velocity of the modal solution will greatly overestimate the initial kinetic energy, and thus overestimate the final deformation. The method of determining the initial velocity in the modal solution will be discussed in the following section.

8.3 Initial Velocity of the Modal Solutions

As shown above, functional $\tilde{\Delta}$, which describes the difference between the modal and the actual solution, decreases with time. Hence, to obtain the "best" approximation to the actual solution, it is reasonable to choose an initial velocity $\dot{w}_*^0$ that $\tilde{\Delta}_0$ renders a minimum at $t = 0$. In fact, at $t = 0$

$$\widetilde{\Delta}_0 = \widetilde{\Delta}(0) = \frac{1}{2}\int_V \rho\left(\dot{u}_i^0 - \dot{w}_*^0\phi_i^*\right)\left(\dot{u}_i^0 - \dot{w}_*^0\phi_i^*\right)\,\mathrm{d}V \tag{8.16}$$

Taking the partial differential of the above relation to initial velocity $\dot{w}_*^0$:

$$\frac{\partial \widetilde{\Delta}_0}{\partial \dot{w}_*^0} = 0 \tag{8.17}$$

leads to

$$\int_V \rho\left(\dot{u}_i^0 - \dot{w}_*^0\phi_i^*\right)\phi_i^*\,\mathrm{d}V = 0 \tag{8.18}$$

That is, the initial velocity of the modal solution can be determined by

$$\dot{w}_*^0 = \frac{\displaystyle\int_V \rho\,\dot{u}_i^0\,\phi_i^*\,\mathrm{d}V}{\displaystyle\int_V \rho\,\phi_j^*\,\phi_j^*\,\mathrm{d}V} \tag{8.19}$$

The above procedure is called "*min* $\widetilde{\Delta}_0$" *technique*, which was proposed by Martin and Symonds (1966). The key point of this technology is to determine the initial velocity of the modal solution by letting the difference in functional $\widetilde{\Delta}$ be at a minimum at time $t = 0$.

Note that the initial velocity $\dot{w}_*^0$ given by Eq. (8.19) results in

$$\widetilde{\Delta}_0 = \frac{1}{2}\int_V \rho\dot{u}_i^0\dot{u}_i^0\,\mathrm{d}V - \frac{1}{2}\left(\dot{w}_*^0\right)^2\int_V \rho\phi_i^*\phi_i^*\,\mathrm{d}V = K_0 - K_0^* \tag{8.20}$$

which is the initial difference between the kinetic energies of the actual solution and the modal solution. The kinetic energy associated with the modal solution is generally smaller than that of the actual solution, but with $\dot{w}_*^0$ chosen in this way the modal solution provides a good approximation to the modal phase of the actual response and to t_f:

$$\widetilde{\Delta}_0 = K_0 - K_0^* \geq 0 \tag{8.21}$$

The velocity histories resulting from the complete and modal solutions are plotted in Figure 8.1. Clearly, the initial velocity of the modal solution, $\dot{w}_*^0$, is smaller than the actual

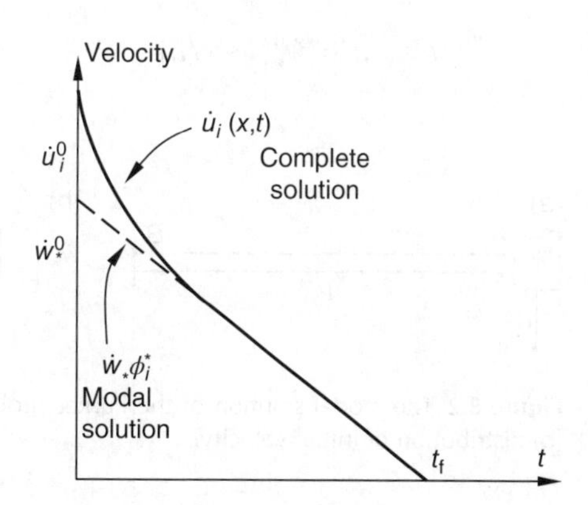

Figure 8.1 History of velocity given by the complete solution and the modal solution.

initial velocity of the structure $\dot{u}_i^0$. In the transient phase, there are obvious differences between velocities of the modal solution and those of the complete solution, but in the modal phase, the curves of velocities tend to coincide, both representing a uniform deceleration. Hereafter, using the initial velocity of the modal solution, $\dot{w}_*^0$, given by the above-mentioned "min $\widetilde{\Delta}_0$" technique will lead to a modal solution that best approximates the actual solution.

8.4 Mode Technique Applications

The procedure for applying the mode technique is as follows:

1) Choose the mode shape, $\phi_i^*(x)$.
2) Determine the initial velocity $\dot{w}_*^0$ using the "min $\widetilde{\Delta}_0$" technique.
3) Calculate the deceleration a_*, the total response time t_f^* and the final displacement w_*^f.
4) Check the bending moment in the beam, in order to ensure that the bending moment distribution within the whole beam satisfies $|M^*(x)| \le M_p$. Violation of this condition implies that the chosen mode shape is incorrect, so we have to start again from the first step.

8.4.1 Modal Solution of the Parkes Problem

The complete solution of the Parkes problem was given in Section 6.3. Here we will discuss the approximate solution given by the modal technique. The Parkes problem is sketched in Figure 8.2(a), in which a cantilever of length L, mass per unit length ρ, and fully plastic bending moment of beam's section M_P is impinged by a concentrated mass m with an initial transverse velocity v_0 at the beam's tip A:

1) First, determine the modal shape. The deformation mechanism of the static collapse is adopted as the modal shape, i.e.

$$\phi_i^*(x) = 1 - x/L \tag{8.22}$$

The distribution of the initial velocity of this modal shape is plotted in Figure 8.2(b). It is

$$\dot{u}_i^{*0}(x,t) = \dot{w}_*^0(1 - x/L) \tag{8.23}$$

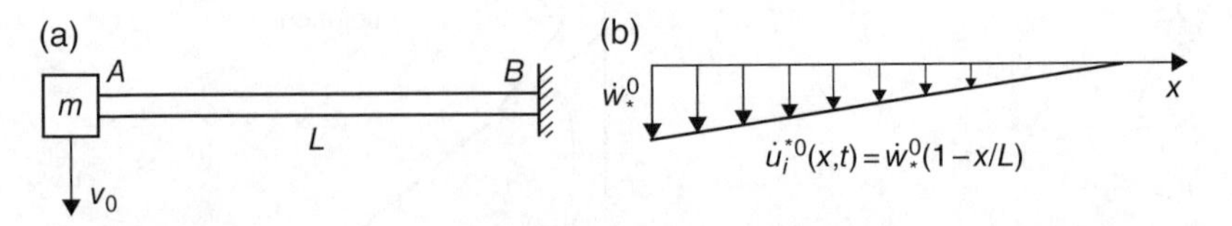

Figure 8.2 The modal solution of the Parkes problem. (a) A cantilever under impact at its tip; (b) distribution of initial velocity.

2) The beam's tip A is adopted as the characteristic point, with initial velocity $\dot{w}_*^0$ determined by the "min $\widetilde{\Delta}_0$" technique:

$$\dot{w}_*^0 = \frac{\int_V \rho\, \dot{u}_i^0\, \phi_i^*\, \mathrm{d}V}{\int_V \rho\, \phi_j^*\, \phi_j^*\, \mathrm{d}V} = \frac{m v_0 \cdot 1}{m + \int_0^L \rho(1 - x/L)^2\, \mathrm{d}x} \tag{8.24}$$

It should be noted that there are two integrations in the numerator and the denominator of Eq. (8.24). From the distribution of the real initial velocity $\dot{u}_i^0$, only the initial velocity of the beam tip A is non-zero, and thus the integration in the numerator results in the initial momentum. To calculate the integration in the denominator, considering both the beam and the concentrated mass will result in an effective mass.

Rearranging Eq. (8.24) gives the initial velocity of the beam tip:

$$\dot{w}_*^0 = \frac{m v_0}{m + \rho L/3} = \frac{3\gamma}{1 + 3\gamma} v_0 < v_0 \tag{8.25}$$

where $\gamma \equiv \dfrac{m}{\rho L}$ is the normalized mass ratio, as defined in Section 6.3. From Eq. (8.25), the initial velocity of the modal solution is obviously smaller than the actual initial velocity, as shown in Figure 8.1.

3) Get the modal response. In this problem, the general stress Q_j^* is taken to be the fully plastic bending moment, $Q_j^* \to M_P$; the general strain rate $\dot{q}_j^*$ is then the angular velocity at the beam's root B, $\dot{q}_j^* \to \dot{\theta}_B = \dot{w}_*(t)/L$. From the relation $\dot{q}_j = \dot{w}_*(t) k_j^*(x)$, we get $k_j^* = 1/L$. Now substitute the above relations into the deceleration in Eq. (8.12):

$$a_* = -\ddot{w}_* = \frac{\int_V Q_j^*\, k_j^*\, \mathrm{d}V}{\int_V \rho\, \phi_i^*\, \phi_i^*\, \mathrm{d}V} = \frac{M_P/L}{m + \rho L/3} \tag{8.26}$$

Then the total response time is found to be

$$t_f^* = \frac{\dot{w}_*^0}{a_*} = \frac{m v_0 L}{M_P} \tag{8.27}$$

which is identical to that given in Section 6.3 from the complete solution.

With uniform deceleration of the beam's tip, the final deflection of the beam's tip is given by

$$w_*^f = \frac{1}{2} \dot{w}_*^0 t_f^* = \frac{m^2 v_0^2 L}{2(m + \rho L/3) M_P} \tag{8.28}$$

Using the energy ratio defined before, $e_0 = K_0/M_P = \frac{1}{2} m v_0^2 / M_P$, the final deflection of the free end given by the modal solution is

$$\Delta_f^* = e_0 \frac{3\gamma L}{1 + 3\gamma} \tag{8.29}$$

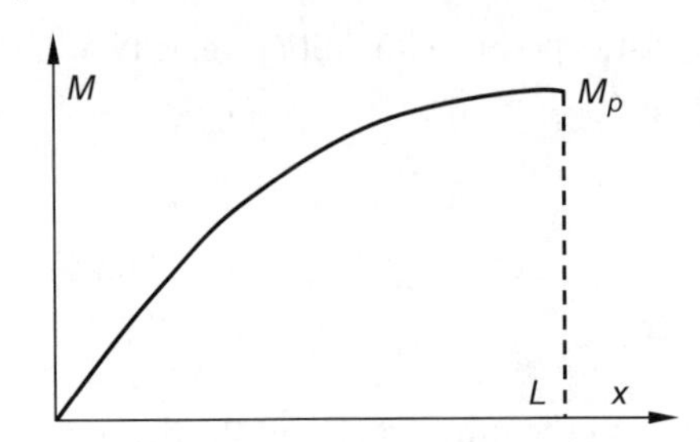

Figure 8.3 Distribution of bending moment by the modal solution.

In comparison, the final deflection of the free end given by the complete solution illustrated in Section 6.3 is

$$\Delta_f = e_0 \frac{2\gamma L}{3} \left\{ \frac{1}{1+2\gamma} + 2\ln\left(1 + \frac{1}{2\gamma}\right) \right\} \tag{8.30}$$

4) Check the bending moment in the beam:

$$M^*(x) = -m\ddot{w}_* x - \int_0^x \rho \ddot{w}_* \frac{L-\chi}{L}(x-\chi)\, \mathrm{d}\chi = a_* \left(mx + \frac{1}{2}\rho x^2 - \frac{1}{6}\rho \frac{x^3}{L} \right) \tag{8.31}$$

The distribution of bending moment given by Eq. (8.31) is plotted in Figure 8.3. Obviously, except at the root of the cantilever, $M^*(L) = M_P$, the bending moment at all the other beam sections is smaller than the fully plastic bending moment, which confirms that the dynamic deformation mode shape assumed in step 1 is appropriate.

After carrying out the four steps, an approximate solution of the Parkes problem is obtained by the modal technique. Figure 8.4 compares the modal solution with the complete solution given in section 6.3. In Figure 8.4(a), the history of the beam tip velocity is plotted for mass ratio $\gamma = 0.2$. Clearly, in the transient phase before the phase transition point, the velocity of the complete solution is higher than that of the modal solution. Due to the traveling plastic hinge, the beam tip velocity varies rapidly and non-linearly with time. Whilst in the modal phase after the phase transition point, the two curves of velocity history completely coincide with each other, and both decrease linearly with time. Also, the total response time of the two solutions is the same.

The final tip deflections $\Delta_f/(e_0 L)$ versus the impact mass ratio γ are plotted in Figure 8.4 (b). It can be seen that the difference in the final tip defections given by the complete solution and the modal solution is quite small for all the mass ratios. Thus, without considering the transient phase that has strong non-linearity, the modal solution can provide a fairly accurate estimation of both the total response time and the final deflection. In view of its simplicity and accuracy, the modal technique can be adopted as a useful approximate method in the analysis of structural dynamic response.

8.4.2 Modal Solution for a Partially Loaded Clamped Beam

The second example of the application of the modal technique is a partially loaded clamped beam. Consider Figure 8.5, which shows a fully clamped beam of length L, mass per unit length ρ, and fully plastic bending moment M_P subjected to an impulsive loading in the region with width b, i.e., at the initial instant the beam sections in this region have suddenly gained an initial velocity v_0.

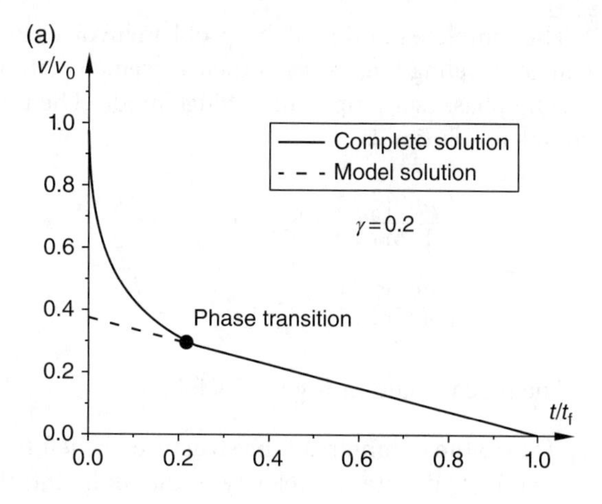

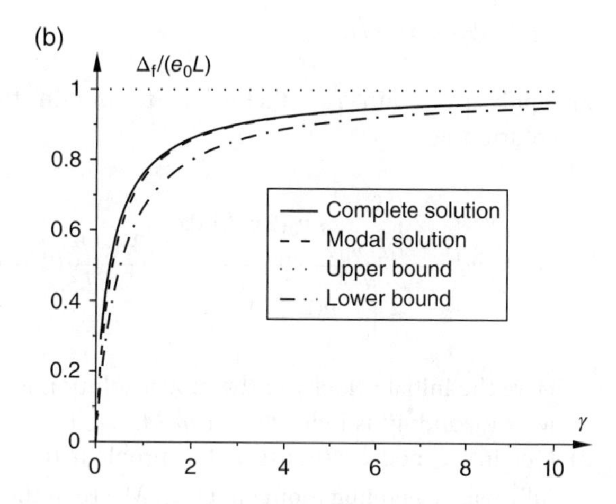

Figure 8.4 Comparison between the complete solution and the modal solution of the Parkes problem. (a) History of the beam tip velocity; (b) final tip deflections versus the mass ratio.

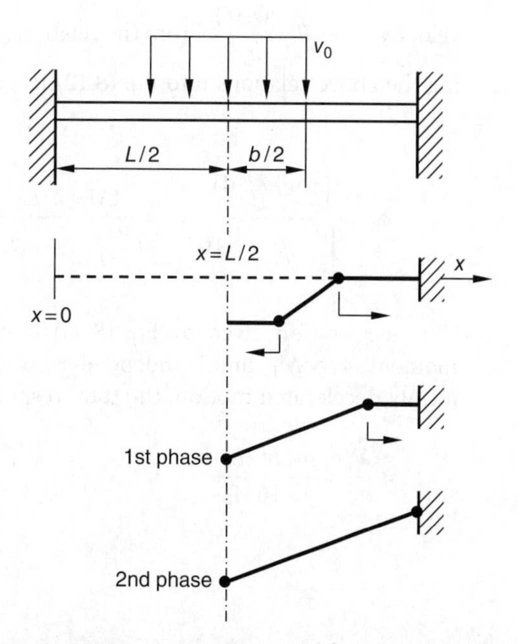

Figure 8.5 Modal solution of a fully clamped beam under local impulsive loading.

The complete solution of this problem involves two phases: the first phase ends when a pair of traveling hinges reach their extreme position (central or ends of the beam); the second phase is a simple three-hinge mode. The final midpoint deflection given by this complete solution is

$$\Delta_f = \begin{cases} \dfrac{\rho b^2 v_0^2}{12 M_P}\left(\dfrac{1}{4} + \ln\dfrac{L}{2b} + 1\right), & b < \dfrac{L}{2} \\ \dfrac{\rho b^2 v_0^2}{48 M_P}\left(\dfrac{6L}{b} - \dfrac{L^2}{b^2} - 3\right), & b \geq \dfrac{L}{2} \end{cases} \tag{8.32}$$

The modal solution is given as follows:

1) Due to the symmetry of the structure, the left half of the beam is used in the analysis. Let $\dot{w}_*(t)$ denote the velocity of the midpoint; then the mode shape is

$$\phi_i^*(x) = 2x/L \tag{8.33}$$

2) Apply the "min $\widetilde{\Delta}_0$" technique to obtain the initial velocity of the modal solution $\dot{w}_*^0$:

$$\dot{w}_*^0 = \frac{2\displaystyle\int_{(L-b)/2}^{L/2} \rho v_0 (2x/L)\, dx}{2\displaystyle\int_0^{L/2} \rho (2x/L)^2\, dx} = \frac{3v_0}{2}\frac{b}{L^2}(2L-b) \tag{8.34}$$

Here the initial velocity of the modal solution is proportional to the impact velocity, $\dot{w}_*^0 \propto v_0$, and $\dot{w}_*^0$ is independent of M_P.

3) Obtain the modal response. In this problem, the general stress Q_j^* is again taken as the fully plastic bending moment, $Q_j^* \to M_P$; then the general strain rate $\dot{q}_j^*$ is the angular velocity, $\dot{q}_j^* \to \dot{\theta} = \dfrac{\dot{w}_*(t)}{L/2}$. From the relation $\dot{q}_j = \dot{w}_*(t) k_j^*(x)$, we get $k_j^* = 2/L$. Substituting the above relations into Eq. (8.12) leads to the deceleration:

$$a_* = \frac{\displaystyle\int_V Q_j^* k_j^*\, dV}{\displaystyle\int_V \rho \phi_i^* \phi_i^*\, dV} = \frac{4M_P(2/L)}{2\displaystyle\int_0^{L/2} \rho (2x/L)^2\, dx} = \frac{24 M_P}{\rho L^2} \tag{8.35}$$

The deceleration given by Eq. (8.35) is proportional to the fully plastic bending moment, $a_* \propto M_P$, and is independent of the initial impact velocity v_0. By the uniformly decelerated motion, the total response time is

$$t_f^* = \frac{\dot{w}_*^0}{a_*} = \frac{\rho v_0 b(2L-b)}{16 M_P} \tag{8.36}$$

Equation (8.36) shows that the total response time is proportional to the initial velocity and is inversely proportional to the fully plastic bending moment, i.e., $t_f^* \propto v_0/M_P$.

The final central deflection of the beam is now calculated as

$$w_*^f = \frac{1}{2}\dot{w}_*^0 t_f^* = \frac{3}{64}\frac{\rho v_0^2}{M_P}\frac{b^2(2L-b)^2}{L^2} \tag{8.37}$$

which can be expressed by the initial energy $K_0 = \rho b v_0^2/2$:

$$w_*^f = \frac{3}{32}\frac{K_0}{M_P}\frac{b(2L-b)^2}{L^2} \tag{8.38}$$

Using the energy ratio, $e_0 = K_0/M_P$, Eq. (8.38) is recast into

$$\Delta_f^* = w_*^f = e_0 \frac{3b}{32}(2-b/L)^2 \tag{8.39}$$

This clearly indicates that the final central deflection of the beam depends on the initial impact velocity and the size of the loading region.

A comparison between the complete solution and the modal solution for a fully clamped beam under local impulsive loading is given in Figure 8.6. From the curves of the beam's central velocity history plotted in Figure 8.6(a), it can be seen that the initial central velocity of the beam in the modal solution is higher than that of the actual initial velocity applied. This does not mean a higher initial kinetic energy in the modal solution, because according to the deformation mode adopted in the modal technique, the distribution of the initial velocity is a triangular one along the beam's length with the maximum velocity at the beam's center, whereas the actual initial velocity is uniformly distributed within a smaller region. From the curves in Figure 8.6(b), it can be seen the central deflection of the modal solution is quite different from that of the complete solution, especially if b/L is small. This is attributed to the large deviation of the modal shape from that of the actual loading condition.

8.4.3 Remarks on the Modal Technique

Following on from the applications of the modal technique, the following points can be noted:

1) The mode solution is much simpler than the complete solution, as the non-linear response during the transient phase is completely ignored.
2) The procedure of finding a modal solution involves three steps: assuming a modal shape; obtaining the initial velocity of the characteristic point; and determining the modal response variables. In addition, sometimes it is safe to check that the relevant moment distribution does not violate the yield condition.
3) The "error" induced by the modal solution depends on the actual initial momentum distribution.

(a)

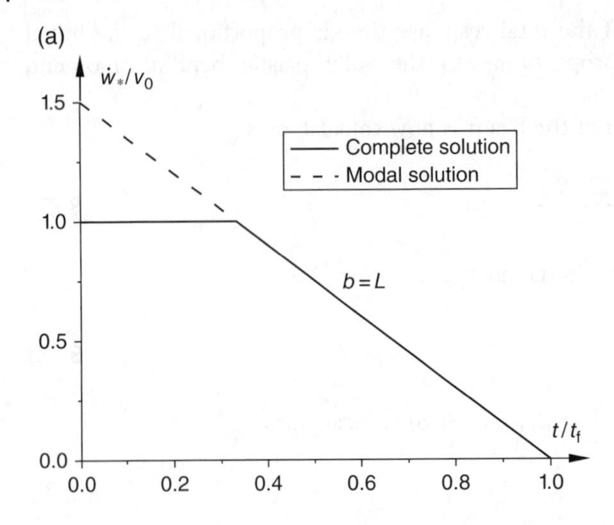

(b)

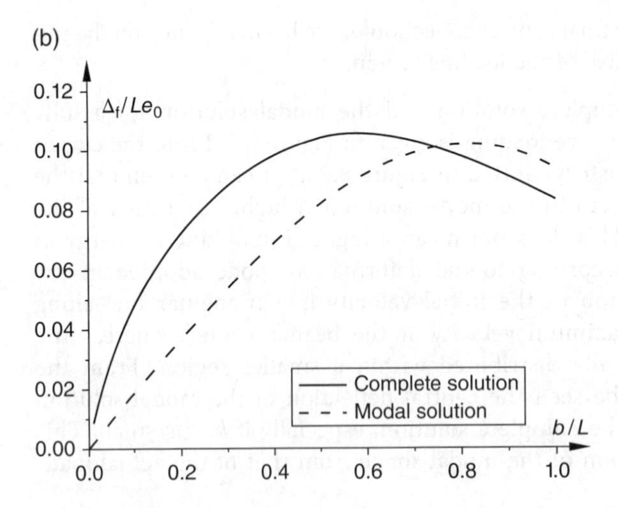

Figure 8.6 Comparison between the complete solution and the modal solution of a fully clamped beam under local impulsive loading. (a) History of beam's central velocity; (b) final central deflections versus the size of the loading region.

8.5 Bound Theorems for RPP Structures

For structures made of RPP material, the bound theorems can be adopted in estimating the final displacement and the respond time (see Stronge and Yu, 1993).

8.5.1 Upper Bound of Final Displacement

If a structure is subjected to impulsive loading only, the final displacement of a characteristic point A in the structure Δ_f should satisfy

$$\Delta_f \leq \frac{K_0}{P_c} \tag{8.40}$$

where K_0 is the initial kinetic energy of the structure due to the impulsive loading, and P_c is the static collapse load when force P applies at point A. The proof of this theorem can be seen in Stronge and Yu (1993).

Here, the Parkes problem is taken as an example again for the application of this upper bound theorem. The beam's tip is taken as the characteristic point A, and the initial kinetic energy carried by the impinging concentrated mass is

$$K_0 = \frac{1}{2} m v_0^2 \tag{8.41}$$

In the static collapse case, if a concentrated force P is applied to the free tip of the cantilever, the static collapse load is

$$P_c = M_P / L \tag{8.42}$$

From the upper bound on the final displacement, i.e., Eq. (8.40), the final deflection of point A should satisfy

$$\Delta_f \leq \frac{K_0}{P_c} = \frac{m v_0^2 L}{2M_P} \tag{8.43}$$

Using the energy ratio e_0, the above relation is normalized as

$$\frac{\Delta_f}{L} \leq \frac{m v_0^2/2}{M_P} \equiv e_0 \tag{8.44}$$

In addition to the complete solution and the modal solution, the maximum displacement of the free tip given by the upper bound theorem is also plotted in Figure 8.4(b). Obviously, the upper bound is higher than both solutions, so it does provide an upper bound of the maximum displacement. Also, in the case of large mass ratio γ, the upper bound of the maximum displacement is quite close to the complete solution.

8.5.2 Lower Bound of Final Displacement

In the structure made of RPP material, the final displacement of a characteristic point A should satisfy

$$\Delta_f \geq \frac{\left(\int_V \rho \dot{u}_0 \dot{u}_c \, dV \right)^2}{2D(\dot{u}_c) \int_V \rho \dot{u}_c \, dV} \tag{8.45}$$

where $\dot{u}_0(x)$ is the initial velocity field, $\dot{u}_c(x)$ is a time-independent kinematically admissible velocity field, and $D(\dot{u}_c)$ is the energy dissipation associated with the field of $\dot{u}_c(x)$.

The *kinematically admissible velocity field* is an important concept in the limit analysis of elastic-plastic or rigid-plastic solids/structures. Here we only briefly explain the conditions it needs to meet:

1) The distribution of velocity/displacement field is continuous in the solid/structure, except that the tangential components may have discontinuities along finite-numbered discrete plastic hinge lines in two-dimensional cases.

2) The velocity/displacement boundary condition of the velocity/displacement field should be satisfied.
3) The associated energy dissipation is non-negative, i.e., the applied external loads should do positive work on this velocity/displacement field.

The derivation of the lower bound theorem is quite complex, so it is not elaborated here. It should be noted that after the concept of bounds of the structural dynamic response was proposed, many researchers constructed different lower bound theorems regarding the final displacement. The lower bound theorem given above is one of those that is generally acknowledged to be effective.

Again the Parkes problem is taken as an example for the application of this lower bound theorem. The initial kinetic velocity field of the Parkes problem is

$$\dot{u}_0(x) = \begin{cases} v_0, & x = 0 \\ 0, & x \neq 0 \end{cases} \tag{8.46}$$

The deformation mechanism of the static collapse is then taken as the time-independent kinematically admissible velocity field:

$$\dot{u}_c(x) = \dot{w}_0(1 - x/L) \tag{8.47}$$

The energy dissipation associated with the field of $\dot{u}_c(x)$ is

$$D(\dot{u}_c) = M_P\,(\dot{w}_0/L) \tag{8.48}$$

The integration in the numerator of Eq. (8.45) is calculated and found to be

$$\int_V \rho \dot{u}_0 \dot{u}_c \, \mathrm{d}V = m v_0 \dot{w}_0 \tag{8.49}$$

And the integration in the denominator of Eq. (8.45) is calculated and found to be

$$\int_V \rho \dot{u}_c \, \mathrm{d}V = m\dot{w}_0 + \int_0^L \rho \dot{w}_0 (1 - x/L)\, \mathrm{d}x = m\dot{w}_0 + \frac{1}{2}\rho L \dot{w}_0 \tag{8.50}$$

Substituting Eqs (8.48)–(8.50) into the lower bound theorem, i.e., Eq. (8.45), leads to

$$\Delta_f \geq \frac{m^2 v_0^2 L}{M_P(2m + \rho L)} \tag{8.51}$$

The above relation could be expressed in terms of the mass ratio $\gamma \equiv m/\rho L$:

$$\frac{\Delta_f}{L} \geq e_0 \frac{2\gamma}{2\gamma + 1} \tag{8.52}$$

Equation (8.52) is the lower bound of the final deflection at the tip of beam by adopting the kinematically admissible velocity field shown in Eq. (8.47). This lower bound of deflection is also plotted in Figure 8.4(b). Clearly, this solution is lower than the complete solution as well as the modal solution.

For the Parkes problem, based on the assumption of the RPP simplification, we have obtained the complete solution, the modal solution, and the estimates of the final deflection at the tip according to the upper and lower bound theorems. The analytical methods in this order become increasingly simple. Which solution should be adopted depends on the required accuracy. If the complete response of the entire dynamic process is required, then the complete solution should be employed; if only the response of the mode phase is of interest, especially in the case of an impact with a large mass ratio, then the accuracy of the modal solution is adequate; if only the estimate of the final displacement at a typical point is needed in engineering applications, then the bound theorems will be the simplest and most efficient method.

The dynamic plastic response of rigid-plastic beams is divided into two phases: the early phase (transient response) and the later phase (modal phase). The total response time of the entire process is much longer than the wave propagation process. It usually takes seconds or even minutes to reach the end of the total dynamic response for large or complex structures.

From the methodological point of view, it is reasonable to separate the timescale and the space scale for complex problems. Thus, we need to establish different models with different accuracy, and adopt different, relevant methods in analysis. Therefore, although the two approximation methods, i.e., the modal solution and the bound theorem estimations, have less accuracy than the complete solution, they are simpler and are useful if only the final displacement is of interest.

8.6 Applicability of an RPP Model

Employing the RPP material model that was introduced in Section 5.3, some theoretical and semi-theoretical analytical results have been discussed regarding the dynamic response of several kinds of structure under various loading conditions in Chapters 5–8.

As introduced in Section 5.3, for a beam made of RPP material, the stress–strain relationship and the bending moment–curvature relationship are plotted in Figure 5.11. When a beam (or another one-dimensional structural component, such as a ring or an arch) is idealized as an RPP structure, its plastic deformation will be concentrated at one or a few cross-sections, where the magnitude of the applied bending moment reaches the fully plastic bending moment of the cross-section, M_P (e.g., $M_P = Ybh^2/4$ in case of rectangular cross-sectional beams). These cross-sections are termed plastic hinges. Any bending moment, whose magnitude is larger than M_P, is not *statically admissible* for an equilibrium configuration. On the other hand, any bending moment with a magnitude smaller than M_P will produce no plastic deformation. Consequently, the deformed configuration of an RPP beam (or a ring or arch) will only contain one or more plastic hinges. Therefore, a finite relative rotation occurs at a plastic hinge as the result of the application of M_P. Away from those plastic hinges all the segments in the beam (or the ring or arch) will remain rigid and their curvatures will remain unchanged (i.e., $\kappa = 0$). However, those rigid segments are allowed to have translation and/or rotation, provided these motions are kinematically admissible in a proposed deformation mechanism. Similarly, for plates or shells, the RPP idealization of materials will mean their plastic deformation is concentrated at discrete plastic hinge lines (this is discussed further in Chapter 9). Along those hinge lines the magnitude of the bending moment per unit length must be equal to M_0, which denotes the fully plastic bending moment per unit length, $M_0 = Yh^2/4$.

Thus, it can be seen that by adopting the RPP idealization for the material, the plastic deformation in a structural component of N dimensions ($N = 1$ for beams, rings, arches etc.; and $N = 2$ for plates, shells etc.) will be concentrated in discrete regions of $(N - 1)$ dimensions. This will greatly simplify the plastic analysis of structures and is very useful for theoretical modeling of the structural components under consideration.

It should be noted that physically the rigid-plastic idealization of materials' behavior is based on the fact that the elastic strain (typically limited by $\varepsilon_Y \approx 0.002$ for structural metals) is much smaller than the plastic strain in structural components under intensive loading and/or used for energy absorption. However, the elastic deformation is always a precursor to subsequent plastic deformation. No matter whether the structure is subjected to a quasi-static load or a dynamic load, the first phase of its deformation is always elastic, and after the plastic deformation is completed, the structure will undergo an elastic springback to reach its final deformed configuration. For these reasons, the validity of rigid-plastic idealization should be thoroughly examined (Lu & Yu, 2003).

The behavior of idealized materials can be demonstrated by mechanical models such as those shown in Figure 8.7. The model shown in Figure 8.7(a) contains a linear elastic spring and a friction pair, so that when a force F is applied along the axis of the spring, the relationship between the force F and axial displacement Δ will be

$$F = \begin{cases} k\Delta, & \Delta \le \Delta_{\mathrm{y}} = F_{\mathrm{y}}/k \\ F_{\mathrm{y}}, & \Delta_{\mathrm{y}} \le \Delta < \Delta_{\mathrm{f}} \end{cases} \tag{8.53}$$

where k is the spring constant, and F_{y} is the critical friction when relative motion begins. If the elastic spring is taken away from the model shown in Figure 8.7(a), which means the elastic deformation of the material is considered negligible, then the mechanical model shown in Figure 8.7(b) demonstrates RPP behavior.

First, assume that a force F is quasi-statically applied to the left end of the elastic-plastic model shown in Figure 8.7(a). Consider a deformation process of the model, in which the displacement at the left end gradually increases from zero to a total displacement $\Delta_{\mathrm{tl}}^{\mathrm{ep}} > \Delta_{\mathrm{y}} = F_{\mathrm{y}}/k$, where the subscript "tl" refers to the total displacement produced, and the superscript "ep" refers to the elastic-plastic model. Correspondingly force F should first increases from zero to F_{y} (as $0 < \Delta \le \Delta_{\mathrm{y}}$) and subsequently remains at F_{y} (as $\Delta > \Delta_{\mathrm{y}}$). In this elastic-plastic deformation process, the total input energy, i.e., the work done by force F, is

$$E_{\mathrm{in}} = \frac{1}{2}F_{\mathrm{y}}\Delta_{\mathrm{y}} + F_{\mathrm{y}}\left(\Delta_{\mathrm{tl}}^{\mathrm{ep}} - \Delta_{\mathrm{y}}\right) = E_{\mathrm{max}}^{\mathrm{e}} + F_{\mathrm{y}}\Delta_{\mathrm{p}}^{\mathrm{ep}} \tag{8.54}$$

where $E_{\mathrm{max}}^{\mathrm{e}} = F_{\mathrm{y}}\Delta_{\mathrm{y}}/2$ is the maximum elastic energy that can be stored in the model, and $\Delta_{\mathrm{p}}^{\mathrm{ep}} = \Delta_{\mathrm{tl}}^{\mathrm{ep}} - \Delta_{\mathrm{y}}$ represents the plastic (permanent) component of the displacement. The

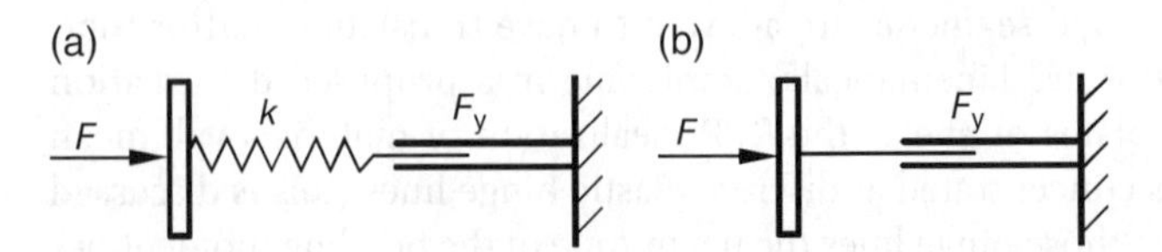

Figure 8.7 Mechanical models of material idealizations. (a) An elastic-perfectly plastic model; (b) a rigid-perfectly plastic model.

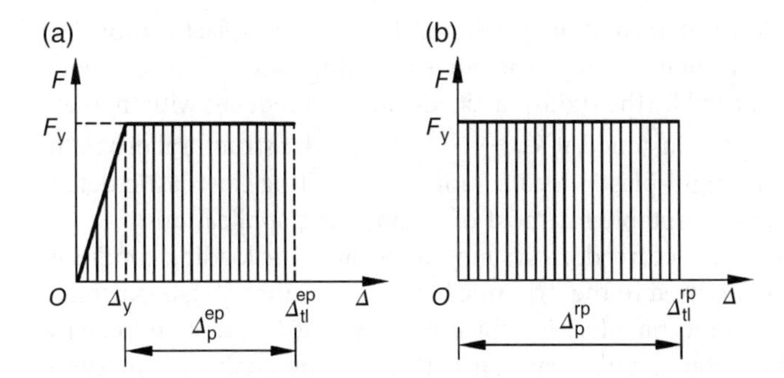

Figure 8.8 Force–displacement relationship for perfectly plastic models. (a) Elastic-plastic model; (b) rigid-plastic model.

total input energy E_{in} given in Eq. (8.54) can be represented by the hatched area in Figure 8.8(a).

Next, consider that a force F is quasi-statically applied to the left end of the rigid-plastic model shown in Figure 8.7(b). In this case, the deformation becomes possible only when $F = F_y$. Assume the final displacement at the left end of the model is Δ_{tl}^{rp}, where the superscript "rp" refers to the rigid-plastic model; then the total input energy (i.e., the work done) in the process is

$$E_{in} = F_y \Delta_{tl}^{rp} = F_y \Delta_p^{rp} \tag{8.55}$$

Here the plastic component of the displacement is equal to the total displacement, $\Delta_p^{rp} = \Delta_{tl}^{rp}$, because the elastic component is ignored. The total input energy E_{in} given in Eq. (8.55) can be represented by the hatched area in Figure 8.8(b).

Suppose the same amount of energy is input into these two models, i.e., the values of E_{in} in Eqs (8.54) and (8.55) are identical, and we define an energy ratio

$$R_{er} \equiv \frac{E_{in}}{E_{max}^{e}} \tag{8.56}$$

then Eqs (8.54) and (8.55) result in

$$E_{in} = F_y \Delta_p^{rp} = F_y \Delta_p^{ep} + E_{max}^{e} \tag{8.57}$$

Dividing Eq. (8.57) by E_{in} leads to

$$1 = \frac{\Delta_p^{ep}}{\Delta_p^{rp}} + \frac{E_{max}^{e}}{E_{in}} = \frac{\Delta_p^{ep}}{\Delta_p^{rp}} + \frac{1}{R_{er}} \tag{8.58}$$

This can be rewritten as

$$\frac{\Delta_p^{rp}}{\Delta_p^{ep}} = \frac{R_{er}}{R_{er} - 1} \tag{8.59}$$

Therefore, the relative "error" of employing a rigid-plastic model in predicting the plastic displacement is found to be

$$\text{"Error"} \equiv \frac{\Delta_p^{rp} - \Delta_p^{ep}}{\Delta_p^{ep}} = \frac{1}{R_{er} - 1} > 0 \tag{8.60}$$

which indicates that the plastic deformation predicted by the rigid-plastic model is always slightly larger than that obtained from the corresponding elastic-plastic model, $\Delta_p^{rp} > \Delta_p^{ep}$, whilst the "error" caused by the rigid-plastic idealization reduces with increasing energy ratio $R_{er} = E_{in}/E_{max}^e$. For instance, if $R_{er} = 10$ in a particular structural problem, then the "error" caused by the rigid-plastic idealization in predicting the plastic deformation is about 11%, which is acceptable for most of engineering applications.

The response of the single degree of freedom system under the static loading condition discussed earlier was then generalized to the dynamic loading condition. If a beam/plate with a mass and corresponding inertia effect is subjected to a load F that is no longer a monotonically increasing force but a pulse varying with time, the analysis (Yu, 1993) indicated that in order to the make the relative "error" between the plastic displacement predicted by a rigid-plastic model and the elastic plastic model small, in addition to the large energy ratio condition (i.e., large R_{er}), it is also required that the period of the applied pulse should be far smaller than the fundamental natural vibration period of the elastic-plastic structure. For example, if the fundamental elastic vibration period of the elastic-plastic structure shown in Figure 8.7(a) is $2\pi\sqrt{m/k}$, the period of the applied pulse should be much smaller than $2\pi\sqrt{m/k}$.

Problems 8

8.1 Consider a fully clamped RPP beam AB of length $2L$, mass per unit length ρ, and fully plastic bending moment M_P, as shown in the figure. A particle of mass m with initial velocity v_0 strikes the beam at point C, located at a distance $L/2$ from one of its ends. Use the mode approximation to calculate:

i) the initial velocity at point C for the modal solution;
ii) the final time at which the modal response ceases;
iii) the final deflection at point C.

Also, discuss in general terms how the actual dynamic response of the beam under impact would be different from that assumed in the modal solution.

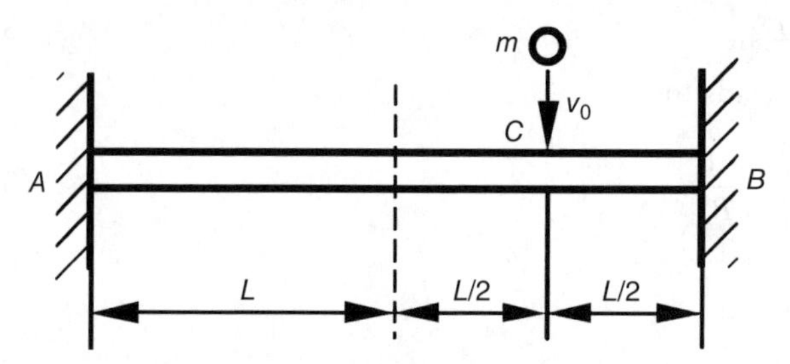

8.2 For Problem 8.1, use the bound theorems to estimate the upper and lower bounds of the final displacement of point C.

8.3 Assume that the two ends in Problem 8.1 are simply supported, the elastic bending stiffness is EI, and the maximum elastic bending moment is M_e for the beam section. Also suppose the bending moment distribution is triangular with the maximum bending moment M_e at the section of point C. Calculate the energy ratio defined in Eq. (8.56), and discuss the error resulting from the RPP model.

9

Response of Rigid-Plastic Plates

To study the dynamic behavior of plates, we first have to learn how to assess their static load-carrying capacity.

For both quasi-static and dynamic loadings on thin plates under small deformation, the following assumptions are adopted:

1) the material is rigid-perfectly plastic and rate-independent;
2) the loading is normal (transverse) to the plates' middle surface;
3) the deflection is small in comparison with the thickness of the plate.

For a rigid-perfectly plastic plate, the *fully plastic bending moment per unit width* is $M_0 = \sigma_0 h^2/4$, where h is the thickness, $\sigma_0 = Y$ for Tresca material or $\sigma_0 = 2Y/\sqrt{3}$ for Mises' material, where Y is the yield stress of material under uniaxial tension/compression.

As a result of assumptions 2 and 3, only the bending response of plates is considered; the membrane stresses and shear stresses are ignored. All the geometrical relations and equations of equilibrium/motion are referred to the initial configuration of the plate. Hence, the following analyses are limited to the *incipient deformation.*

9.1 Static Load-Carrying Capacity of Rigid-Plastic Plates

In this section, we will apply an upper bound method to estimate the load-carrying capacity of plates (see, e.g., Johnson and Mellor, 1973 or Johnson, 1972).

As we have seen, the static plastic collapse mechanisms of beams usually contain rigid segments. Knowing this, we may construct *plastic collapse mechanisms* for plates, which contain *rigid segments* connected by *plastic hinge lines,* along which the bending moment is equal to the fully plastic bending moment M_0, as shown in Figure 9.1.

According to the theorem of limit analysis (discussed in Chapter 8), if a displacement/velocity field related to this kind of mechanism is kinematically admissible, i.e., it satisfies the boundary conditions on displacement/velocity and the continuity conditions at the hinge lines, then *the energy balance between the external work (or its rate) and the plastic dissipation (or its rate) provides an upper bound* estimate to the plastic collapse load of the plate.

Introduction to Impact Dynamics, First Edition. T.X. Yu and XinMing Qiu.

Published 2018 by John Wiley & Sons Singapore Pte. Ltd.

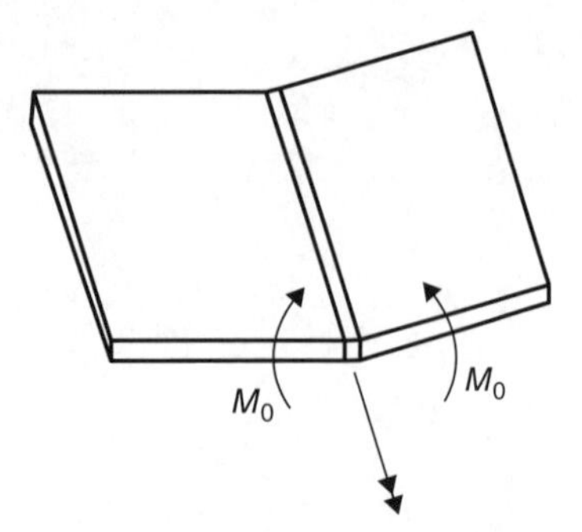

Figure 9.1 Rigid segments connected by a plastic hinge line.

In general, the collapse mechanisms are not unique for a particular configuration, so the "best" mechanism should be that one that is associated with the lowest upper bound and which gives the best estimate for the actual load-carrying capacity of the plate.

9.1.1 Load Capacity of Square Plates

Suppose a square plate has sides of length $2a$ and is of thickness h, with different boundary and loading conditions. Its load-carrying capacity will be analyzed in this subsection.

Case A: Simply Supported, Loaded by a Concentrated Force *F* at its Center

As shown in Figure 9.2, a simply supported square plate is loaded by a concentrated force F at its center. Due to the small deformation assumption, the deflection at the plate center is small compared with the plate thickness. As illustrated in Figure 9.3(a), suppose that plastic hinge lines form along two diagonals. These lines divide the plate into four segments (quadrants) A, B, C, and D; each segment rotates about an edge with angular velocity ω. By using a conventional right-hand screw rule, the angular velocities are represented by vectors within the plane of plate. In Figure 9.3(b), these vectors are redrawn together with a common pole O. This diagram is called a *hodograph*. For instance, $\overrightarrow{Oa}$ represents the angular velocity of segment A, etc. Note that $\overrightarrow{ab}$ represents the relative angular velocity between segment A and segment B.

That rate of work $\dot{W}$ done by force F on this velocity field is equal to the force times the velocity of the plate center, v_0, i.e.

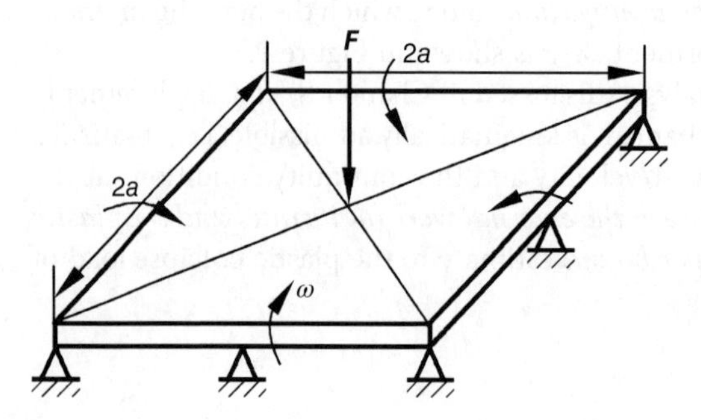

Figure 9.2 Simply supported square plate loaded by a concentrated force F at its center.

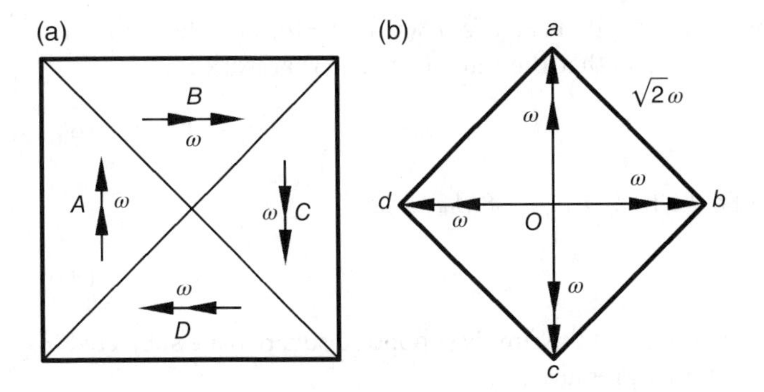

Figure 9.3 The deformation mechanism of a square plate. (a) Angular velocities of rigid segments; (b) hodograph of angular velocity.

$$\dot{W} = Fv_0 = F\omega a \tag{9.1}$$

The rate of plastic dissipation alone each line is equal to the fully plastic bending moment, M_0, times the length of the plastic hinge line, and the relative angular velocity between the segments connected by the plastic hinge line. There are four plastic hinge lines of length $\sqrt{2}a$ in the deformation mechanism shown in Figure 9.3 (a), so the total rate of plastic dissipation should take account of all four. By using the hodograph shown in Figure 9.3(b), the relative angular velocities are all equal to $\sqrt{2}\omega$. Hence

$$\dot{D}_{\text{p}} = M_0\sqrt{2}\omega\ 4\sqrt{2}a = 8M_0\omega a \tag{9.2}$$

From the energy balance, the rate of external work should be equal to the rate of total plastic dissipation

$$\dot{W} = \dot{D}_{\text{p}} \tag{9.3}$$

Thus, $F\omega a = 8M_0\omega a$, i.e.

$$F_{\text{s}} = 8M_0 \tag{9.4}$$

where the subscript "s" denotes the static collapse load. Eq. (9.4) provides an upper bound of the simply supported square plate subjected to a quasi-static, concentrated force F at its center. It should be noted that the critical force is independent of the side length of the plate, a. As M_0 is the fully plastic bending moment per unit width, it has the same magnitude with the force.

Case B: Simply Supported, Loaded by a Uniformly Distributed Pressure *p*

A simply supported square plate, as shown in Figure 9.2, is loaded by a uniformly distributed pressure p. Suppose the deformation mechanism is the same as that in Case A; then the hodograph remains the same (see Figure 9.3b). Thus, the rate of plastic dissipation is the same as that given in Eq. (9.2). The only difference is that the rate of the external work. The area of each triangle in Figure 9.3(a) is $A = a^2$, and the transverse velocity of its centroid is $v_c = a\omega/3$, so that the rate of the external work is

$$\dot{W} = 4pAv_c = \frac{4}{3}pa^3\omega \tag{9.5}$$

From the energy balance Eq. (9.3), $4pa^3\omega/3 = 8M_0\omega a$, i.e.

$$p_s = \frac{6M_0}{a^2} \tag{9.6}$$

Equation (9.6) gives an upper bound of the simply supported square plate subjected to a quasi-static, uniformly distributed pressure p.

Case C: Fully Clamped, Loaded by a Uniformly Distributed Pressure *p*

A fully clamped square plate, as shown in Figure 9.2, is loaded by a uniformly distributed pressure p. Suppose the deformation mechanism is the same as that in Case A and Case B, so the hodograph remains the same (see Figure 9.3b). At the same time, the rate of external work is the same as that in Case B, i.e., it is given by Eq. (9.5). However, in addition to the four hinge lines along the diagonals, *four edges* have to become hinge lines in order to make the plate a mechanism. Hence, the plastic dissipation rate now is

$$\dot{D}_p = 8M_0\omega a + M_0\omega\ 8a = 16M_0\omega a \tag{9.7}$$

Clearly the plastic dissipation rate is doubled under the fully clamped boundary condition. Consequently, the static collapse pressure in this case is

$$p_s = \frac{12M_0}{a^2} \tag{9.8}$$

Similarly, it is easy to confirm that for the fully clamped square plate subjected to a concentrated force at its center, the upper limit of the collapse force is doubled in comparison to the simply supported case, i.e., Eq. (9.4).

9.1.2 Load Capacity of Rectangular Plates

Now suppose a simply supported rectangular plate with sides of length $2a$ and $2b$, where $a \geq b$, is loaded by uniformly distributed pressure p. The deformation mechanism is assumed to be the one depicted in Figure 9.4(a). The rectangular plate is divided into four rigid segments by five plastic hinge lines, with c being a distance to be determined.

Along the middle line AC, all the points have the same velocity, i.e.

$$\omega_1 c = \omega_2 b = v \quad (b \leq c) \tag{9.9}$$

To calculate the rate of external work, three kinds of segments should be included in the deformation mechanism shown in Figure 9.4(a). First, the rate of external work on two triangles on the left and right sides, as represented by segments $\boxed{1}$, is

$$\dot{W}_1 = p\left(\frac{1}{2}2bc\right)\frac{1}{3}v \tag{9.10}$$

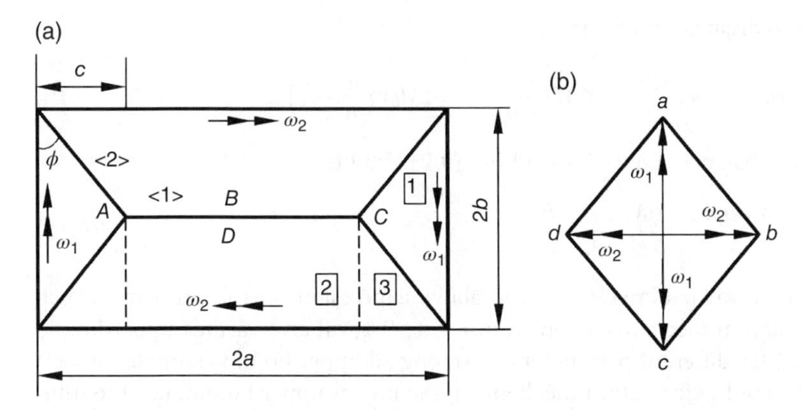

Figure 9.4 A rectangular plate under transverse load. (a) Deformation mechanism; (b) hodograph of angular velocity.

Then the rate of external work on the rectangular part of the trapezoid, as represented by segments 2, is

$$\dot{W}_2 = pb(2a - 2c)\frac{1}{2}v \tag{9.11}$$

Finally, the work rate of the triangular part of the trapezoid, as represented by segments 3, is

$$\dot{W}_3 = p\frac{1}{2}bc\frac{1}{3}v \tag{9.12}$$

Hence, the total rate of external work is obtained by the summary of Eqs (9.10)–(9.12), i.e.

$$\dot{W} = 2\dot{W}_1 + 2\dot{W}_2 + 4\dot{W}_3 = p\frac{2b}{3}(3a - c)v \tag{9.13}$$

Then to calculate the plastic dissipation rate, we note that among the five plastic hinge lines as shown in Figure 9.4(a), the horizontal line is marked by a <1>, while the other four inclined lines are marked by a <2>. The length of the plastic hinge line <1> is $(2a - 2c)$, and the relative angular velocity between the two rigid segments adjacent to this plastic hinge line is $2\omega_2$. Therefore, the plastic dissipation on this line is

$$\dot{D}_1 = M_0\ (2a - 2c)\ 2\omega_2 \tag{9.14}$$

The length of the plastic hinge line <2> is $\sqrt{b^2 + c^2}$, and according to the hodograph of angular velocity given in Figure 9.4(b), the relative angular velocity between two rigid segments adjacent to the plastic hinge line is $\sqrt{\omega_1^2 + \omega_2^2}$. Thus the plastic dissipation on one of the plastic hinge lines <2> is

$$\dot{D}_2 = M_0\sqrt{b^2 + c^2}\sqrt{\omega_1^2 + \omega_2^2} \tag{9.15}$$

The total plastic dissipation rate is

$$\dot{D}_p = \dot{D}_1 + 4\dot{D}_2 = 4M_0\frac{a-c}{b}v + 4M_0\frac{b^2+c^2}{bc}v = 4M_0 v\left(\frac{a}{b}+\frac{b}{c}\right) \quad (9.16)$$

From the energy balance, Eq. (9.16) and Eq. (9.13) lead to

$$p = \frac{6M_0(a/b+b/c)}{b(3a-c)} = \frac{6M_0(ac+b^2)}{b^2c(3a-c)} \quad (9.17)$$

There is an unknown parameter c in the above expression on the pressure, which means the deformation mechanism shown in Figure 9.4 will give various upper bound values of pressure for different parameters c. Among all upper bound estimates associated with the assumed deformation mechanism, the lowest upper bound, i.e., the minimum value of Eq. (9.17), should be closest to the actual collapse pressure. Hereby, the unknown distance c should be determined by $\partial p/\partial c = 0$, which leads to

$$ac^2 + 2b^2c - 3ab^2 = 0 \quad (9.18)$$

Solving Eq. (9.18) with respect to parameter c, we obtain

$$c/b = \sqrt{3+(b/a)^2} - b/a, \quad (b \le a) \quad (9.19)$$

Equation (9.19) is also expressed in the form of angle ϕ, as indicated in Figure 9.4(a):

$$\phi = \tan^{-1}(c/b) = \tan^{-1}\left[\sqrt{3+(b/a)^2} - b/a\right] \quad (9.20)$$

Equation (9.18) also gives a relation

$$\frac{ac+b^2}{3a-c} = \frac{b^2}{c} \quad (9.21)$$

By substituting Eq. (9.21) into Eq. (9.17), the critical pressure in the static case, p_s, is found to be

$$p_s = \frac{6M_0}{c^2} = \frac{6M_0}{(b\tan\phi)^2} \quad (9.22)$$

For example:

1) If $a = b$, then $c = b$, $\phi = 45°$, and $p_s = 6M_0/a^2 = 6M_0/b^2$. That is, the result for a rectangular plate given by Eq. (9.22) reduces to that of a square plate.
2) If $a = 3b$, then $c = 1.43b$, $\phi = 55°$, and $p_s = 2.93M_0/b^2$.
3) If $a \gg b$, then $c \to \sqrt{3}b$, $\phi \to 60°$, and $p_s \to 2M_0/b^2$. In the case of a large difference between the length and width of the rectangular plate, the pressure-carrying capacity p_s is significantly reduced to only one-third of that of the square plate.

9.1.3 Load-Carrying Capacity of Regular Polygonal Plates

As shown in Figure 9.5, in a regular polygonal plate with n sides, the distance from its center to one of its sides is a and its fully plastic bending moment per unit width is M_0.

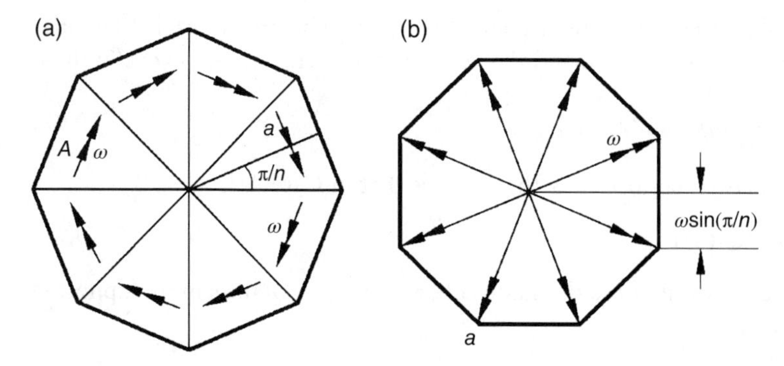

Figure 9.5 The deformation mechanism of a regular polygonal plate with n sides. (a) Angular velocities of rigid segments; (b) hodograph of angular velocity.

The load-carrying capacity of this plate with different boundary and loading conditions is analyzed in this subsection.

Case A: Simply Supported, Loaded by a Concentrated Force F at its Center

The deformation mechanism of a simply supported, regular polygonal plate with n sides subjected to a concentrated force at its center is shown in Figure 9.5(a). It is composed of n isosceles triangles as rigid segments, connected by radial plastic hinge lines. Each rigid isosceles triangle rotates about its bottom edge. Based on the deformation compatibility, the magnitude of the angular velocity of each segment must be the same, i.e., ω, while the direction of angular velocity is parallel to the bottom edge of each isosceles triangle.

By noting that the velocity of the plate center is $v_0 = \omega a$, the rate of the external force is

$$\dot{W} = F\omega a \tag{9.23}$$

Each of n plastic hinge lines shown in Figure 9.5(a) is of length $a/\cos(\pi/n)$. The relative angular velocity between two adjacent rigid segments is $2\omega\sin(\pi/n)$, as shown in Figure 9.5(b). Thus, the total plastic dissipation rate on the plastic hinges is

$$\dot{D}_{\mathrm{p}} = nM_0 a/\cos(\pi/n) 2\omega\sin(\pi/n) = 2M_0\omega a n\tan(\pi/n) \tag{9.24}$$

From the energy balance, Eqs (9.23) and (9.24) lead to the static collapse force

$$F_{\mathrm{s}} = 2M_0 n\tan(\pi/n) \tag{9.25}$$

When the side number tends to infinity, i.e., $n \to \infty$, the above analysis approaches the collapse load of a circular plate. Therefore, Eq. (9.25) leads to the static collapse load of a circular plate subjected to a concentrated force at its center:

$$F_{\mathrm{s}} = 2\pi M_0 \tag{9.26}$$

Case B: Simply Supported, Loaded by a Uniformly Distributed Pressure p

When a simply supported regular polygon plate with n sides as shown in Figure 9.5 is subjected to a uniformly distributed pressure p, the deformation mechanism is assumed to be the same as that in Case A, as shown in Figure 9.5(a).

First, let us calculate the work rate of applied pressure. The area of each triangular rigid segment is $A = a \times a\tan(\pi/n)$, and the velocity of its centroid is $v_c = \omega a/3$; thus

$$\dot{W} = npAv_c = \frac{1}{3}\omega npa^3 \tan(\pi/n) \tag{9.27}$$

The total rate of plastic dissipation is the same as that in Case A:

$$\dot{D}_p = 2M_0\omega an \tan(\pi/n) \tag{9.28}$$

From the energy balance, $\dot{W} = \dot{D}_p$, the static collapse pressure of the simply supported regular polygonal plate is

$$p_s = 6M_0/a^2 \tag{9.29}$$

It is noted that the collapse pressure given in Eq. (9.29) is independent of the side number of the regular polygon. Hence, this expression is also applicable to simply supported circular plates, if $n \to \infty$.

Case C: Fully Clamped, Loaded by a Uniformly Distributed Pressure *p*

When a fully clamped regular polygon plate with n sides as shown in Figure 9.5 is subjected to a uniformly distributed pressure p, the deformation mechanism is assumed to be the same as that in Case A and Case B, so the hodograph remains the same too (see Figure 9.5b). The rate of external work is the same as that in Case B, i.e., Eq. (9.5). However, the plastic dissipation should consider the additional energy dissipation along the boundary, i.e.

$$\dot{D}_p = 2M_0\omega an \tan(\pi/n) + M_0 n\omega \times 2a\tan(\pi/n) = 4M_0\omega an \tan(\pi/n) \tag{9.30}$$

From the energy balance, $\dot{W} = \dot{D}_p$, the static collapse pressure of the fully clamped regular polygon plate is

$$p_s = 12M_0/a^2 \tag{9.31}$$

Obviously, due to the plastic dissipation on the boundary, the static collapse pressure of the fully clamped regular polygon plate is double that of the simply supported plate. Similarly, Eq. (9.31) is independent of the side number of the regular polygon, and therefore it is also applicable to fully clamped circular plates.

9.1.4 Load-Carrying Capacity of Annular Plate Clamped at its Outer Boundary

An annular plate with radius of inner circle b and radius of outer circle a, as shown in Figure 9.6(a), is analyzed in this subsection. It is clamped at its outer boundary and subjected to a distributed force along its inner boundary, with the total force being P.

In the deformation mechanism shown in Figure 9.6(a), this annular plate is divided into n sectors by n plastic hinges along the radial direction, with $n \to \infty$. The plastic hinge lines along the radial direction are marked as <2>, while the plastic hinge lines along the outer boundary are noted as <1>. When the total force P is applied uniformly along the inner circle $r = b$, the pressure per unit length is $p = P/(2\pi b)$. Each of the vectors of angular velocity of the rigid segments shown in Figure 9.6(b) is parallel to the bottom line of the respective rigid segment. If the magnitude of angular velocity is ω, then

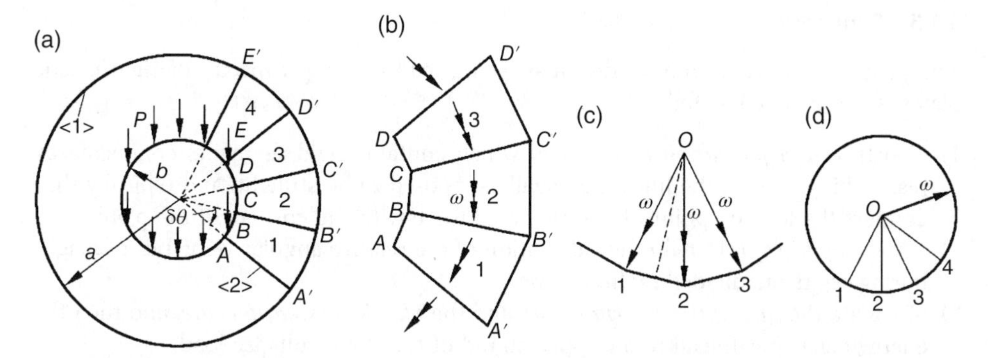

Figure 9.6 An annual plate subjected to distributed load on its inner boundary. (a) Deformation mechanism; (b) vectors of angular velocity of rigid segments; (c, d) hodograph of angular velocity.

the transverse velocity at the inner circle is $v = (a-b)\omega$. Thus, the work rate of the applied force is

$$\dot{W} = p \times 2\pi b v = p \times 2\pi b(a-b)\omega \tag{9.32}$$

The plastic dissipation rate on plastic hinge lines <1> is

$$\dot{D}_1 = M_0 \times 2\pi a\omega \tag{9.33}$$

There are in total n plastic hinge lines like <2>, each having length $(a-b)$. The relative angular velocity between the adjacent rigid segments is $2\omega\sin(\pi/n)$. Therefore, the plastic dissipation rate on plastic hinge lines like <2> is

$$\dot{D}_2 = M_0 n(a-b) \times 2\omega\sin(\pi/n) \tag{9.34}$$

when $n \to \infty$, the above relation approaches

$$\dot{D}_2 = 2\pi\omega M_0(a-b) \tag{9.35}$$

Hence, the total plastic dissipation rate is

$$\dot{D}_\mathrm{p} = \dot{D}_1 + \dot{D}_2 = 2\pi(2a-b)\omega M_0 \tag{9.36}$$

From the energy balance, $\dot{W} = \dot{D}_\mathrm{p}$, Eqs (9.32) and (9.36) lead to the static collapse force per unit length of this annular plate:

$$p_\mathrm{s} = \frac{2a-b}{a-b}\frac{M_0}{b} \tag{9.37}$$

And the corresponding total collapse force is given by

$$P_\mathrm{s} = 2\pi b p_\mathrm{s} = 2\pi M_0 \frac{2a-b}{a-b} \tag{9.38}$$

In a special case, $b = 0$, the above relation becomes $P_\mathrm{s} = 4\pi M_0$, i.e. Eq. (9.38) is reduced to the case of a fully clamped circular plate under central concentrated force.

9.1.5 Summary

The procedure of determining the quasi-static load-carrying capacity of rigid-plastic plates is summarized as follows:

1) Construct a *deformation mechanism* which contains rigid segments connected by plastic hinge lines. The lines are usually supposed to be straight, and typically they connect the loading point of concentrated force with the corners of the plate.
2) Draw a *hodograph* to help the calculation of the relative angular velocities of neighboring segments divided by hinge lines.
3) Calculate the *rate of the external work* and the *plastic dissipation rate*, and then the energy balance will lead to an upper bound of the static collapse load.
4) For complex configurations, unknown geometric parameter(s) may be adopted in the assumed deformation mechanism and then determined by the minimization of the upper bounds of the collapse load.

9.2 Dynamic Deformation of Pulse-Loaded Plates

In this section, the dynamic response of plates under pulse loading will be discussed. In general, the pulses applied to structures can be very different in magnitude, shape, and duration. So, the pulse approximation method is introduced first.

9.2.1 The Pulse Approximation Method

As shown in Figure 9.7, Youngdahl (1970) first proposed that the effect of a general loading pulse on a structure's response is approximately equivalent to that of a rectangular pulse impulse I_e with an effective magnitude P_e and pulse duration $2t_{mean}$, which are defined by

$$I_e = \int_{t_y}^{t_f} P(t)\,\mathrm{d}t \tag{9.39}$$

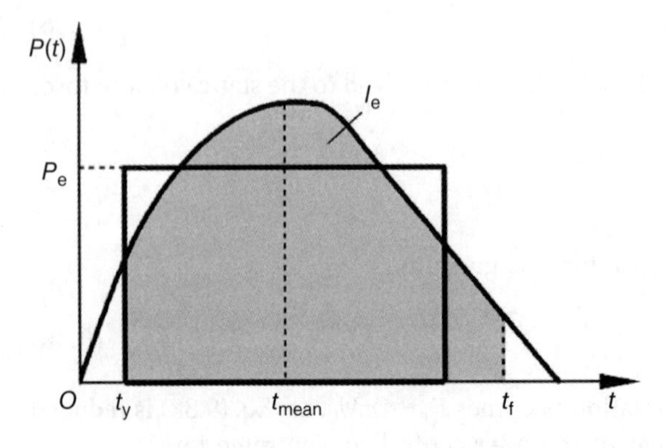

Figure 9.7 The equivalent of an arbitrary pulse.

$$P_e = I_e/(2t_{mean}) \tag{9.40}$$

$$t_{mean} = \frac{1}{I_e}\int_{t_y}^{t_f} (t - t_y)P(t)\ dt \tag{9.41}$$

where $P(t)$ is the non-negative pulse magnitude varying with time t, and t_y and t_f are the times at which plastic deformation begins and ends, respectively. The final deflection of the structure in concern is determined by I_e and P_e for a general pulse load. As both I_e and P_e depend on the integral of the general pulse, they are not very sensitive to the actual pulse shape. The method reduces the influence of pulse shape and allows a general pulse to be approximately represented by only two parameters. However, in actual applications, it is still difficult to determine t_y and t_f. Hence, Youngdahl also suggested the following relation:

$$P_y\left(t_f - t_y\right) = \int_{t_y}^{t_f} P(t)\ dt \tag{9.42}$$

where P_y is the static collapse load of the structure.

Using Eqs (9.39)–(9.41), we are able to eliminate pulse shape effects on the dynamic plastic bending response of widely used structural members, such as beams, circular plates, and a cylindrical shell. It was verified that for typical structures, including beams and plates, their dynamic response under the linear decreasing pulse, triangular pulse, cosine pulse, and so on, were very close to the equivalent rectangular pulse obtained by the pulse approximation method, i.e. Eq. (9.39). By comparing the final deflections obtained from this approximate method and from the known rigid-plastic solutions, it was found that the deviations caused by the pulse approximation method are within 5%. Thus, the dynamic response of the structure under pulses of various shape could be equivalent to that under the rectangular pulses.

9.2.2 Square Plate Loaded by Rectangular Pulse

Here we will only illustrate a simple solution of plates subjected to pulse loading. This was proposed by Kaliszky (1970), with the following assumptions:

1) The pulse is *rectangular*, which refers to the equivalent replacement of pulses proposed by Youngdahl.
2) The intensity of the pulse is "*moderate*", i.e., $p_s \le p_0 \le (2-3)p_s$.
3) The *deformation mechanism is the same* as that in the quasi-static case.

The previous three assumptions at the beginning of this chapter are still required: (i) the material is rigid-perfectly plastic and rate-independent; (ii) the loading is normal (transverse) to the plate's middle surface; and (iii) the deflection is small in comparison to the thickness of the plate.

Now let us consider a simply supported square plate with sides of length $2a$ and with thickness h, which is subjected to uniformly distributed pressure from a rectangular pulse (p_0, t_d), as shown in Figure 9.8(a).

In a study made by Cox and Morland (1959), it was found that the quasi-static mechanism remains correct only for *moderate* pulse, $p_s \le p_0 \le 2p_s$. For intense loading $(p > 2p_s)$, the deformation mechanism will be changed and contains travel hinge lines.

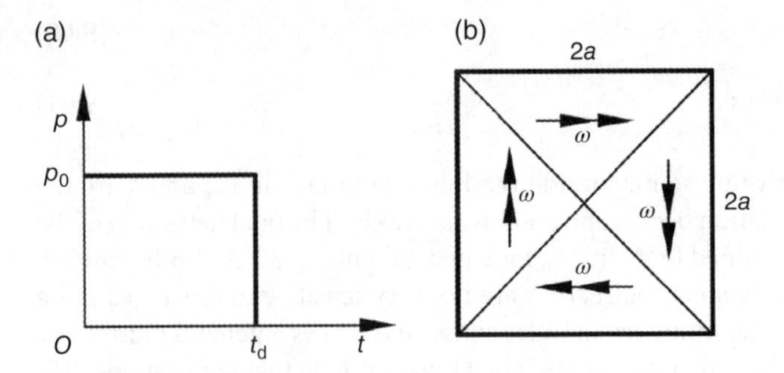

Figure 9.8 Square plate under uniformly distributed pressure of a rectangular pulse shape. (a) Rectangular pulse; (b) deformation mechanism.

From the analysis of Section 9.1, for the square plate subjected to uniformly distributed pressure, the static collapse load is

$$p_s = \frac{6M_0}{a^2}, \text{for a simply supported plate} \tag{9.43}$$

$$p_s = \frac{12M_0}{a^2}, \text{for a fully clamped plate} \tag{9.44}$$

Now we consider only the case of moderate pulse loading. The response consists of two phases: *phase I*, accelerated deformation during the time period of $0 \le t \le t_d$; and *phase II*, decelerated deformation in the time period of $t > t_d$.

Phase I

The *resistance to deformation* comes from plastic hinge lines, whether in the static deformation or the dynamic response. Therefore, in view of the energy balance used for analyzing static deformation cases, the resistance to deformation can be represented by the static collapse pressure p_s. Thus, the *"net" pressure*, or the *"excess" pressure* applied to the plate, is $(p - p_s) = (p_0 - p_s)$, which is a constant, and will cause the accelerated deformation of the plate segments.

It is assumed that the angular acceleration of each quadrant is a constant α; then the angular velocity of each quadrant is

$$\omega = \alpha t \tag{9.45}$$

and the velocity at the center is

$$v = \alpha a t \tag{9.46}$$

It is easy to calculate the displacement at the center:

$$\Delta = vt/2 \tag{9.47}$$

In order to determine the magnitude of the velocity at the plate center, or the angular velocity of each quadrant, one of the four rigid segments is taken into analysis. From the energy balance, the work done by the *net pressure* on top of the plate during the period 0–t should be equal to the increase in the kinetic energy of the plate:

$$W = \Delta K \tag{9.48}$$

where the work done by net pressure is equal to the product of the net pressure $(p_0 - p_s)$, the area of the triangle a^2, and the deflection at the triangle center $\Delta/3 = vt/6$, i.e.

$$W = (p_0 - p_s)a^2 vt/6 \tag{9.49}$$

while the kinetic energy increase of the triangle plate is

$$\Delta K = \frac{1}{2} J\omega^2 \tag{9.50}$$

where J is the *moment of inertia* of the quadrant about the edge. As shown in Figure 9.9, for a plate with the mass per unit area μ,

$$J = 2\int_0^a \int_0^x \mu y^2 \, dy \, dx = 2\int_0^a \frac{1}{3}\mu x^3 \, dx = \frac{1}{6}\mu a^4 \tag{9.51}$$

As $\omega = v/a$, the energy equation (Eq. 9.48) leads to the velocity of the plate center:

$$v = \frac{2(p_0 - p_s)}{\mu} t \tag{9.52}$$

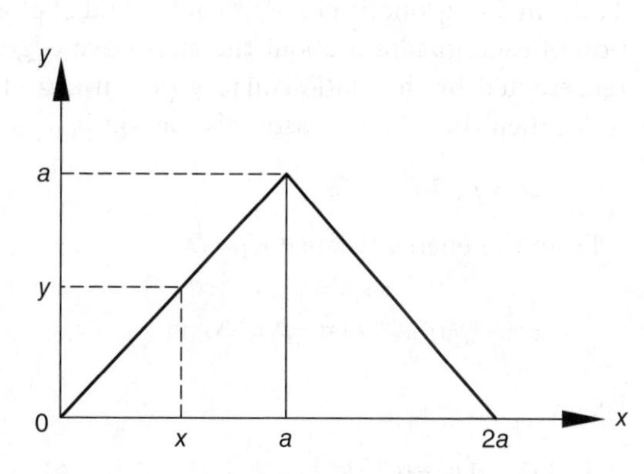

Figure 9.9 Triangular segment rotating with respect to its bottom edge.

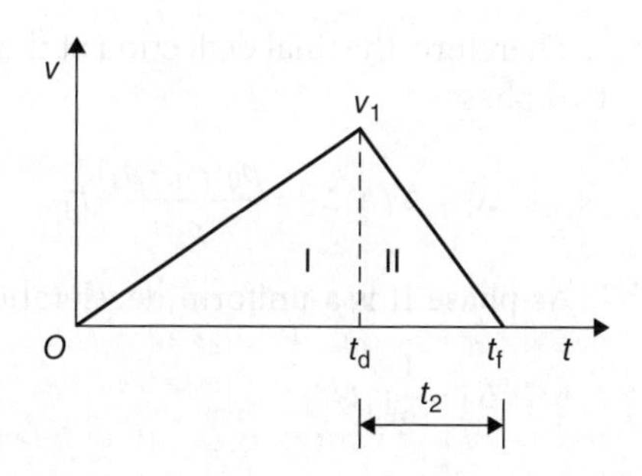

Figure 9.10 Central velocity of a square plate versus time.

From the above relation, clearly the velocity of the central point is proportional to the excess pressure $(p_0 - p_s)$, and it is also in proportional to time t, which implies a uniform acceleration.

At the end of phase I when $t = t_d$, the velocity of the plate center is

$$v_1 = \frac{2(p_0 - p_s)}{\mu} t_d \tag{9.53}$$

and the deflection of the plate center is

$$\Delta_1 = \frac{1}{2} v_1 t_d = \frac{(p_0 - p_s)}{\mu} t_d^2 \tag{9.54}$$

The kinetic energy at this time instant is

$$K_1 = 4 \cdot \frac{1}{2} J \omega_1^2 = 2 \cdot \frac{1}{6} \mu a^4 \cdot \left[\frac{2(p_0 - p_s) t_d}{\mu} \frac{}{a} \right]^2 = \frac{4a^2}{3\mu} (p_0 - p_s)^2 t_d^2 \tag{9.55}$$

By using Eq. (9.54), the kinetic energy at this instant could be also expressed by the current deflection,

$$K_1 = (p_0 - p_s)\, 4a^2\, \Delta_1 / 3 \tag{9.56}$$

Phase II

The remaining kinetic energy K_1 at the end of phase I will be dissipated by a further rotation of each quadrant about the respective edge. And the resistance to deformation is represented by the static collapse pressure p_s. If the additional deflection in phase II is Δ_2, then the related plastic dissipation is

$$D_2 = p_s\, 4a^2\, \Delta_2 / 3 \tag{9.57}$$

From the energy balance $K_1 = D_2$

$$\frac{4}{3} \frac{a^2}{\mu} (p_0 - p_s)^2 t_d^2 = \frac{4}{3} p_s a^2 \Delta_2$$

i.e.

$$\Delta_2 = \frac{(p_0 - p_s)^2 t_d^2}{\mu p_s} \tag{9.58}$$

Therefore, the final deflection at the plate center is the sum of the displacements in the two phases

$$\Delta_f = \Delta_1 + \Delta_2 = \frac{p_0 (p_0 - p_s)}{p_s \mu} t_d^2 \tag{9.59}$$

As phase II is a uniform deceleration

$$\Delta_2 = \frac{1}{2} v_1 t_2 \tag{9.60}$$

the duration of phase II, t_2, can be found by

$$t_2 = \frac{2\Delta_2}{v_1} = \frac{(p_0 - p_s)t_d}{p_s} \tag{9.61}$$

The velocity of the plate center changing with time is given in Figure 9.10. Obviously, in Phase I and Phase II, the motion of the plate center is uniformly accelerated and uniformly decelerated, respectively.

Hence, the total response time of the plate is found as

$$t_f = t_1 + t_2 = t_d + \frac{1}{p_s}(p_0 - p_s)t_d \tag{9.62}$$

Or it can be rewritten as

$$t_f = \frac{p_0}{p_s}t_d$$

i.e.

$$p_s t_f = p_0 t_d \tag{9.63}$$

The physical meaning of Eq. (9.63) is that the total input impulse of the applied load $p_0 t_d$ must be equal to the resistance pressure multiplied by the total response time, $p_s t_f$. In fact, the same result is obtained from the impulse–momentum relation, i.e., the net input impulse in phase I is

$$I_{in} = (p_0 - p_s)t_d\, 4a^2 \tag{9.64}$$

while the impulse absorbed in phase II is

$$I_{out} = p_s(t_f - t_d)4a^2 \tag{9.65}$$

Equation (9.63) could be resulted from $I_{in} = I_{out}$.

The following should be noted:

1) The use of "net" load allows us to treat the simply supported and fully clamped plates simultaneously, provided the relevant p_s is used.
2) Can you visualize the dynamic deformation mechanism of a square plate in the case where $p_0 > 2p_s$? You need to construct a mechanism for the transient response and another one for the modal phase.

9.2.3 Annular Circular Plate Loaded by Rectangular Pulse Applied on its Inner Boundary

Consider an annular circular plate with radius of inner circle b, radius of outer circle a, and mass per unit area μ which is clamped at its outer edge and loaded by a rectangular pulse (P_0, t_d) at its inner edge (Figure 9.11). In Section 9.1.4 the total static collapse force was found:

$$P_s = 2\pi M_0 \frac{2a - b}{a - b} \tag{9.66}$$

Or the static collapse force per unit length is

$$p_s = \frac{P_s}{2\pi b} = \frac{M_0}{b}\frac{2a - b}{a - b} \tag{9.67}$$

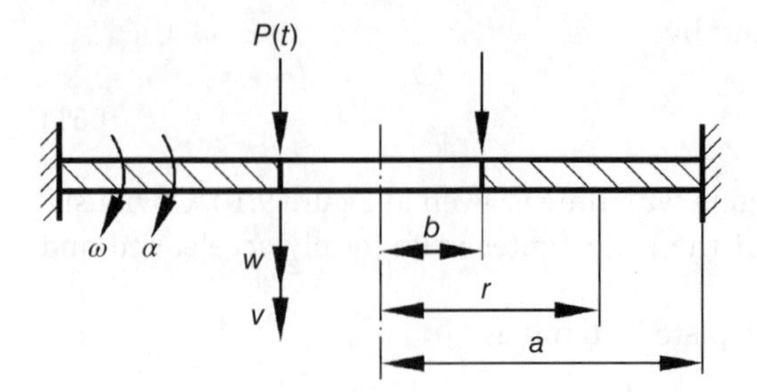

Figure 9.11 Annular circular plate subjected to a rectangular pulse on its inner boundary.

For a rectangular pulse $P = P_0$ $(0 \le t \le t_d)$, as shown in Figure 9.8(a), suppose the load is moderate, i.e., $P_s < P_0 \le 3P_s$. Similar to a square plate under rectangular pulse, the dynamic response of the annular circular plate is also divided into two phases. In *phase I*, accelerated deformation occurs during the time period $0 \le t \le t_d$; in *phase II*, decelerated deformation occurs when $t > t_d$.

Phase I

In phase I, the angular acceleration α of the rotation about the outer edge is constant, so that

$$\omega = \alpha t \tag{9.68}$$

The transverse velocity of the inner edge is

$$v = (a-b)\omega = (a-b)\alpha t \tag{9.69}$$

and the transverse displacement of the inner edge is obtained by integrating Eq. (9.69) with respect to time:

$$\Delta = \frac{1}{2}vt = \frac{1}{2}(a-b)\alpha t^2 \tag{9.70}$$

Therefore, the velocity and displacement of the inner edge at the end of phase I, $t = t_d$, are

$$v_1 = (a-b)\alpha t_d \tag{9.71}$$

$$\Delta_1 = \frac{1}{2}(a-b)\alpha t_d^2 \tag{9.72}$$

In the above equations, the angular acceleration α is an unknown constant to be determined. According to the conservation of energy principle, during the deformation process in phase I, the work rate of the net load $(P_0 - P_s)$ is equal to the rate of the increase of the kinetic energy of the plate:

$$(P_0 - P_s)\Delta_1 = \frac{1}{2}J(\alpha t_d)^2 \tag{9.73}$$

where J is the moment of inertia of the annular circular plate about the outer edge:

$$J = \iint_A \mu(a-r)^2 \, \mathrm{d}A = \int_b^a \mu(a-r)^2 \, 2\pi r \, \mathrm{d}r = \frac{\pi}{6}\mu(a-b)^3(a+3b) \tag{9.74}$$

Thus, the angular acceleration α in phase I is obtained from Eq. (9.73):

$$\alpha = \frac{1}{J}(a-b)(P_0 - P_s) \tag{9.75}$$

At the end of phase I, i.e., $t = t_d$, the transverse velocity of the inner edge is

$$v_1 = \frac{1}{J}(a-b)^2(P_0 - P_s)t_d \tag{9.76}$$

whilst the displacement, i.e., deflection, of the inner edge is

$$\Delta_1 = \frac{1}{2J}(a-b)^2(P_0 - P_s)t_d^2 \tag{9.77}$$

The kinetic energy of the plate at this instant is

$$K_1 = (P_0 - P_s)\Delta_1 = \frac{1}{2J}(a-b)^2(P_0 - P_s)^2 t_d^2 \tag{9.78}$$

Phase II

In phase II, the remaining kinetic energy K_1 at the end of phase I is dissipated by a further rotation of all the rigid segments shown in Figure 9.6(a) about the outer edge, while the resistance to deformation is represented by the static collapse total force, P_s. If the additional deflection in phase II is Δ_2, then the related plastic dissipation is equal to the kinetic energy at $t = t_d$:

$$K_1 = P_s\Delta_2 \tag{9.79}$$

Thus, the additional deflection gained in phase II is

$$\Delta_2 = \frac{2}{JP_s}(a-b)^2(P_0 - P_s)^2 t_d^2 \tag{9.80}$$

Therefore, the final deflection of the inner edge of the annular circular plate is

$$\Delta_f = \Delta_1 + \Delta_2 = \frac{(a-b)^2(P_0 - P_s)}{2J} t_d^2 \left[1 + \frac{1}{P_s}(P_0 - P_s)\right] \tag{9.81}$$

Or it can be rewritten as

$$\Delta_f = \frac{(a-b)^2}{2J}\frac{P_0}{P_s}(P_0 - P_s)t_d^2 = \frac{3t_d^2}{\pi\mu(a-b)(a+3b)}\frac{P_0}{P_s}(P_0 - P_s) \tag{9.82}$$

Also, due to the uniform decelerated motion in phase II

$$\Delta_2 = v_1 t_2/2 \tag{9.83}$$

Equations (9.80) and (9.83) give the time duration of phase II as

$$t_2 = \frac{2\Delta_2}{v_1} = \frac{1}{P_s}(P_0 - P_s)t_d \tag{9.84}$$

Hence, the total time of the dynamic response is

$$t_f = t_1 + t_2 = \frac{P_0}{P_s} t_d$$

i.e.,

$$P_s t_f = P_0 t_d \tag{9.85}$$

Again, the physical meaning of Eq. (9.85) is that the total input impulse of the applied load $P_0 t_d$ must be equal to the resistance force multiplied by the total response time, $P_s t_f$, which is the same as that we have seen in the response of a square plate under rectangular pulse loading, i.e., Eq. (9.63).

9.2.4 Summary

The analysis of the dynamic response of pulse-loaded plates can be summarized as follows:

1) The Kaliszky method employed here is applicable only for *moderate* dynamic pulse loading, which slightly exceeds the quasi-static collapse load.
2) In phase I, the plate's deformation is accelerated, with the acceleration being proportional to the "excess" load, or net load, $(P_0 - P_s)$; in phase II, the deceleration is proportional to the static collapse load P_s.
3) All the examples show that the final deflection of the plate possesses the property of $\Delta_f \propto [P_0(P_0 - P_s)/P_s] t_d^2$, i.e., it is determined by the input impulse $P_0 t_d$ and the excess impulse $(P_0 - P_s)t_d$. In addition, the total response time satisfies $P_s t_f = P_0 t_d$.

9.3 Effect of Large Deflection

If the deformation of the plate is relatively large, for example, the maximum deflection is close to or exceeds the plate's thickness, i.e., $\Delta \approx h$ or $\Delta > h$, then the membrane force induced by the large deflection will strengthen the plate and make its load-carrying capacity increase with deflection. This strengthening effect of the membrane force has been seen in elastic plates with large deformation and has attracted numerous studies. For plates that are not very thin, the membrane force will have a significant effect and will usually be coupled with the plastic deformation, resulting in more difficulties in the analysis.

The essential differences between plate and beam should be emphasized here. For straight beams without axial constraint at the supports, as discussed in Section 7.1, no tensile force or corresponding strengthening effect will be produced. For plates, on the other hand, regardless of whether the boundary condition provides in-plane constraint, the membrane deformation and membrane strain must occur in the mid-surface of the plate, if the mid-surface deforms into a curved surface with non-zero Gauss curvature, such as a circular plate deforming into a rotary shell. Consequently, the strengthening effect of membrane force will significantly increase the static and dynamic load-carrying capacity of plates as long as the deflection reaches the order of the thickness of the plate. In other words, compared with beams, the effect of large deflection is much more important for the behavior of plates.

9.3.1 Static Load-Carrying Capacity of Circular Plates in Large Deflection

A simple method proposed by Calladine (1968) can be used to assess the load-carrying capacity of plates under large deflection in the static loading condition. Employing the rigid-perfectly plastic material idealization, when the circular plate collapses, at any point on a section along a diameter (see Figure 9.12), the circumferential stress satisfies $\sigma_\theta = \pm Y$, where Y is the yield stress under uniaxial tension. By assuming the deflection shape (i.e., the deformation mechanism) of the plate under bending, and using the energy conservation to balance the internal plastic dissipation rate with the external work rate, the optimum upper bound of the collapse load will be obtained for the circular plate under large deformation.

For instance, based on the initial configuration without considering the large deformation effect, the "initial" static collapse load for a simply supported circular plate of radius a subjected to a concentrated force F is (refer to Eq. 9.26)

$$F_0 = 2\pi M_0 \tag{9.86}$$

Suppose there is no axial constraint on the outer boundary of the circular plate, then the above initial collapse load $F_0 = 2\pi M_0$ leads to a conically shaped velocity distribution in the deformation mechanism. Thus, at the cross-section along a diameter of a simply supported circular plate (Figure 9.12), the two halves of the section divided by the middle line will rotate about their own instantaneous centers of velocity, I. As the transverse displacement and velocity are constrained at the support, each of the instantaneous centers of velocity, I, is directly above the respective support point, at a to-be-determined distance from the mid-surface. If the rotation angle with respect to I of half of the cross-section along the diameter is ϕ, then the transverse displacement of an arbitrary point $B(x, y)$ is $\phi(a-x)$, and the displacement in the radial direction is ϕy (refer to Figure 9.12). The circumferential strain of point $B(x, y)$ is $\varepsilon_\theta = \phi y/x$. Consider a small area element $\mathrm{d}A$ around point B; the volume of the rotating body of this area element $\mathrm{d}A$ is $\mathrm{d}V = 2\pi x\ \mathrm{d}A$. It is already known that the circumferential stress is $\sigma_\theta = \pm Y$, and thus the plastic dissipation on this volume is $Y|\varepsilon_\theta|\ \mathrm{d}V = 2\pi Y\phi|y|\ \mathrm{d}A$. The absolute value of ε_θ is adopted here as the plastic dissipation is always positive. It should also be noted that the

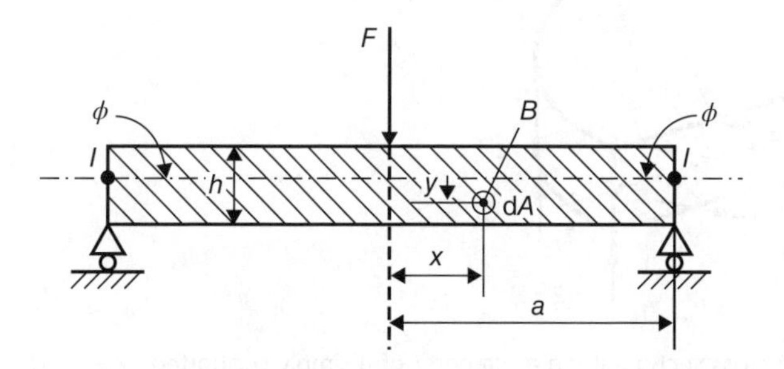

Figure 9.12 Cross-section along a diameter of a simply supported circular plate.

expression of the plastic dissipation is independent of the position x, indicating the uniform distribution of energy absorption along the radial direction.

For the deformation mechanism shown in Figure 9.12, from the energy conservation we obtain

$$F\phi a = 2\pi Y\phi \int_A |y| \, dA \tag{9.87}$$

where A is the area of the cross-section along a diameter. By eliminating the rotation angle ϕ from the Eq. (9.87), Eq. (9.88) provides an upper bound of the collapse load:

$$F = \frac{2\pi Y}{a} \int_A |y| \, dA \tag{9.88}$$

In order to obtain the best estimate of the collapse load from Eq. (9.88), the position of I–I must be determined first. From the expression of circumferential strain, i.e., $\varepsilon_\theta = \phi y / x$, it is clear that at the cross-section along a diameter, an arbitrary point below the I–I line is under tension along the circumferential direction, i.e., $\sigma_\theta = +Y$, whilst an arbitrary point above the I–I line is under compression, i.e., $\sigma_\theta = -Y$. For the half-circular plate shown in Figure 9.13, the reaction force along the radial direction is zero as there is no axial constraint on the outer boundary. Hence, from the overall equilibrium of the half-circular plate, it is clear that the area with $\sigma_\theta = +Y$ should be equal to the area with $\sigma_\theta = -Y$, which is termed the *principle of equal areas*. It is easy to prove that according to the principle of equal areas, the determination of the I–I line should make the integration $\int_A |y| dA$ a minimum. Thus, the principle of equal areas will result in a minimum upper bound of the collapse load given in Eq. (9.88). It should be emphasized that the deformation mechanism may not satisfy the equilibrium equations everywhere in the plate, so this "minimum upper bound" is usually not equal to the lower bound, i.e., that coming from the complete solution; it is just an optimized upper bound.

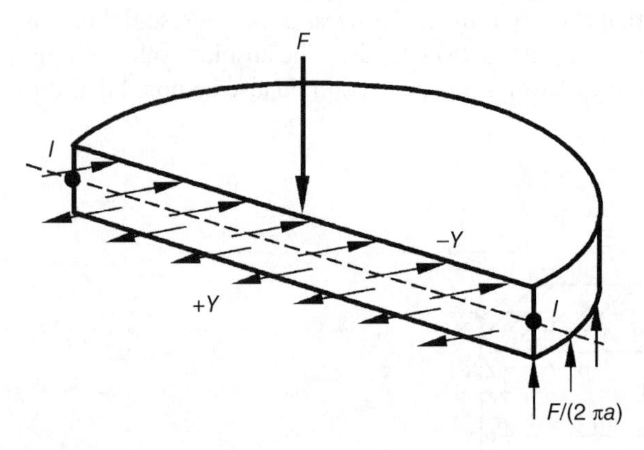

Figure 9.13 Stress distribution at the cross-section along a diameter of a simply supported circular plate.

For the initial collapse state of the circular plate, the I–I line determined by the principle of equal areas is coincident with the mid-plane of the plate. Therefore

$$\int_A |y| \mathrm{d}A = 2a\int_0^{h/2} y\,\mathrm{d}y = ah^2/4 \tag{9.89}$$

By substituting Eq. (9.89) into Eq. (9.88), the initial collapse load of the simply supported circular plate is

$$F_0 = \frac{1}{2}\pi Y h^2 = 2\pi M_0 \tag{9.90}$$

which is the same as that obtained in previous analysis (see Eq. 9.86). However, Calladine's method is not limited to obtaining the initial collapse load F_0; rather, it can be used to assess the load-carrying capacity of plates in large deflection.

If the central deflection of the plate is smaller than the beam thickness, i.e., $\Delta \le h$, the position of the I–I line is determined on the basis of Figure 9.14(a). By integrating $\int_A |y|\mathrm{d}A$ with this configuration, the static collapse load is estimated as

$$\frac{F_s}{F_0} = 1 + \frac{1}{3}\left(\frac{\Delta}{h}\right)^2 = 1 + \frac{1}{3}\bar{\Delta}^2, \bar{\Delta} \equiv \Delta/h \le 1 \tag{9.91}$$

If the central deflection of the plate is greater than the beam thickness, i.e., $\Delta > h$, the position of the I–I line is determined by the principle of equal areas based on Figure 9.14 (b). As a result, the static collapse load is estimated as

$$\frac{F_s}{F_0} = \frac{\Delta}{h} + \frac{1}{3}\frac{h}{\Delta} = \bar{\Delta} + \frac{1}{3\bar{\Delta}}, \bar{\Delta} \equiv \Delta/h \ge 1 \tag{9.92}$$

The load-carrying capacity of a simply supported circular plate without axial constraint, given by Eqs (9.91) and (9.92), is plotted as solid line in Figure 9.15. Obviously, as the deflection increases, the load-carrying capacity of the plate, F_s, will increase rapidly. When plate's central deflection Δ is much greater than the plate thickness h, the load-carrying capacity F_s can be regarded as approximately linearly increasing with the central deflection Δ (dashed line in Figure 9.15).

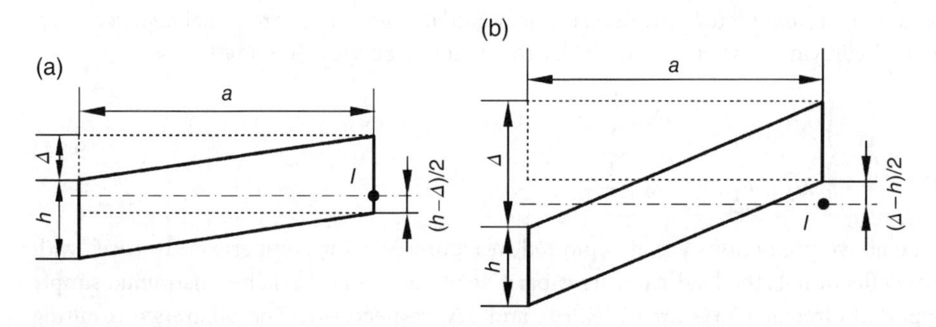

Figure 9.14 Determination of the position of the I–I line. (a) $\Delta \le h$; (b) $\Delta > h$.

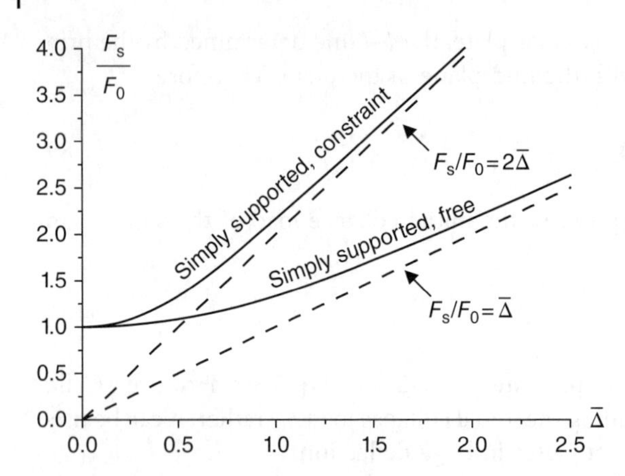

Figure 9.15 Load-carrying capacity of circular plates versus the central deflection.

The load-carrying capacity F_s given by Eqs (9.91) and (9.92) can be integrated with respect to the normalized deflection $\bar{\Delta}$, i.e., the area under the solid curves in Figure 9.15. This area represents the plastic energy dissipated until a specified normalized deflection. The normalized integration gives

$$\frac{1}{F_0 h}\int_0^{\Delta} F_s \, d\Delta = \int_0^{\bar{\Delta}} \frac{F_s}{F_0} \, d\bar{\Delta} = \begin{cases} \bar{\Delta} + \dfrac{1}{9}\bar{\Delta}^3, & \bar{\Delta} \le 1 \\ \dfrac{11}{18} + \dfrac{\bar{\Delta}^2}{2} + \dfrac{1}{3}\ln \bar{\Delta}, & \bar{\Delta} \ge 1 \end{cases} \tag{9.93}$$

The Calladine's method can be also applied in the large deflection analysis of other structures, such as fully clamped circular plates, sandwich plates, reinforced concrete plates and so on.

The large deflections of rigid-plastic circular plates were also analyzed by Kondo and Pian (1981), using the *generalized yield line method* which takes into account the changes in geometry of the structures. The circular plates were assumed to deform into a number of right circular cones separated by concentric hinge circles with no radial strains in each cone. Then, the general equation to obtain the load-deflection relations was derived from the principle of virtual velocity. Simply supported circular plates under circular loading were investigated, with the boundary conditions with and without axial constraint. In the case of no axial constraint, the results predicted by the generalized yield line method are exactly the same as those obtained by the Calladine's method (see Eqs 9.91 and 9.92). In the case of circular plates restrained at the boundary against inward radial displacement, the load-carrying capacity predicted by the generalized yield line method is

$$\frac{F_s}{F_0} = \begin{cases} 1 + \dfrac{4}{3}\bar{\Delta}^2, & \bar{\Delta} \le \dfrac{1}{2} \\ 2\bar{\Delta} + \dfrac{1}{6\bar{\Delta}}, & \bar{\Delta} \ge \dfrac{1}{2} \end{cases} \tag{9.94}$$

The above predictions are also plotted in Figure 9.15 for comparison. In cases with large deflection $\bar{\Delta}$, the load-carrying capacities of the free and axially constrained simply supported circular plates are $F_s/F_0 = \bar{\Delta}$ and $2\bar{\Delta}$, respectively. The difference resulting from the boundary conditions is clearly important.

9.3.2 Dynamic Response of Circular Plates with Large Deflection

If the dynamic load applied to a circular plate is slightly higher than its static collapse load, the deformation mechanism can be usually assumed to be the same as that occurring in the static case. Thus, the deflection of plates under the dynamic loading could be estimated directly from their static behavior.

For example, if a circular plate is subjected to an impact at its center by mass m with initial velocity v_0, then the initial kinetic energy $K_0 = mv_0^2/2$ should be dissipated by its large deflection mechanism, which has been discussed in Section 9.3.1. Thus

$$\int_0^{\Delta_f} F_s \, d\Delta = K_0 \tag{9.95}$$

Then, combining Eqs (9.90)–(9.92) and (9.95) leads to

$$\frac{K_0}{F_0 h} = \frac{mv_0^2/2}{2\pi M_0 h} = \begin{cases} \bar{\Delta}_f + \dfrac{1}{9}\bar{\Delta}_f^3, & \bar{\Delta}_f \le 1 \\ \dfrac{11}{18} + \dfrac{\bar{\Delta}_f^2}{2} + \dfrac{1}{3}\ln \bar{\Delta}_f, & \bar{\Delta}_f \ge 1 \end{cases} \tag{9.96}$$

where $\bar{\Delta}_f \equiv \Delta_f/h$ is the normalized final central deflection. Solving $\bar{\Delta}_f$ from Eq. (9.96) could give an estimate of the final deflection of the plate after impact, as plotted in Figure 9.16.

This approximate method can also be applied to other dynamic loading cases. For example, consider a circular plate of radius a and mass per unit area μ subjected to an impulsive loading so it gains a uniformly distributed initial velocity v_0. The final deflection Δ_f can be found in a similar way, i.e., the initial kinetic energy $K_0 = \pi a^2 \mu v_0^2/2$ should be dissipated by this large deflection mechanism.

As discussed in Section 7.3, the membrane factor method is used in analyzing the dynamic response of beams with large deflections. Similarly, the membrane factor method could also be used in analyzing the dynamic response of circular plates, rectangular plates and other plates with large deflections (see Yu & Chen, 1992).

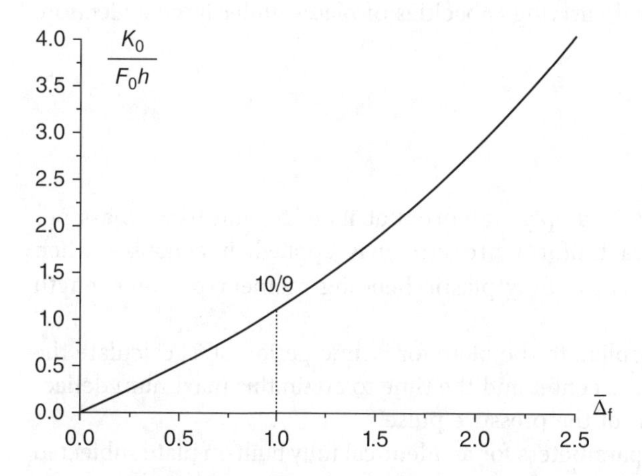

Figure 9.16 Initial kinetic energy versus the normalized maximum central deflection.

For example, in the analysis of the large deformation of a simply supported circular plate made of rigid-perfectly plastic material, by comparing the energy dissipation rate due solely to bending, J_m, with that due to both bending moment and membrane force J_{mn}, Yu and Chen (1990) derived the membrane factor in the radial direction, f_n^r, and the circumferential direction, f_n^θ,

$$f_n^r \equiv \frac{J_{mn}^r}{J_m^r} = \begin{cases} 1 + 4\bar{\Delta}^2, & \bar{\Delta} \equiv \dfrac{\Delta}{h} \leq \dfrac{1}{2} \\ 4\bar{\Delta}, & \bar{\Delta} \equiv \dfrac{\Delta}{h} \geq \dfrac{1}{2} \end{cases} \tag{9.97}$$

$$f_n^\theta \equiv \frac{J_{mn}^\theta}{J_m^\theta} = \begin{cases} 1 + \dfrac{4}{3}\bar{\Delta}^2, & \bar{\Delta} \equiv \dfrac{\Delta}{h} \leq \dfrac{1}{2} \\ 2\bar{\Delta} + \dfrac{1}{6\bar{\Delta}}, & \bar{\Delta} \equiv \dfrac{\Delta}{h} \geq \dfrac{1}{2} \end{cases} \tag{9.98}$$

When a simply supported circular plate is subjected to uniform impulsive loading, the whole plate has an uniform initial velocity v_0. Similar to the simply supported beam discussed in Section 7.1, there are two phases during the dynamic response of the circular plate. In phase I, i.e., the transient phase, the deformation mechanism has a plastic hinge circle, traveling from the outer edge towards the center of the plate, and inside this plastic hinge circle the velocity remains at v_0. At the moment this plastic hinge circle shrinks to the plate center, phase II starts, which is the modal phase of the response. The load-carrying capacity will increase due to the strengthening effect by the membrane force associated with large deflection. The plate could possibly become a plastic membrane, similar to the case of a beam becoming a plastic string, as illustrated in Section 7.2. Employing the membrane factor method, Yu and Chen (1992) derived the relation between the final deflection of the circular plate and the initial kinetic energy, exhibiting good agreement with experimental results.

If the deformation of the plate is relatively large, e.g., the maximum deflection is close to or exceeds the maximum deflection of the plate thickness (i.e., $\Delta \approx h$ or $\Delta > h$), then the strengthening effect caused by the membrane force must be considered. As discussed earlier in this section, the Calladine method and the membrane factor method are applicable when analyzing the load-carrying capacities of plates under large deflection.

Problems 9

9.1 A circular plate of radius a is simply supported at its edge, and has a mass per unit area μ. Show that if a uniform pressure p is applied, it collapses when $p = p_s = 6M_0/a^2$, where M_0 is the fully plastic bending moment per unit length along plastic hinge lines.

a) If a pressure $p = 2p_s$ is applied to the plate for a time period of T, calculate the maximum deflection of the center and the time to attain this maximum deflection after the application of the pressure pulse.

b) Find the corresponding parameters for an identical fully built-in plate subjected to the same pressure pulse.

9.2 Consider a simply supported circular plate undergoing large deflection under a uniformly distributed impulsive pressure that imposes a uniform initial velocity. The deformation mechanism of the plate may vary according to the loading velocity. Considering the transition of the deformation mechanism, and also the large deflection effect, explain how to analyze its dynamic response.

10

Case Studies

Through four typical cases, this chapter will demonstrate how to apply the theoretical models and analytical/numerical methods of structural impact dynamics illustrated in previous chapters to investigate engineering problems.

10.1 Theoretical Analysis of Tensor Skin

In this section, the analysis of a special sandwich structure called "tensor skin" will be illustrated.

10.1.1 Introduction to Tensor Skin

A special kind of sandwich structure comprising cover plies, tensor plies and carrying plies, tensor skin (see Figure 10.1) was originally developed to improve the crashworthiness of composite helicopter structures subjected to water impact. The cover ply is the loading face; the carrying ply provides the structural stiffness; and the tensor ply provides the capability to unfold through the formation of *plastic hinges*, before it stretches and fails, leading to an increase in the load-bearing capability and energy absorption of the structure.

It is noted that most of the published studies have focused on the static/dynamic responses of tensor skin under concentrated loads, which are different from the pressure-loading condition under water impact. In order to understand the fundamental behavior of tensor skin under pressure loading and its sensitivities to the pulse shapes, and also the fluid–structure interaction between the water and the structure, Ren *et al.* (2014a,b) established a theoretical model to analyze the static and dynamic responses of a kind of tensor skin.

10.1.2 Static Response to Uniform Pressure Loading

Assumptions

The total response of tensor skin is divided into three stages: an elastic deformation stage of the whole structure; an unfolding stage of the tensor ply; and a stretching stage of the tensor ply, as shown in Figure 10.2. In the first stage, elastic deformation occurs

Introduction to Impact Dynamics, First Edition. T.X. Yu and XinMing Qiu.

Published 2018 by John Wiley & Sons Singapore Pte. Ltd.

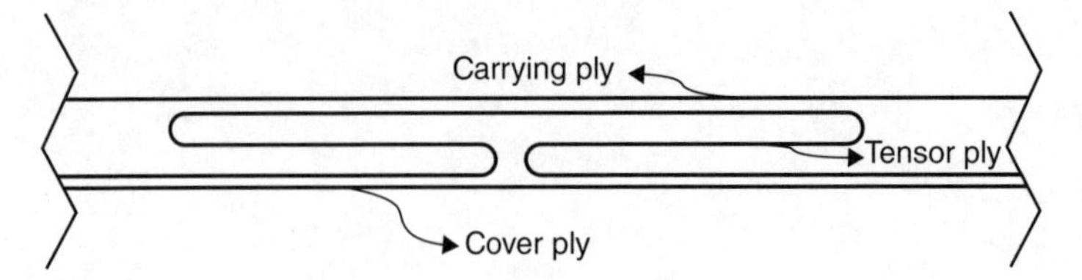

Figure 10.1 The schematic of a tensor skin. *Source*: Ren *et al.* (2014b). Reproduced with permission of Elsevier.

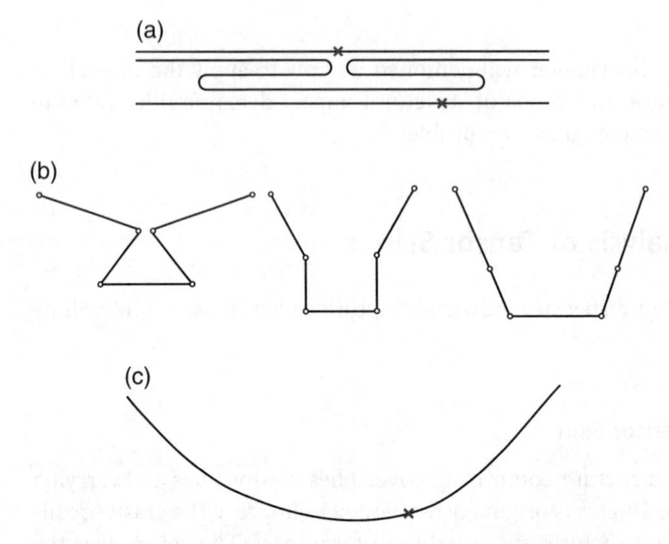

Figure 10.2 Three response stages of a tensor skin. (a) Elastic deformation stage; (b) unfolding stage of tensor ply; (c) stretching stage of tensor ply. *Source*: Ren *et al.* (2014a). Reproduced with permission of Elsevier.

throughout the whole structure, including the cover ply, carrying ply, and the tensor ply, and is ended by the fracture of the cover and carrying plies made of brittle composite. The second stage is an unfolding process of tensor ply, which can be simplified as three rigid segments connected by plastic hinges. The third stage is characterized by the stretching of the tensor ply, which may be terminated by tensile failure of the ply.

As revealed by finite element simulations, the energy dissipation in the first and third stages are small compared with that in the second stage. In addition, as there is no adhesion and friction between different plies, after the brittle failure of the cover and carrying plies, their load-carrying contributions can be ignored. Thereby, in the theoretical modeling, only the tensor ply is of concern.

Deformation Mechanism

As observed from the finite element simulations, there is localization of deformation i.e., plastic hinges clearly form near the corners. The elastic energy is found to be less than

10% of the total deformation energy when the tensor skin experiences large bending. Therefore, the rigid-perfectly plastic assumption is adopted in the theoretical analysis.

The deformation mechanisms during the unfolding process are depicted in Figure 10.3. Only half of the structure is illustrated due to symmetry. Point D represents the fixed boundary and point C is the midpoint of the tensor ply. Three rigid segments, DA, AB, and BC, are connected by plastic hinges A, B, and D. The lengths of different segments are $\overline{DA} = l$ and $\overline{AB} = \overline{BC} = kl$, where k $(0 < k < 1)$ is a geometric parameter. The normalized central deflection is defined as $\bar{u} \equiv u/l$, where u is the vertical displacement of point C:

$$\bar{u} = \sin\alpha + k\sin\beta \tag{10.1}$$

where

$$\beta = \arccos\left(1 - \frac{1-\cos\alpha}{k}\right) \tag{10.2}$$

The variation of angle, $\delta\alpha$, gives rise to the normalized central deflection and angle β:

$$\delta\bar{u} = \left(\cos\alpha + \frac{\sin\alpha}{\tan\beta}\right)\delta\alpha,\ \delta\beta = \frac{\sin\alpha}{k\sin\beta}\delta\alpha, \tag{10.3}$$

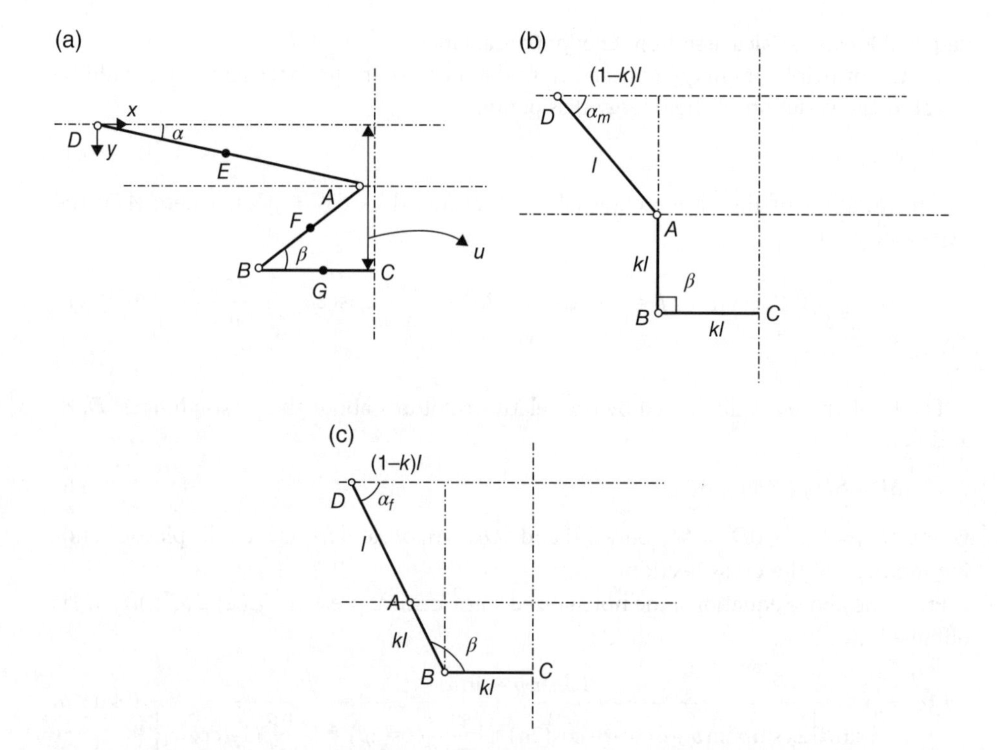

Figure 10.3 Deformation mechanisms of unfolding process of tensor ply, (a) $\beta < \pi/2$, (b) $\beta = \pi/2$, (c) $\beta > \pi/2$. *Source:* Ren *et al.* (2014a). Reproduced with permission of Elsevier.

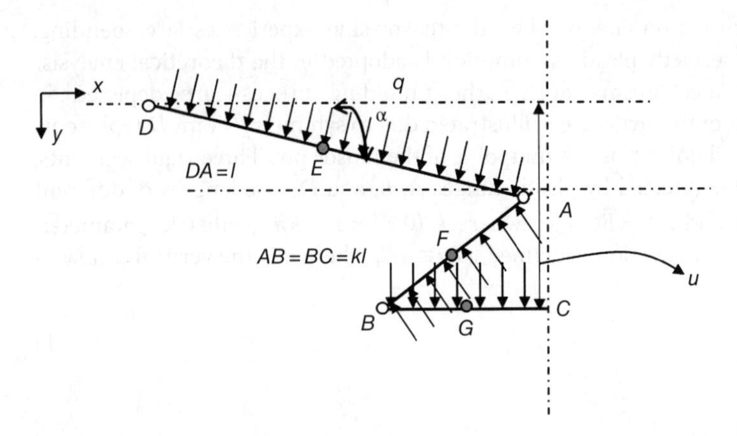

Figure 10.4 Tensor ply under uniform distributed pressure. *Source*: Ren *et al.* (2014a). Reproduced with permission of Elsevier.

Suppose a uniform distributed pressure q is slowly applied on the top surface of the tensor ply, as shown in Figure 10.4; then the mechanism undergoes quasi-static deformation. Thus, as angle α increases, the quasi-static critical pressure $q_s(\alpha)$ will vary.

Required Pressure Calculated from Energy Conservation

From the principle of energy conservation, the increase in the external work should be equal to the variation in the energy dissipation:

$$\delta W = \delta D \tag{10.4}$$

The variation of the above external work is found as $\delta W = q\,\delta A$, where A is the expanded area:

$$\delta W = \frac{1}{2}ql^2\left[2k\sin\alpha\sin\beta + 2\cos\alpha - \cos(2\alpha) + k\frac{\sin\alpha}{\sin\beta}\cos(2\beta) + 2\frac{\sin\alpha}{\tan\beta}(1-\cos\alpha)\right]\delta\alpha \tag{10.5}$$

The total energy is dissipated by the relative rotations about the plastic hinges, D, A, and B, as

$$\delta D = \delta D_D + \delta D_A + \delta D_B \tag{10.6}$$

where $\delta D_D = M_p\delta\alpha$, $\delta D_A = M_p(\delta\alpha + \delta\beta)$ and $\delta D_B = M_p\delta\beta$, and M_p is the fully plastic bending moment of the cross-section.

From the above equations, the normalized static critical pressure $\bar{q}_s(\alpha) \equiv q_s(\alpha)l^2/M_p$ is obtained as,

$$\bar{q}_s(\alpha) = \frac{4(k\sin\beta + \sin\alpha)}{k\sin\beta\left[2k\sin\alpha\sin\beta + 2\cos\alpha - \cos(2\alpha) + k\frac{\sin\alpha}{\sin\beta}\cos(2\beta) + 2\frac{\sin\alpha}{\tan\beta}(1-\cos\alpha)\right]}, 0 \le \alpha \le \alpha_f \tag{10.7}$$

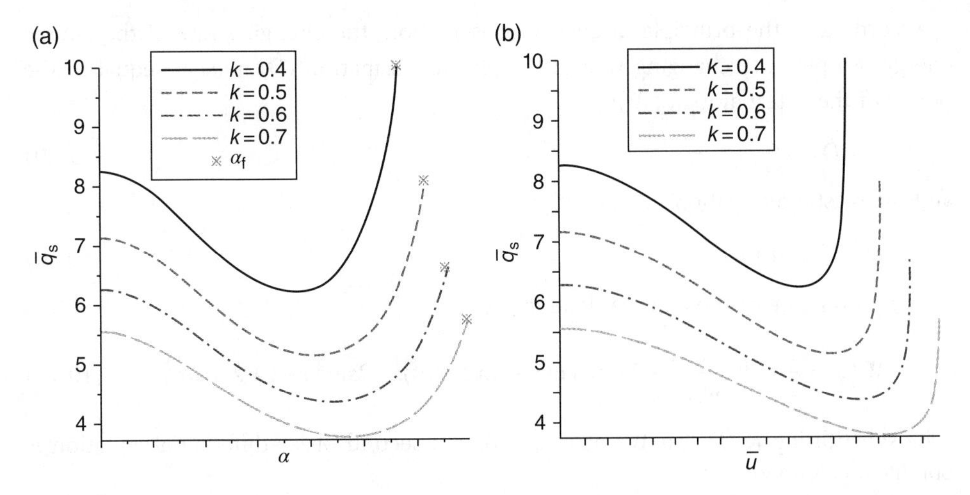

Figure 10.5 The normalized critical static pressure versus angle (a) and normalized central deflection (b). *Source*: Ren *et al.* (2014a). Reproduced with permission of Elsevier

As demonstrated in Figure 10.5, for increasing angle α or normalized central deflection $\bar{u}$, the static critical pressure $\bar{q}_s$ first decreases and then increases. The curves' tendency indicates a kind of "softening effect", i.e., the pressure to produce the initial deformation is high, but the resistance of the structure reduces as its deformation becomes larger. It is worth emphasizing that in the above analysis, an upper bound of critical pressure is calculated based on the assumed kinematically admissible field, which is not necessary to satisfy the equilibrium equations.

10.1.3 Dynamic Response of Tensor Skin

As found earlier, the static critical pressure $\bar{q}_s$ decreases first and then increases with the central deflection $\bar{u}$ (see Figure 10.5). Thus, the static critical pressure $\bar{q}_s$ can be regarded as a kind of "structural friction", and $\bar{q}_e \equiv \bar{q} - \bar{q}_s$ is defined as the *excess pressure.* It has been seen in Chapter 9 that the excess pressure will lead to the acceleration (or deceleration, if it is negative) of the structure.

As seen in Chapters 5–9, the dynamic response of a structure under a rectangular pulse can represent its fundamental dynamic characteristics. Suppose a constant pressure q is applied to the tensor ply; here q is a moderate pressure that is larger than the static critical pressure q_s. Owing to the unstable character of $\bar{q}_s$ shown in Figure 10.5, the tensor ply will be accelerated first and then decelerated.

As shown in Figure 10.3, the coordinates of each characteristic point is a function of α. Thus, the total kinetic energy of the tensor ply is given by:

$$K = \frac{1}{2}\rho l^3 \left\{ \frac{1}{3} + \frac{k}{\sin^2\beta}\left[\frac{1}{3}\sin^2\alpha + \cos\alpha \sin\beta \sin(\alpha+\beta) + \sin^2(\alpha+\beta) \right] \right\} \dot{\alpha}^2 \tag{10.8}$$

where ρ is the density per unit length of beam. The changing rate of the kinetic energy rate is $\dot{K}$.

According to the principle of energy conservation, the changing rate of the kinetic energy ($\dot{K}$) plus the changing rate of the plastic dissipation ($\dot{D}$) must be equal to the power of the external load ($\dot{W}$):

$$\dot{K} + \dot{D} = \dot{W}(q) \tag{10.9}$$

and the plastic dissipation rate is

$$\dot{D} = 2M_P(\dot{\alpha} + \dot{\beta}) \tag{10.10}$$

The power rate of the external load is

$$\dot{W}(q) = \frac{1}{2}ql^2\left(\frac{k\sin\alpha}{\sin\beta} + 2\sin\alpha\cot\beta - \sin 2\alpha\cot\beta + 2\sin^2\alpha - 1 + 2\cos\alpha\right)\dot{\alpha} \tag{10.11}$$

By substituting $\dot{K}$, $\dot{D}$ and $\dot{W}$ into Eq. (10.9), a second-order differential equation is obtained as follows:

$$a_1\ddot{\alpha} + a_2\dot{\alpha}^2 + a_3 = 0 \tag{10.12}$$

The coefficients of the above differential equation are omitted here but can be found in Ren *et al.* (2014a). By solving Eq. (10.12) with initial conditions, the dynamic response of the tensor ply can be obtained.

10.1.4 Pulse Shape

As illustrated in Chapter 9, Youngdahl (1970) proposed that the effect of a general loading pulse could be approximately equivalent to that of a rectangular pulse impulse. The method reduces the influence of pulse shape and allows a general pulse to be approximately represented by just two parameters. By comparing the final deflections obtained from this approximate method and from the known rigid-perfectly plastic solutions, it is found that the deviations caused by the pulse approximation method are small for stable structures, such as beams and plates. However, tensor skin is a typical geometrically unstable structure, for which the pressure to initiate its plastic deformation is high, but then the resistance will decrease as deformation proceeds. Hence, the applicability of the pulse approximation method for unstable structures should be investigated by examining the tensor skin.

Considering a tensor skin with geometric parameter k=0.5, unit width and half-length l=1, the initial static critical pressure is $\bar{q}_s{}^0$. A reference rectangular impulse with normalized pressure $\bar{q}_e = 1.3\bar{q}_s^0$ and normalized time duration $\bar{t}_{de} \equiv t_{de}/t_0 = 1$ is adopted in the following analysis. Here the reference time duration is taken as $t_0 = 0.20$ s. The corresponding normalized impulse is defined as $\bar{I} = \bar{q}_e\bar{t}_{de}$.

According to the pulse approximation method, various pulses that are approximated by the reference rectangular pulse, $\bar{I} = 1.3\bar{q}_s^0$, $\bar{t}_{de} = 1$, are depicted in Figure 10.6. For the triangular pulse, the pressure first increases from $\bar{q}_s^0$ and then decreases. The triangular pulse is regarded to be the closest one to the actual pulse produced by a flat body impinging on a water surface.

Figure 10.7 depicts the dynamic response of the tensor skin under various pulse loadings shown in Figure 10.6. Clearly, the deformations all increase with time, while the kinetic energy first increases and then decreases with time, which implies that the tensor

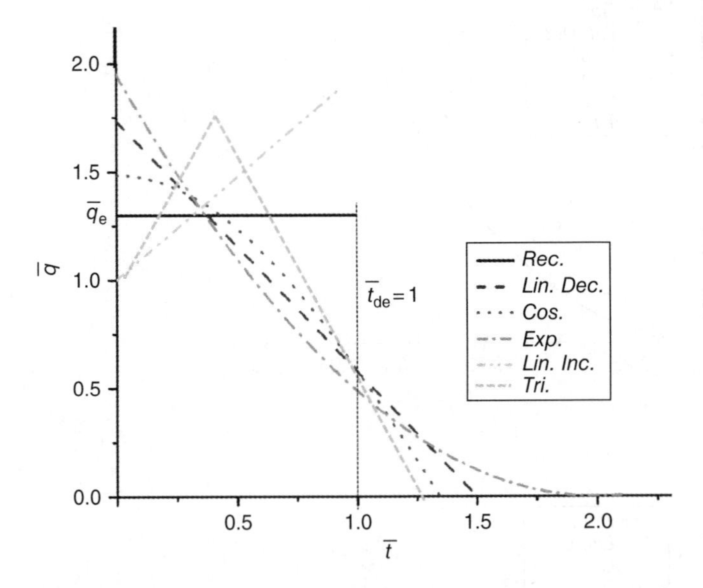

Figure 10.6 Various pulses that approximate to a rectangular pulse $\bar{I}=1.3\bar{q}_s^0$, $\bar{t}_{de}=1$. Rec., rectangular pulse; Lin. Dec., linearly decreased pulse; Cos., cosine pulse; Exp., explosive pulse; Lin. Inc., linearly increased pulse; Tri., triangular pulse. *Source*: Ren *et al.* (2014b). Reproduced with permission of Elsevier.

skin is accelerated first and then decelerated. Also, the differences between the curves of the monotonically decreasing pulses, i.e., explosive, cosine and linear decreasing pulses, are all small. The deviations of final deflection and total response time are within 10%, which is consistent with the previous remarks on stable structures. Although the pulse approximation method cannot follow the detailed deformation history of the structure exactly, it is applicable for estimating the overall dynamic response of a geometrically unstable structure, such as tensor skin.

It should be emphasized that the dynamic response under a triangular pulse is quite close to that under the equivalent rectangular pulse, as shown in Figure 10.7(a)–(c). The relative differences in the final deflection and in the total response time are only 4% and 1%, respectively. Hence, it does make sense to adopt the pulse approximation method when studying the tensor skin impinging on water.

10.2 Static and Dynamic Behavior of Cellular Structures

Cellular solids mean an assembly of cells with solid edges or faces, packed together so that they fill space, e.g., wood, cork, sponge, and coral are all cellular solids existing in nature. Recently a lot of synthetic cellular solids have also appeared. At the simplest level, honeycomb-like materials, made of parallel, prismatic cells, have been used to produce lightweight structural components (Gibson & Ashby, 1997). Techniques now exist for foaming not only polymers, but also metals, ceramics and glasses, which are increasingly used to absorb the kinetic energy resulting from impacts.

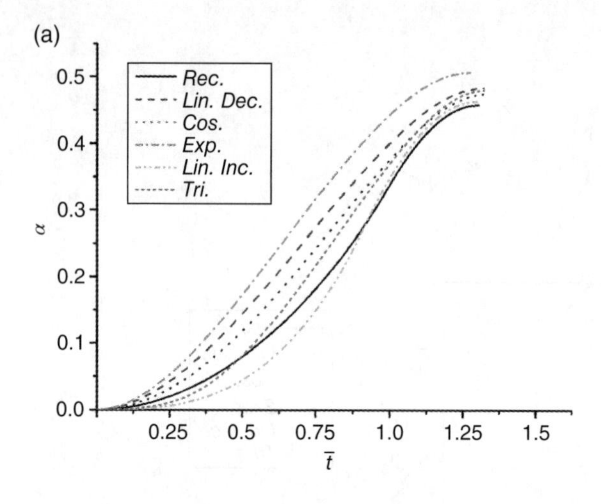

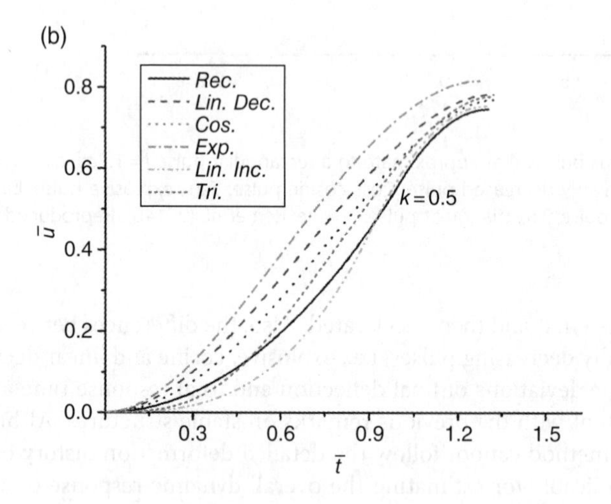

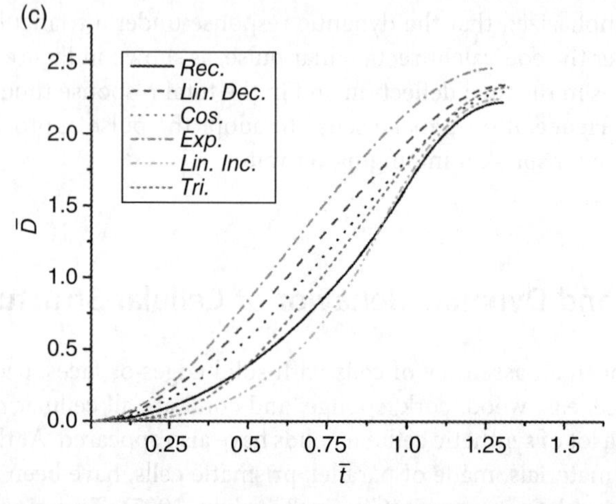

Figure 10.7 The dynamic response of the tensor skin (k = 0.5) under pulses approximated by the reference rectangular pulse $\bar{I} = 1.3\bar{q}^0{}_s$, $\bar{t}_{de} = 1$. (a) Angle; (b) normalized central deflection; (c) normalized energy dissipation; (d) normalized kinetic energy. Rec., rectangular pulse; Lin. Dec., linearly decreased pulse; Cos., cosine pulse; Exp., explosive pulse; Lin. Inc., linearly increased pulse; Tri., triangular pulse. *Source*: Ren *et al.* (2014b). Reproduced with permission of Elsevier.

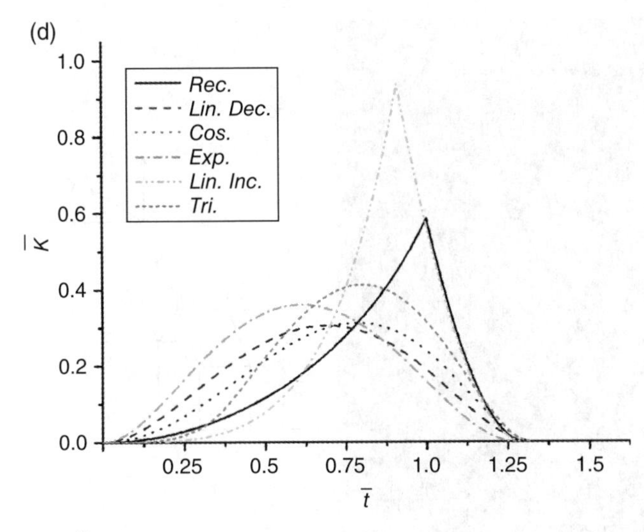

Figure 10.7 (Continued)

As a typical cellular material, honeycomb structures are widely adopted as energy absorbers, such as the core of sandwich structures. As the microstructures of honeycomb are essentially two-dimensional and regular, they are easier to analyze than foams, which have three-dimensional cell structures. In a narrow sense, honeycomb usually refers to the traditional hexagonal honeycomb, but the definition can also be generalized to prismatic cells with other shapes, and these are also called lattice materials. In this section, the equivalent stress–strain curves of generalized honeycomb and their dynamic behavior are illustrated.

10.2.1 Static Response of Hexagonal Honeycomb

As shown in Figure 10.8, a typical hexagonal honeycomb consists of a series of hexagonal cells whose dimensions are defined by cell wall lengths l and c, an angle between two cell walls of θ and a cell wall thickness h. Deformation caused by loading in the global plane of the honeycomb X_1X_2, is known as in-plane response, while that caused by loading in the X_3 direction is an out-of-plane response. Here, only the in-plane response will be discussed.

One of the most important parameters is *relative density* which is defined as $\bar{\rho} \equiv \rho / \rho_s$, where ρ is the overall (nominal) density of the cellular material, ρ_s is the density of the solid material. For the hexagonal honeycomb shown in Figure 10.8, where $h << l$

$$\bar{\rho} \equiv \frac{\rho}{\rho_s} = C_1 \frac{h}{l} \tag{10.13}$$

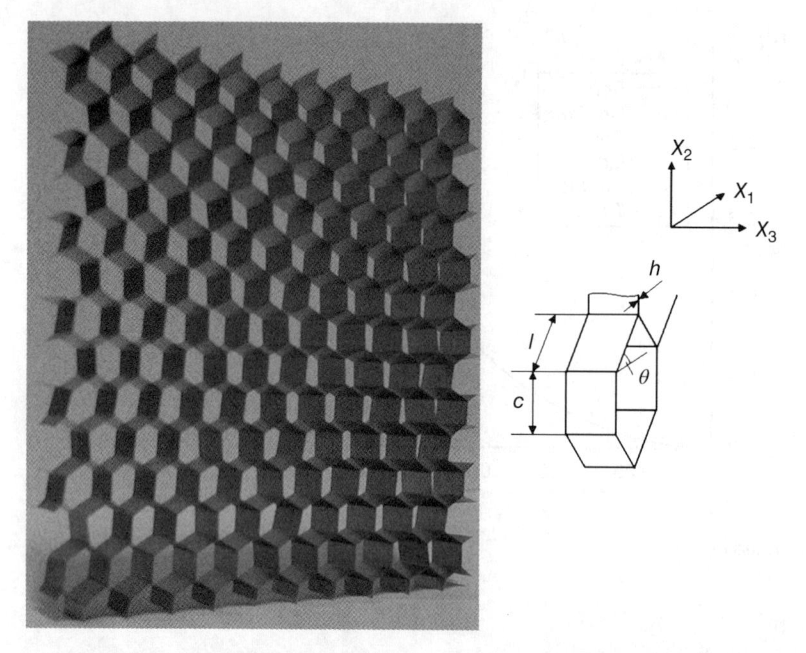

Figure 10.8 A piece of hexagonal honeycomb and its representative cell. *Source*: Lu and Yu (2003) Reproduced with permission of Elsevier.

where C_1 is a numerical constant and is dependent upon the details of cell geometry (Gibson & Ashby, 1997). For regular hexagonal cells, $l = c$ and $\theta = 30°$, and hence

$$\bar{\rho} = \frac{2}{\sqrt{3}}\frac{h}{l} \tag{10.14}$$

Typical stress–strain curves for hexagonal honeycombs under uniaxial compression in either the X_1 or X_2 direction are sketched in Figure 10.9.

The deformation at the cell level is reflected as strain at the macro-level. Hence, the structural response of hexagonal honeycomb cells indicates the global stress–strain curve of the equivalent material. First, the cell walls bend elastically with small deflections in the linear elastic stage. Then, the second stage is governed by one of three possible failure mechanisms of cell walls, i.e., elastic buckling, plastic collapse, or brittle fracture. This third mechanisms is often accompanied by considerable fluctuations in the plateau stress.

The most relevant properties to energy absorption are *plateau stress* and *densification strain* (or *locking strain*), ε_D. Theoretically, the densification strain should be equal to the porosity $(1 - \bar{\rho})$, but in practice, it was found that $\varepsilon_D < (1 - \bar{\rho})$, as the cellular material cannot be fully densified to a piece of solid.

The plateau stress is governed by the cell failure mechanism. For small value of h/l, elastic buckling of cell walls occurs. In this case the vertical walls behave very much like columns under compression loading. For equivalent external stress σ_2, the corresponding column force P is given by vertical force equilibrium:

$$P = 2\sigma_2 bl\cos\theta \tag{10.15}$$

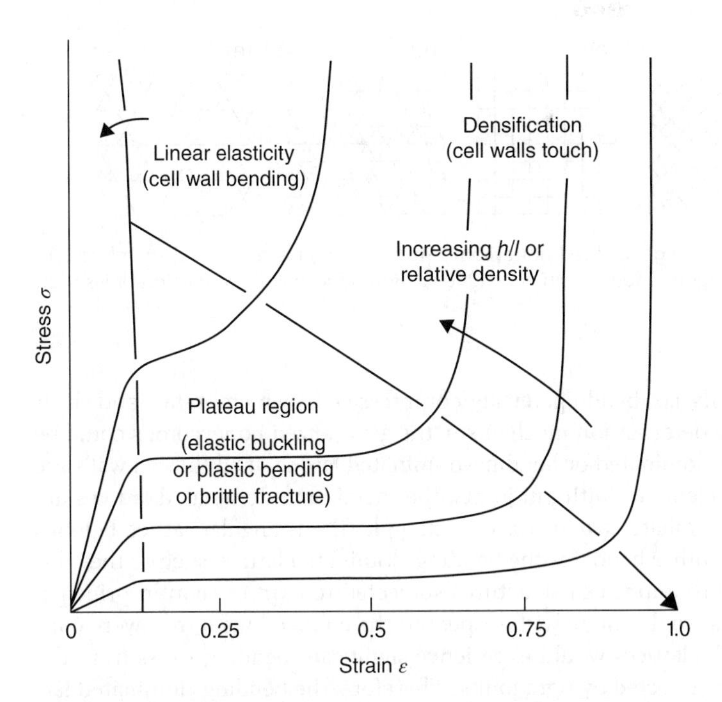

Figure 10.9 A typical stress–strain curve for a hexagonal honeycomb under in-plane loading. *Source*: Lu and Yu (2003). Reproduced with permission of Elsevier.

where b is the breadth of a cell. Gibson and Ashby (1997) gave the elastic buckling stress in the X_2 direction for regular hexagonal honeycomb (l=c and $\theta = 30°$):

$$\frac{\sigma_{e2}}{E_s} = 0.22\left(\frac{h}{l}\right)^3 \tag{10.16}$$

This indicates that the non-dimensional critical bucking stress proportional to h/l or $\bar{\rho}$ to the power of 3.

For hexagonal honeycomb with relative thick cell walls, plastic collapse of these walls will be the mechanism governing the plateau stress. Employing the collapse mechanism including six plastic hinges, Gibson and Ashby (1997) gave the plastic collapse for regular hexagonal honeycomb as

$$\frac{\sigma_p}{Y_s} = \frac{2}{3}\left(\frac{h}{l}\right)^2 \tag{10.17}$$

where Y_s is the yield stress of the solid.

10.2.2 Static Response of Generalized Honeycombs

In addition to the traditional hexagonal honeycomb, the generalized honeycombs have other cross-sectional topologies, which are also called the lattice materials. Five kinds of periodic planar lattices are shown in Figure 10.10: hexagonal, rhombus, square, triangular and Kagome lattices. Under different loading conditions, three types of stress states

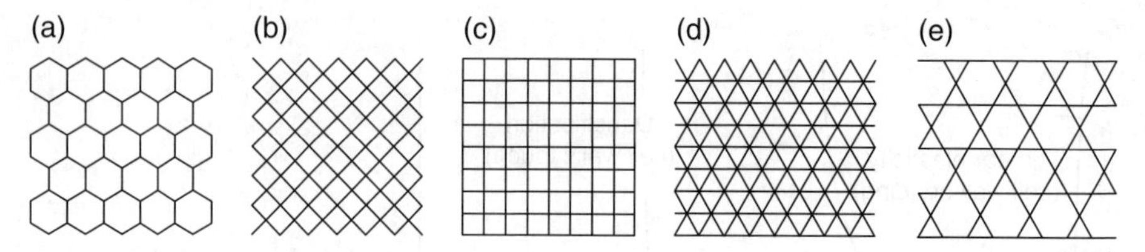

Figure 10.10 Schemes of some generalized honeycomb structures. (a) Hexagonal; (b) rhombus; (c) square; (d) triangular; (e) Kagome. *Source*: Qiu *et al.* (2009a). Reproduced with permission of Elsevier.

may exist in the cell walls, i.e., bending, tensile/compressive (i.e., membrane), and shear stress. According to the deformation mechanism, the generalized honeycombs could be classified as membrane-dominated or bending-dominated lattice. If all the cell walls are replaced by rigid truss elements with pin joints, the membrane-dominated lattices are usually statically indeterminate structures. For example, the triangular lattice belongs to this category. On the other hand, for the bending-dominated lattice, such as the conventional hexagonal or rhombus cell structures subjected to compression in principal directions, the cell walls would rotate with respect to the connections if they were connected by pin joints. The lattices would experience significant bending stress from the joints if the walls were connected by rigid joints. Therefore, the bending-dominated lattices are much more compliant and have lower collapse strength. The Kagome lattice is an intermediate case with a connectivity of 4, and thus the perfect Kagome deformed as membrane-dominated under uniform macroscopic stress states, but missing bars or misplaced nodes lead to a bending response near the imperfection. Considering the deformation mechanism, a bending-dominated structure is usually simplified as an assembly of beams, while a membrane-dominated structure is considered as an assembly of trusses (Qiu *et al.*, 2009a).

A representative unit cell is usually adopted in the analysis of equivalent properties. Table 10.1 lists the in-plane stiffnesses and yield strengths of periodic honeycombs as functions of relative density $\bar{\rho}$, supposing the base material has density ρ_s, Young's modulus E_s, Poisson ratio ν_s, and yield strength σ_{ys}. For honeycombs with $0.1 \le \bar{\rho} \le 0.3$, the initial yield associated with short column compression or bending occurs prior to elastic buckling. Table 10.1 clearly shows that the equivalent stiffness and collapse strength of

Table 10.1 The equivalent material properties of some planar lattices

	Hexagonal	Rhombus	Square	Triangular	Kagome
$\bar{\rho}$	$2h/(\sqrt{3}l)$	$2h/l$	$2h/l$	$2\sqrt{3}h/l$	$\sqrt{3}h/l$
E/E_s	$3\bar{\rho}^3/2$	$\bar{\rho}^3/4$	$\bar{\rho}/2$	$\bar{\rho}/3$	$\bar{\rho}/3$
σ_1^y/σ_{ys}	$\bar{\rho}^2/2$	$\bar{\rho}^2/4$	$\bar{\rho}/2$	$\bar{\rho}/3$	$\bar{\rho}/3$
σ_2^y/σ_{ys}	$\bar{\rho}^2/2$	$\bar{\rho}^2/4$	$\bar{\rho}/2$	$\bar{\rho}/2$	$\bar{\rho}/2$

Source: Qiu *et al.* (2009a). Reproduced with permission of Elsevier.

membrane-dominated structures are both higher than those of bending-dominated structures of the same relative density.

In Table 10.1, the normalized yield strengths σ_1^y/σ_{ys} and σ_2^y/σ_{ys} are derived from the yielding of the base material under the statically admissible stress field. Therefore, they are the lower bound of the initial collapse strength. For the elastic-perfectly plastic base material, the relation between equivalent stress σ and strain ε is given from the static analysis:

$$\sigma^{l}(\varepsilon)/\sigma_{ys} = \min\left(E\varepsilon/\sigma_{ys}, \sigma^{y}/\sigma_{ys}\right) \tag{10.18}$$

It should be emphasized that the above relation is accurate for the small deformation case.

By contrast, the upper bound of the initial collapse strength should be given by the kinematically admissible field. The collapse mechanisms of lattices under the X_2 uniaxial compression are sketched in Figure 10.11. By considering the elastic deformation at the small deformation stage, and also the large deformation after the initial collapse, the post-collapse response of representative block of the lattice was given by Qiu *et al.* (2009a).

The post-collapse equivalent stress will drop after the initial collapses of the representative block. As sketched in Figure 10.12, the curve of OAB represents the equivalent stress–stain relation $\sigma^{l}(\varepsilon)$, while the curve of CAD represents the stress $\sigma^{u}(\varepsilon)$ in the

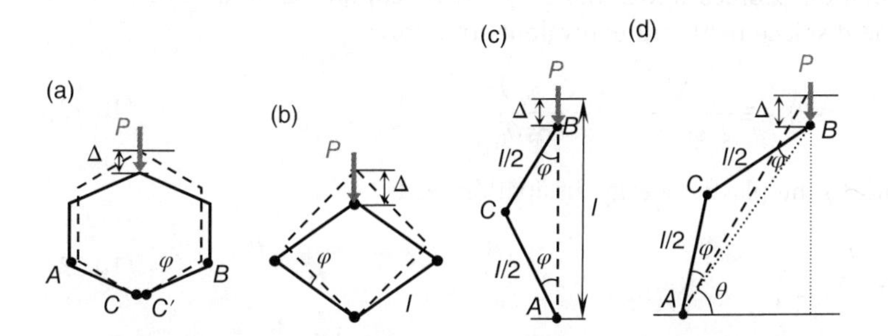

Figure 10.11 Collapse mechanisms of lattices under the X_2 uniaxial compression. (a) Hexagonal; (b) rhombus; (c) square; (d) triangular and Kagome. *Source*: Qiu *et al.* (2009a). Reproduced with permission of Elsevier.

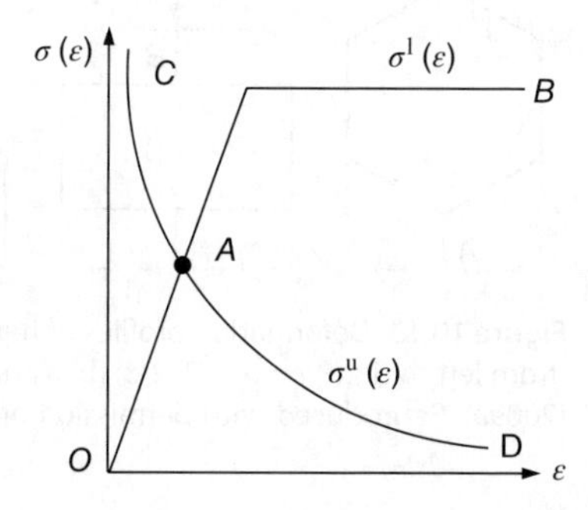

Figure 10.12 Equivalent compression stress versus the equivalent strain. *Source*: Qiu *et al.* (2009a). Reproduced with permission of Elsevier.

post-collapse stage. The lower branches of the two curves, *OAD*, indicate the actual stress–stain relation,

$$\sigma(\varepsilon) = \min\left[\sigma^{l}(\varepsilon), \sigma^{u}(\varepsilon)\right] \tag{10.19}$$

Here the intersection of the two curves, as marked by *A* in Figure 10.12, provides the initial collapse load. Then, based on the analysis of collapse mechanisms, the load of each cell is obtained by considering the large deformation effect.

Here the hexagonal lattice is still adopted as an example of a bending-dominated lattice. As plotted in Figure 10.13(a), the dotted-line square that envelops the hexagonal cell is regarded as the equivalent material unit. From the deformation of this dotted-line square, the equivalent stress–strain curve of hexagonal lattice can be obtained. Under the X_2 compression, the distance between points *A* and *B* is a function of the angle θ:

$$\Delta = 2l(\sin\theta_0 - \sin\theta),\ \varepsilon = \frac{\sin\theta_0 - \sin\theta}{1 + \sin\theta_0} \tag{10.20}$$

where $\theta_0 = \pi/6$ is the initial value of θ. Equating the work done by the force to the plastic energy dissipation required by this collapse mechanism leads to $8M_p\ d\theta = P\ d\Delta$, that is,

$$\frac{\sigma_2^{u}(\varepsilon)}{\sigma_{ys}} = \frac{\bar{\rho}^2}{2\sqrt{1 + 2\varepsilon - 3\varepsilon^2}},\ \text{for}\ \varepsilon \le 2/3 \tag{10.21}$$

The finite deformation profiles under the X_1 uniaxial compression are sketched in Figure 10.13(b). The displacement and equivalent strain are

$$\Delta = 2l(\cos\theta_0 - \cos\theta),\ \varepsilon \equiv \frac{\Delta}{2l\cos\theta_0} = 1 - \frac{\cos\theta}{\cos\theta_0} \tag{10.22}$$

Equating the work to the plastic energy dissipation leads to

$$\frac{\sigma_1^{u}(\varepsilon)}{\sigma_{ys}} = \frac{\bar{\rho}^2}{2\sqrt{1 + 6\varepsilon - 3\varepsilon^2}} \tag{10.23}$$

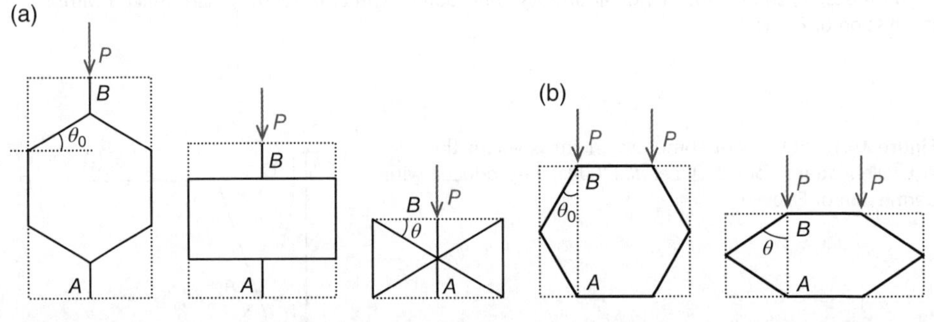

Figure 10.13 Deformation profiles of hexagonal under compression. (a) X_2 direction: equivalent strains (from left to right) $\varepsilon = 0$, 1/3, 2/3; (b) X_1 direction: strains (from left to right) $\varepsilon = 0$, $\varepsilon > 0$. *Source*: Qiu *et al.* (2009a). Reproduced with permission of Elsevier.

The square lattice is adopted as an example of the membrane-dominated lattice. By referring to Figure 10.11(c), the equivalent compression strain is found as $\varepsilon = (1 - \cos\varphi)$. Considering the plastic work done due to the rotations at three plastic hinges:

$$4M\varphi = P\Delta \tag{10.24}$$

As the cell wall is now subjected to both bending and compression, the interaction between bending moment and axial force should be taken into account. For the cell walls of rectangular section, the bending moment will be reduced by the compression force

$$M = M_{\mathrm{p}}\left[1 - (P/N_{\mathrm{p}})^2\right] \tag{10.25}$$

where $N_{\mathrm{p}} = \sigma_{\mathrm{ys}}bh$ is the fully plastic membrane force. Defining $\lambda(\varepsilon) = [\arccos(1-\varepsilon)]/\varepsilon$ gives

$$\lambda\bar{\rho}^2\bar{P}^2 + 16\bar{P} - 64\lambda = 0 \tag{10.26}$$

where $\bar{P} = Pl/M_{\mathrm{p}}$ is denoted as non-dimensional load. The equivalent stress–strain relation of the square lattice under uniaxial compression is obtained by solving Eq. (10.26):

$$\frac{\sigma^{\mathrm{u}}(\varepsilon)}{\sigma_{\mathrm{ys}}} = \frac{1}{16}\bar{\rho}^2\bar{P}(\varepsilon) = \frac{1}{2}\frac{1}{\lambda(\varepsilon)}\left(\sqrt{1 + \lambda(\varepsilon)^2\bar{\rho}^2} - 1\right) \tag{10.27}$$

Considering both the static approach and the collapse mechanism, the membrane-dominated structures are found typically to have a stress–strain curve with an initial peak followed by a rapid drop. This character is similar to the type II energy absorption structures, as described by Calladine and English (1984).

Calladine and English (1984) pointed out that most impact energy-absorbing structures can be classified into one of two types in terms of the shape of the overall quasi-static load-displacement curve in the early stages of deformation. Type I has a relatively "flat-topped" curve, and type II has an initial peak load followed by a "steeply falling" curve (see Figure 10.14a). Figures 10.14(b) and (c) show examples of typical type

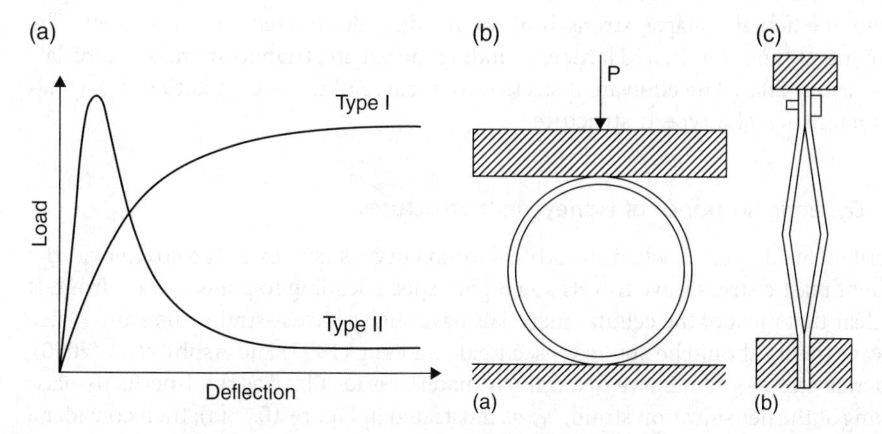

Figure 10.14 Two types of structure. (a) Response curves; (b) type I: a circular ring; (c) type II: pre-bent plates. *Source*: Gao *et al.* (2005). Reproduced with permission of Elsevier.

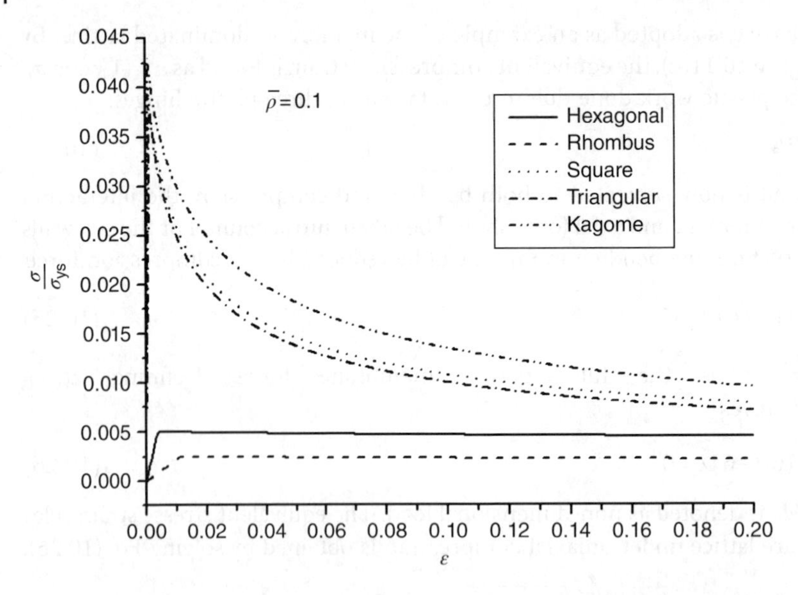

Figure 10.15 The equivalent stress–strain curves of some lattices under compression in small strain stage. *Source*: Qiu *et al.* (2009a). Reproduced with permission of Elsevier.

I and type II structures, respectively: a circular ring under compression is a type I structure, as its force–displacement curve is a "flat-topped" curve, and a pair of pre-bent plates under compression is a type II structure.

In Figure 10.15, the equivalent stress–strain curves of various lattices with the same relative density $\bar{\rho} = 0.1$, under compression in a small strain range, are compared with each other. The behavior of bending-dominated lattices is found to be similar to type I structures. From the two examples of bending-dominated structures, the "plateau stress", which is close to both upper and lower bounds of the initial collapse strength, represents the actual collapse strength of the bending-dominated structure itself. The curves of membrane-dominated lattice, including the square, triangular and Kagome lattices, are also similar. The equivalent stress–strain curve of the square lattice clearly has the characteristics of a type II structure.

10.2.3 Dynamic Response of Honeycomb Structures

As described in Chapter 2, when the stress–strain curve is convex to the strain axis, the subsequent plastic stress wave travels at a higher speed, leading to a shock wave front. It is now clear that most of the cellular materials have such a stress–strain curve and hence shock wave theory should be applied – see Reid and Peng (1997) and Ashby *et al.* (2000).

The actual stress–strain curve of a cellular material is idealized as rigid-perfectly plastic, locking at the densification strain, ε_D, as illustrated in Figure 10.16(a); then consider a mass G with velocity v_0 impinging an initially stationary cylindrical bar made of this material (see Figure 10.16b). A plastic wave front develops, traveling at velocity c_p towards the right. Ahead of this wave front, the material is stationary with stress σ_p, while

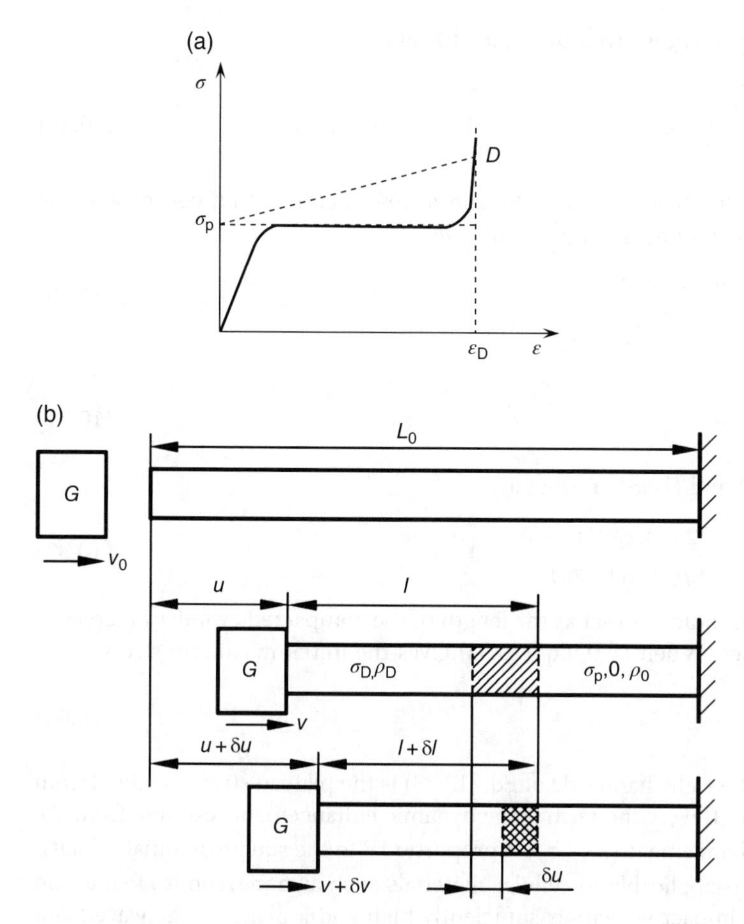

Figure 10.16 Rigid-perfectly plastic shock theory. (a) A typical stress–strain curve for cellular material with rigid-perfectly plastic idealization. (b) Sketch of a mass G impacting an initially stationary cylinder. *Source*: Lu and Yu (2003). Reproduced with permission of Elsevier.

behind the wave front, the material is compacted at the densification strain ε_D with a density $\rho_D = \rho_0/(1-\varepsilon_D)$. The material also moves at the same velocity as the instantaneous velocity of the mass G, which decreases with time. The stress level for the densified material at the wave front jumps to σ_D, which varies with the instantaneous velocity v_D, as will be seen later. Consider the instant when the current length of the compacted cylinder is l. The corresponding initial length is $l_0 = l/(1-\varepsilon_D)$. The plastic work done in compressing this length is $\sigma_p \varepsilon_D A l/(1-\varepsilon_D)$. Here A is the cross-sectional area, which is assumed to be constant. Consideration of energy balance gives

$$\frac{1}{2}\left(G + \frac{\rho_0}{1-\varepsilon_D} A l\right) v_D^2 + \sigma_p \varepsilon_D A \frac{l}{1-\varepsilon_D} = \frac{1}{2} G v_0^2 \tag{10.28}$$

The plastic wave speed c_p is given by Figure 10.16(a):

$$c_\mathrm{p} = \sqrt{\frac{(\sigma_\mathrm{D} - \sigma_\mathrm{p})/\varepsilon_\mathrm{D}}{\rho_0}} \tag{10.29}$$

And the particle velocity is $v_\mathrm{D} = c_\mathrm{p}\varepsilon_\mathrm{D}$. Within a time increment δt, conservation of momentum for an element at the wave front gives

$$(\sigma_\mathrm{D} - \sigma_\mathrm{p})A\ \delta t = \frac{\rho_0 A c_\mathrm{p}\ \delta t\ v_\mathrm{D}}{1 - \varepsilon_\mathrm{D}} \tag{10.30}$$

Hence

$$\sigma_\mathrm{D} = \sigma_\mathrm{p} + \frac{\rho_0 c_\mathrm{p} v_\mathrm{D}}{1 - \varepsilon_\mathrm{D}} \tag{10.31}$$

Solving Eqs (10.28) and (10.31) results in

$$\sigma_\mathrm{D} = \sigma_\mathrm{p} + \frac{\rho_0}{\varepsilon_\mathrm{D}} \frac{G v_0^2 - 2\sigma_\mathrm{p} A l \varepsilon_\mathrm{D}/(1 - \varepsilon_\mathrm{D})}{G + \rho_0 A l/(1 - \varepsilon_\mathrm{D})} \tag{10.32}$$

The above equation indicates that as the length of the compacted cylinder l increases, the stress σ_D decreases. When $l = 0$, Eq. (10.32) gives the initial maximum stress:

$$\sigma_\mathrm{D} = \sigma_\mathrm{p} + \frac{\rho_0 v_0^2}{\varepsilon_\mathrm{D}} \tag{10.33}$$

The first term on the right-hand side of Eq. (10.33) is the plateau stress obtained from the static analysis, and the second term is the dynamic enhancement resulting from the inertia effect. The enhancement $(\sigma_\mathrm{D} - \sigma_\mathrm{p})$ is proportional to the square of initial velocity. Plastic shock theory is applicable to cellular materials such as honeycombs, foams and woods, provided the impact velocity is sufficiently high and a plane plastic wave front develops.

For example, a plastic wave front will be created when a hexagonal honeycomb specimen is subjected to impact with high speed, while deformation localization will be produced in the case of lower speed impact. In the finite element simulations given by Ruan *et al.* (2003), the hexagonal honeycomb specimen is compressed by a rigid plate at different velocities. It was found that the deformation modes vary with the impact velocity. Under a lower velocity impact (see Figure 10.17a), the "V"-shaped initial localized bands occur firstly at the impact end and then at the remote fixed end, until the cell walls contacted each other which enhanced stiffness. In a higher impact velocity case (see Figure 10.17b), no obvious "V"-shaped deformation is observed. The localized band displays a transition from the "V" shape to the "I" shape, and the cells are crushed row by row in a manner similar to the plastic wave propagation.

As seen, the energy absorption is proportional to the plateau stress. From the finite element simulations of hexagonal honeycomb under impact loading, an empirical equation of plateau stress was proposed by Ruan *et al.* (2003) as

$$\frac{\sigma_\mathrm{D}}{Y_\mathrm{s}} = 0.8\left(\frac{h}{l}\right)^2 + \left[62\left(\frac{h}{l}\right)^2 + 41\left(\frac{h}{l}\right) + 0.01\right] \times 10^{-6} v^2 \tag{10.34}$$

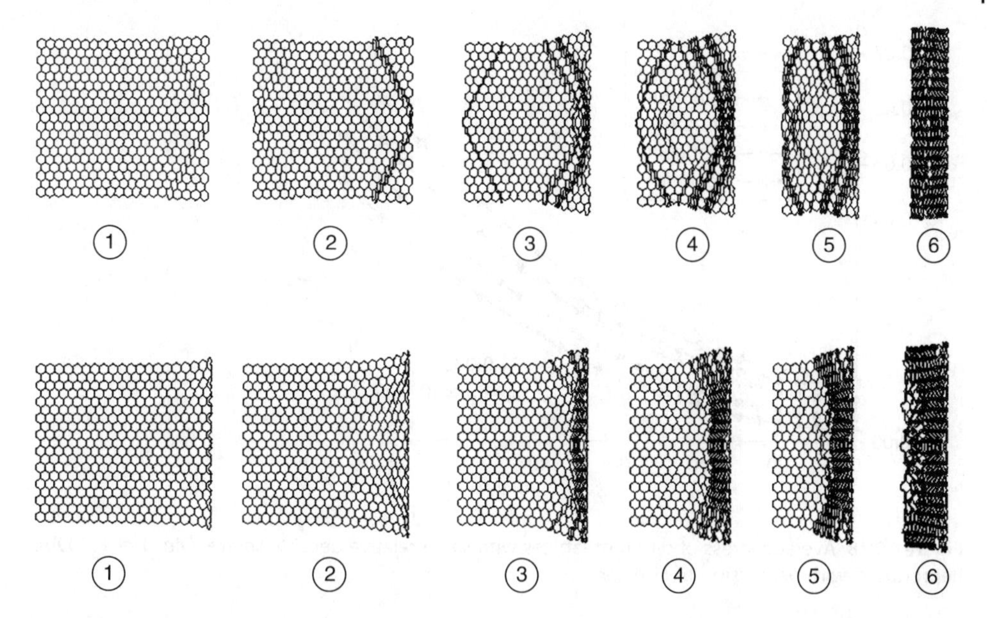

Figure 10.17 A hexagonal lattice under dynamic compression. (a) Velocity 3.5 m/s; (b) velocity 28 m/s. ① $\varepsilon = 0.05$, ② $\varepsilon = 0.11$, ③ $\varepsilon = 0.30$, ④ $\varepsilon = 0.46$, ⑤ $\varepsilon = 0.57$, ⑥ $\varepsilon = 0.80$. *Source*: Qiu et al. (2009b). Reproduced with permission of Elsevier.

Clearly, the dynamic plateau stress depends on the thickness to length ratio h/l, and the impact velocity v.

For the generalized honeycombs (or lattices) shown in Figure 10.10, Qiu *et al.* (2009b) investigated their dynamic crushing behaviors by finite element simulations. Under different impact velocities, the deformation modes and the average stress of these generalized honeycombs of different relative densities were studied. It is found that the average stress increases with the impact velocity, the density of the base material, and the relative density of the lattice. As shown in Figure 10.18, the dynamic average stresses of different honeycombs with the same relative density all increase with the impact velocity.

Consequently, by employing the least-squares fitting of the average stress, two simplified expressions of the dynamic average stress were given by Qiu *et al.* (2009b):

$$\frac{\sigma_D}{Y_s} = A_1\bar{\rho}^2 + B_1\bar{\rho}\frac{\rho_s v^2}{Y_s} \tag{10.35}$$

and

$$\frac{\sigma_D}{Y_s} = A\bar{\rho}^2 + \frac{\bar{\rho}}{1-B\bar{\rho}}\frac{\rho_s v^2}{Y_s} \tag{10.36}$$

where A, B, A_1, and B_1 are fitting parameters. The physical meaning of the term $(1-B\bar{\rho})$ in Eq. (10.36) is the densification strain, $\varepsilon_D = 1-\lambda\bar{\rho}$.

In both the expressions of Eqs (10.35) and (10.36), the dynamic average stress σ_D is increased with the relative density $\bar{\rho}$ and impact velocity v. Further, Eqs (10.35) and (10.36) are both similar to the equation given by shock wave theory, i.e., Eq. (10.33). The first term of the right-hand side of Eqs (10.35) and (10.36) represents the average

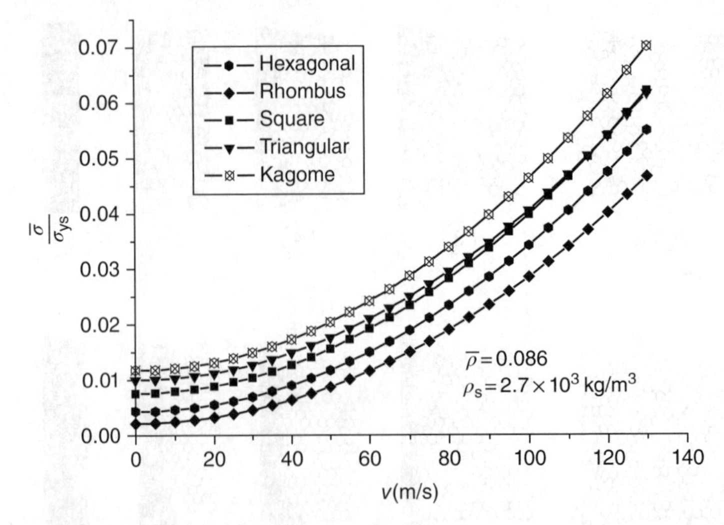

Figure 10.18 Average stress of different lattices with same relative density. *Source*: Qiu *et al.* (2009b). Reproduced with permission of Elsevier.

static stress, $\bar{\sigma}_p = A\bar{\rho}^2$, being in the same order of the yield strength of a bending-dominated lattice. The fitting parameters are $A_1 = 0.575$, $A_1 = 0.2875$, $A = 1.02$, $A = 1.35$, and $A = 1.59$, for the hexagonal, rhombus, square, triangular, and Kagome lattices, respectively. These values illustrate that the actual collapse stress of membrane-dominated honeycomb is not as high as in the order of $\bar{\rho}$, although its static collapse stress $\bar{\sigma}_p = A\bar{\rho}^2$ is higher than those of the bending-dominated honeycombs, which can be verified by the curves plotted in Figure 10.15.

Equation (10.35) is employed to compare the contribution of the inertia effect of each honeycomb. The second term of the right-hand side of equations represents the dynamic enhancement due to the inertia effect σ_d. In Eq. (10.35) where $\sigma_d = B_1\bar{\rho}(\rho_s v^2)$, the fitting parameters are $B_1 = 0.98, 0.862, 1.056, 1.00,$ and 1.13, for the hexagonal, rhombus, square, triangular, and Kagome lattices, respectively. Evidently, all the values of B_1 are around 1.0. The magnitudes of B_1 of the membrane-dominated honeycombs are only slightly larger than those of the bending-dominated ones.

In summary, the average dynamic stress can be expressed as the sum of the average static stress and the dynamic enhancement caused by the inertia effect. The average dynamic stress for membrane-dominated honeycombs is higher than that of bending-dominated honeycombs, but the difference in the dynamic enhancement is not as significant as that in the static collapse stress.

Qiu *et al.* (2009b) gave an institutive explanation of the dynamic enhancement stress. A representative cell of each lattice under impact in the X_2 direction is sketched in Figure 10.19. The dashed-line square represents a cell of height H and width W. A block of lattice could be generated by repeating the representative cell at regular intervals in both the horizontal and vertical directions. Under the uniform acceleration assumption, the total mass of the unit cell, m, is accelerated from a stationary state to a moving state with velocity v; hereafter the relation between the acceleration a and the average moving

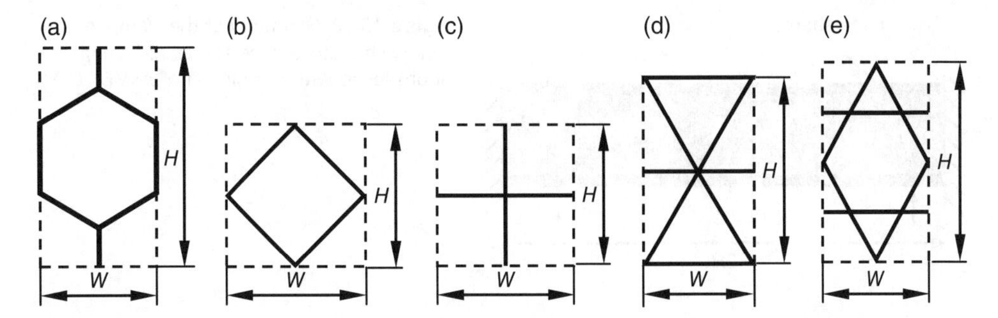

Figure 10.19 The representative cell of the generalized honeycombs, (a) Hexagonal; (b) rhombus; (c) square; (d) triangular; (e) Kagome. *Source*: Qiu *et al.* (2009b). Reproduced with permission of Elsevier.

distance $H/2$ is given by $v^2 = aH$. From Newton's second law, the reaction force F of this unit cell is

$$F = ma = \rho_s Sbv^2/H \tag{10.37}$$

where S is the area filled by the base material, and b is the cell dimension in the X_3 direction. Denote $S_0 = HW$ as the nominal area covered by the unit cell. By considering the definition of the relative density, $\bar{\rho} = \rho/\rho_s = S/S_0$, the dynamic enhancement stress caused by the inertia effect is

$$\sigma_d = F/Wb = \rho_s Sv^2/S_0 = \bar{\rho}\rho_s v^2 \tag{10.38}$$

Hence, Eq. (10.38) gives an estimation of the dynamic enhancement stress during unidirectional crushing. It is found that the dynamic enhancement stress only depends on the density, but not the configuration of the lattice. Eq. (10.38) is validated by the fitting parameters; $B_1 \approx 1$ is found for all different honeycombs. Furthermore, it should be noted that the simple model described here becomes more accurate in the case of high impact velocity, in which the unit cells are crushed row by row (corresponding to the "I" mode of the hexagonal and rhombus lattice). As the localized deformation of a membrane-dominated lattice is much more significant than a bending-dominated lattice, especially under low-velocity impact, the estimation of dynamic enhancement (Eq. 10.38) is more accurate for the membrane-dominated lattices.

In conclusion, the physical phenomenon "shock wave propagation" described at the macroscale, where the generalized honeycombs are usually equivalent to continuous media, is, in fact, the overall result of the collapse of the cell wall structures at the microscale. Therefore, the "strain rate effect" of the cellular material discussed at the macroscale is actually the same physical phenomenon of the "inertia effect" of the dynamic crushing.

10.3 Dynamic Response of a Clamped Circular Sandwich Plate Subject to Shock Loading

Sandwich structures are composed of two strong face-sheets and a weak core, as shown in Figure 10.20. Sandwich structures are widely adopted in the design of commercial and military vehicles owing to their high bending stiffness relative to mass. For

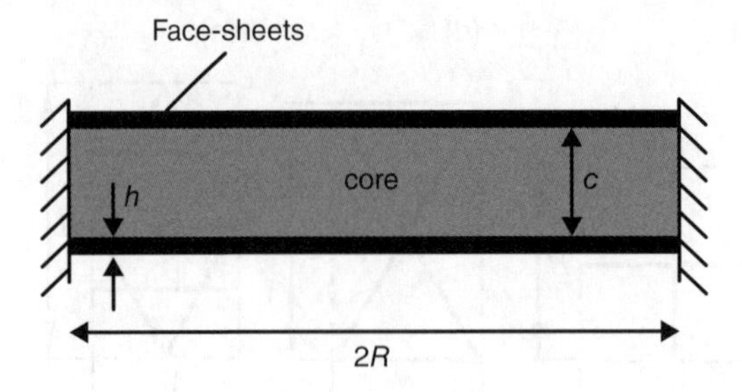

Figure 10.20 Geometry of the clamped sandwich plate. *Source*: Qiu *et al.* (2004). Reproduced with permission of ASME.

example, the outermost structure of a ship comprises plates welded to an array of stiffeners. The superior performance of sandwich plates relative to monolithic solid plates is well known for applications requiring high quasi-static strength. The resistance of sandwich plates to dynamic loads has also been investigated in order to quantify the advantages of sandwich design over monolithic design for application in shock-resistant structures. In the study of Qiu *et al.* (2004), an analytical model was developed for the deformation response of clamped circular sandwich plates subjected to shock loading in air and water.

10.3.1 An Analytical Model for the Shock Resistance of Clamped Sandwich Plates

Fleck and Deshpande (2004) developed an analytical model for the response of clamped sandwich beams subjected to shock loading. This model was extended to analyze the response of clamped axisymmetric sandwich plates to a spatially uniform air or underwater shock. Consider a clamped circular sandwich plate of radius R with identical face-sheets of thickness h and a core of thickness c, as shown in Figure 10.20. The face-sheets are made from a rigid-perfectly plastic solid of yield strength σ_{fy}, density ρ_f, and tensile failure strain ε_f. The core is taken to be a compressible isotropic solid of density ρ_c, and in uniaxial compression the core deforms at a constant strength σ_c with no lateral expansion up to a densification strain ε_D; beyond densification the core is treated as rigid. Fleck and Deshpande (2004) split the response of the sandwich structure into three sequential stages:

i) stage I – fluid–structure interaction phase;
ii) stage II – core compression phase;
iii) stage III – plate bending and stretching phase.

Stage I – The Initial Fluid–Structure Interaction Phase

The momentum is transmitted to the sandwich beam by treating the outer face of the sandwich beam as a free-standing plate. The pressure p at any point in the fluid of density ρ_w engulfed by the pressure wave traveling at a velocity c_w is taken to be (starting at time $t = 0$)

$$p = p_0 e^{-t/\theta} \tag{10.39}$$

where p_0 is the peak pressure and θ is the decay constant of the wave. When this pressure wave hits a stationary rigid plate at normal incidence it imparts an impulse I to the plate:

$$I = 2\int_0^{\infty} p_0 e^{-t/\theta} \mathrm{d}t = 2p_0\theta \tag{10.40}$$

The factor of 2 arises in Eq. (10.40) due to full reflection of the pressure wave.

If, instead, the pressure wave impinges a free-standing plate, the imparted impulse is less than I and can be estimated as follows. When the pressure wave strikes a free-standing plate of thickness h made from a material of density ρ_f, it sets the plate in motion and is partly reflected. At the instant the plate achieves its maximum velocity, the pressure at the interface between the plate and the fluid is zero and cavitation sets in shortly thereafter. The momentum per unit area I_{trans} transmitted into the structure is then given by

$$I_{\text{trans}} = \zeta I \tag{10.41}$$

where

$$\zeta \equiv \psi^{\psi(1-\psi)} \tag{10.42}$$

and $\psi \equiv \rho_w c_w \theta / (\rho_f h)$. It is assumed that this transmitted impulse imparts a uniform velocity $v_0 = I_{\text{trans}} / (\rho_f h)$ to the outer front face of the sandwich plate.

Stage II – Core Compression Phase

At the start of this phase, the outer face has a velocity v_0 while the core and inner face are stationary. The outer face compresses the core, while the core with compressive strength σ_c decelerates the outer face and simultaneously accelerates the inner face. The final common velocity of the faces and the core is dictated by momentum conservation, and the ratio ϕ of the energy lost U_{lost} in this phase to the initial kinetic energy $I^2\zeta^2/(2\rho_f h)$ of the outer face is given by

$$\phi \equiv \frac{U_{\text{lost}}}{I^2\zeta^2/(2\rho_f h)} = \frac{1+\overline{m}}{2+\overline{m}} \tag{10.43}$$

where $\overline{m} = \rho_c c / (\rho_f h)$ is the ratio of the mass of the core to the mass of a face-sheet. This energy lost is dissipated by plastic dissipation in compressing the core and thus the average through-thickness strain ε_c in the core is given by

$$\varepsilon_c = \frac{\bar{I}^2 \zeta^2}{2\bar{\sigma}\bar{c}^2 \bar{h}} \frac{\bar{h}+\bar{\rho}}{2\bar{h}+\bar{\rho}} \tag{10.44}$$

where $\bar{h} \equiv h/c$, $\bar{c} \equiv c/R$, $\bar{\rho} \equiv \rho_c/\rho_f$, $\bar{I} \equiv I / \left(R\sqrt{\sigma_{fy}\rho_f}\right)$, and $\bar{\sigma} \equiv \sigma_c/\sigma_{fy}$. However, if U_{lost} is too high such that ε_c as given by Eq. (10.44) exceeds the densification strain ε_D, then ε_c is set equal to ε_D and the model does not account explicitly for the additional dissipation mechanisms required to conserve energy. Rather it is assumed that inelastic impact of the outer face against the combined core and inner face leads to the additional dissipation. After the core has been compressed by a strain of ε_c, the core height is reduced to $(1-\varepsilon_c)c$. An approximate estimate of the time T_c for this second stage of motion (calculated by neglecting the mass of the core) is given by

$$\bar{T}_c \equiv \frac{T_c}{R\sqrt{\rho_f/\sigma_{fy}}} = \begin{cases} \dfrac{\bar{I}\zeta}{2\bar{\sigma}}, & \text{if } \bar{I}^2\zeta^2 < 4\bar{\sigma}\bar{c}^2\bar{h}\varepsilon_D \\ \dfrac{\bar{I}\zeta}{2\bar{\sigma}}\left(1-\sqrt{1-\dfrac{4\bar{\sigma}\bar{c}^2\bar{h}\varepsilon_D}{\bar{I}^2\zeta^2}}\right), & \text{otherwise} \end{cases} \tag{10.45}$$

This time T_c is typically small compared with the structural response time and thus the transverse deflection of the inner face of the sandwich plate in this stage can be neglected.

Stage III – Plate Bending and Stretching Phase

At the end of stage II, the sandwich plate has a uniform velocity except for a boundary layer near the supports. The plate is brought to rest by plastic bending and stretching. The problem under consideration is a classical one: the dynamic response of a clamped plate of radius R with an initial uniform transverse velocity v. First, stationary plastic hinges form at the supports while plastic hinges travel inwards from each clamped support. Then the moving hinges coalesce at the center of the plate, and continued rotation occurs about the central hinge until the plate is brought to rest. This dynamic response process is very similar to that analyzed for beams in Chapter 7. Note that the plastic bending moment M_0 of the circular sandwich plate is given by

$$M_0 = \sigma_c\frac{(1-\varepsilon_c)^2c^2}{4} + \sigma_{fy}h[(1-\varepsilon_c)c+h] \tag{10.46}$$

In the small deflection regime, the maximum central deflection w of the inner face of the sandwich plate and the structural response time T are given by

$$\bar{w} \equiv \frac{w}{R} = 0.28\frac{\bar{I}^2\zeta^2}{\bar{c}\hat{c}^2\alpha_1(2\bar{h}+\bar{\rho})} \tag{10.47}$$

and

$$\bar{T} \equiv \frac{T}{R}\sqrt{\frac{\sigma_{fy}}{\rho_f}} = 0.36\frac{\bar{I}\zeta}{\hat{c}^2\alpha_1} \tag{10.48}$$

where $\alpha_1 = (1+2\bar{h})^2 - 1 - \bar{\sigma}$ and $\alpha_2 = \sqrt{\frac{2\bar{h}+\bar{\rho}}{2\bar{h}+\bar{\sigma}}}$, with $\hat{c} \equiv \bar{c}(1-\varepsilon_c)$ and $\hat{h} \equiv h/(1-\varepsilon_c)$, respectively.

If the displacement is large, the coupling effect of bending moment and membrane force should be considered. The yield locus of an axisymmetric sandwich element subjected to a circumferential membrane force N_θ and a circumferential bending moment M_θ is well approximated by

$$\frac{M_\theta}{M_0} + \frac{N_\theta}{N_0} = 1 \tag{10.49}$$

where M_0 is the plastic bending moment specified by Eq. (10.46) and N_0 the circumferential plastic membrane force given by

$$N_0 = 2h\sigma_{fy} + (1-\varepsilon_c)c\sigma_c \tag{10.50}$$

Analytical formulas for the deflection and structural response time of the circular plate can be obtained by approximating the above yield locus by either inscribing or circumscribing squares as sketched in Figure 10.21. The maximum central deflection w of the inner face and structural response time T of a clamped circular sandwich plate are given by

$$\bar{w} \equiv \frac{\bar{c}\alpha_1}{2\hat{h}+\bar{\sigma}}\left(\sqrt{1+\frac{2}{3\bar{c}\hat{c}^3}\frac{\bar{I}^2\zeta^2}{\alpha_1^2\alpha_2^2}}-1\right) \tag{10.51}$$

and

$$\bar{T} = \alpha_2\sqrt{\frac{\bar{c}}{6\hat{c}}}\arctan\left(\sqrt{\frac{2}{3\bar{c}\hat{c}}}\frac{\bar{I}\zeta}{\hat{c}\alpha_1\alpha_2}\right) \tag{10.52}$$

respectively, for the choice of a circumscribing yield locus, and by

$$\bar{w} \equiv \frac{\bar{c}\alpha_1}{2\hat{h}+\bar{\sigma}}\left(\sqrt{1+\frac{4}{3\bar{c}\hat{c}^3}\frac{\bar{I}^2\zeta^2}{\alpha_1^2\alpha_2^2}}-1\right) \tag{10.53}$$

and

$$\bar{T} = \alpha_2\sqrt{\frac{\bar{c}}{3\hat{c}}}\arctan\left(\frac{2}{\sqrt{3\bar{c}\hat{c}}}\frac{\bar{I}\zeta}{\hat{c}\alpha_1\alpha_2}\right) \tag{10.54}$$

for an inscribing yield locus.

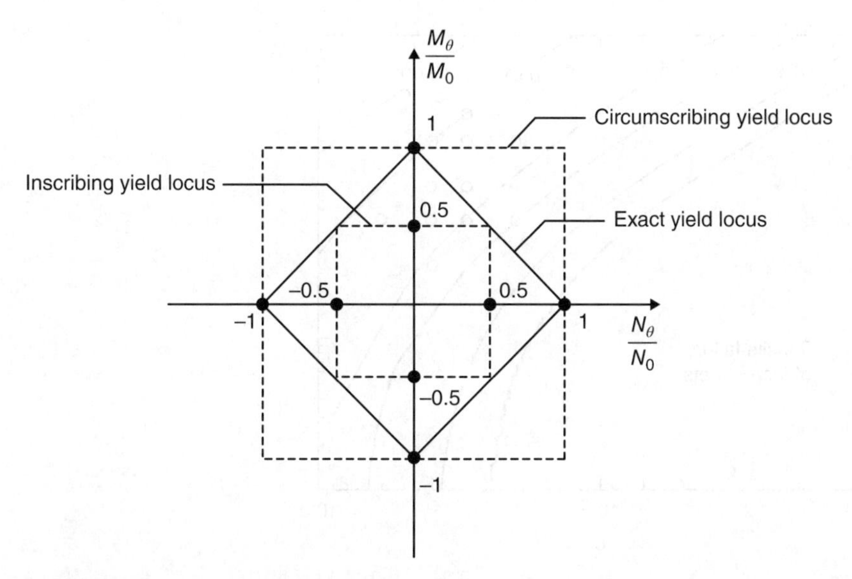

Figure 10.21 The exact, inscribing, and circumscribing yield loci of the sandwich plate. *Source*: Qiu *et al.* (2004). Reproduced with permission of ASME.

The circumferential tensile strain ε_m in the face-sheets due to stretching is approximately equal to

$$\varepsilon_m = \frac{1}{2}\bar{w}^2 \tag{10.55}$$

Neglecting the strains due to bending, an approximate failure criterion for the sandwich plates is given by setting the face-sheet tensile strain ε_m equal to the tensile ductility ε_f of the face-sheet material.

10.3.2 Comparison of Finite Element and Analytical Predictions

In order to validate the analytical model, finite element studies were conducted with the effects of fluid–structure interaction neglected. In the limit of no fluid–structure interaction ($\psi = 0$ and $\zeta = 1$), it is assumed that the entire shock impulse I is transferred uniformly to the outer face of the sandwich plate. In all the finite element calculations presented here, loading corresponding to a non-dimensional impulse $\bar{I}$ is specified by imparting an initial uniform velocity v_0 to the outer face-sheet of the sandwich plate:

$$v_0 = \frac{\bar{I}}{\bar{c}\bar{h}}\sqrt{\frac{\sigma_{fy}}{\rho_f}} \tag{10.56}$$

First, for a fixed impulse, the response of the sandwich plate is investigated as a function of the plate geometry and, secondly, the response of a sandwich plate with a representative geometry is studied for varying levels of impulse. The design chart for a plate with $\bar{\sigma} = 0.05$, $\varepsilon_D = 0.5$ and $\varepsilon_f = 0.2$ is shown in Figure 10.22. Contours of the maximum normalized central deflection $\bar{w}$ of the inner face-sheet subject to a normalized impulse

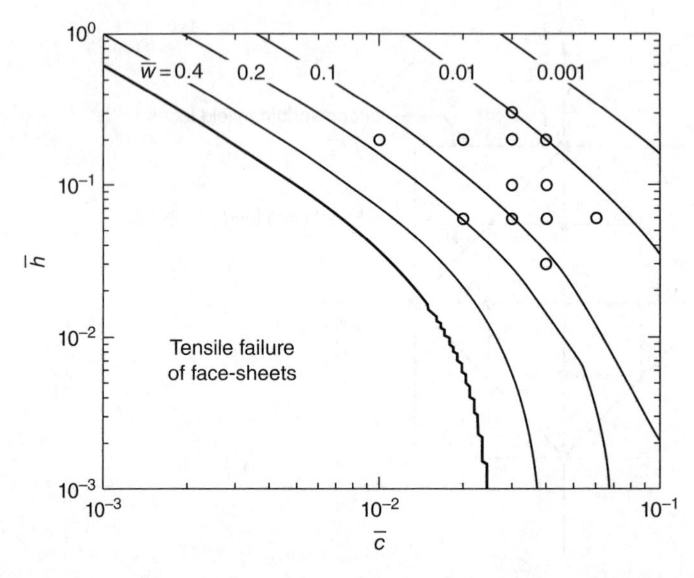

Figure 10.22 Design chart for a clamped sandwich plate with $\bar{\sigma} = 0.05$, $\varepsilon_D = 0.5$ and $\varepsilon_f = 0.2$ under $\bar{I} = 10^{-3}$. *Source*: Qiu *et al.* (2004). Reproduced with permission of ASME.

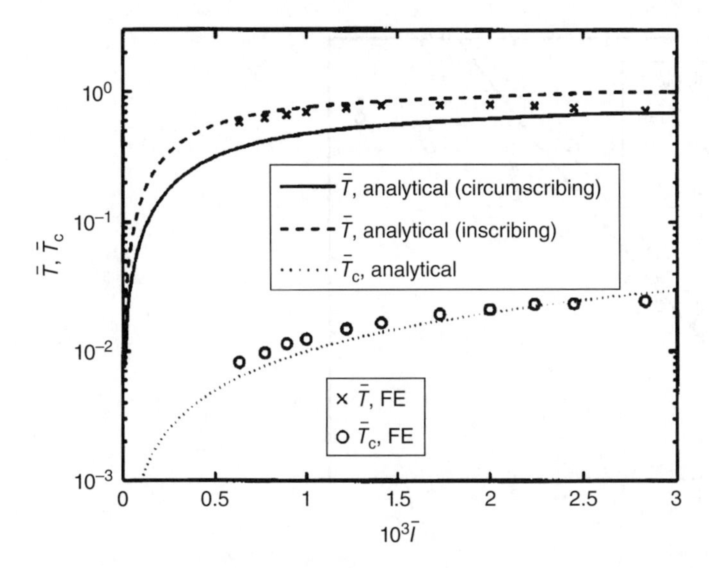

Figure 10.23 The structural response time $\bar{T}$ and core compression time $\bar{T}_c$ as a function of applied impulse $\bar{I}$ ($\bar{c} = 0.03$ and $\bar{h} = 0.1$). FE, finite element analysis. *Source*: Qiu *et al.* (2004). Reproduced with permission of ASME.

$\bar{I} = 10^{-3}$ are included. The symbols denote the sandwich plate geometries selected for the finite element calculations. The design chart was constructed using the analytical model with the circumscribing yield locus.

The normalized structural response time $\bar{T}$ and the core compression time $\bar{T}_c$ are shown in Figure 10.23 as functions of $\bar{I}$. Good agreement between the analytical and finite element predictions is seen for both $\bar{T}$ and $\bar{T}_c$. Further, $\bar{T}_c$ is at least an order of magnitude smaller than $\bar{T}$, which supports the timescale separation assumption in the analytical model.

It is noted that the finite element predictions of the maximum deflections of the clamped sandwich beams and plates as functions of the applied normalized impulse are approximately equal when the half-span L of the sandwich beam is equated to the radius R of the sandwich plate. The analytical predictions (employing the inscribing yield locus) of the deflections of the beams and plates are included in Figure 10.24. In line with the finite element predictions, the analytical predictions for the beams and plates are approximately equal.

10.3.3 Optimal Design of Sandwich Plates

The analytical formulas was employed to determine the optimal designs of sandwich plates that maximize the resistance of a sandwich plate of given mass to shock loading subject to the constraint of a maximum allowable inner face deflection. The optimization is conducted by assuming that the entire shock impulse is transmitted to the sandwich plate.

A design chart relating the sandwich plate geometry to the shock impulse for a specified deflection is given in Figure 10.25, with axes $\bar{c}$ and $\bar{h}$ for a normalized deflection

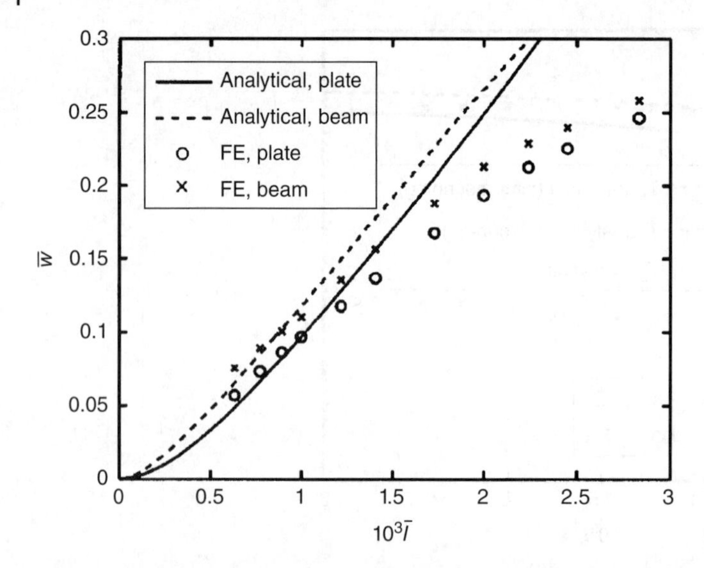

Figure 10.24 The maximum central deflections $\bar{w}$ for clamped sandwich plates and beams, as a function of applied impulse $\bar{I}$. ($\bar{c}=0.03$ and $\bar{h}=0.1$). FE, finite element analysis. *Source*: Qiu *et al.* (2004). Reproduced with permission of ASME.

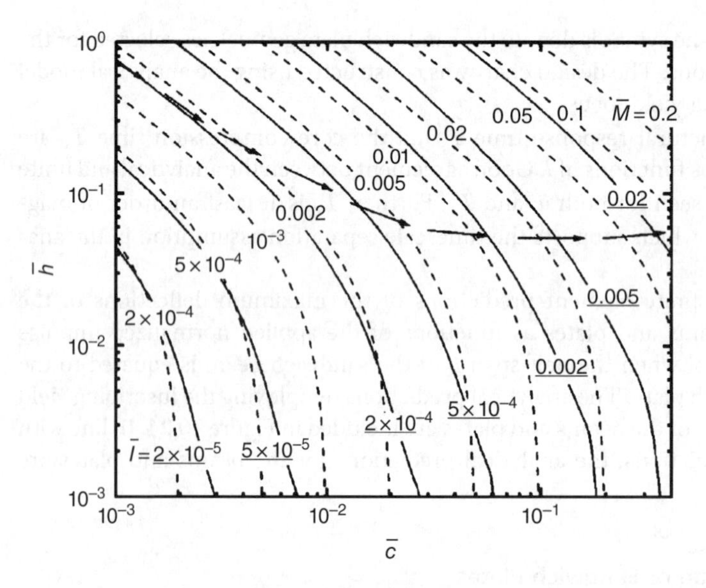

Figure 10.25 Design chart for a clamped sandwich plate with a fixed $\bar{w}=0.1$. *Source*: Qiu *et al.* (2004). Reproduced with permission of ASME.

$\bar{w}=0.1$, by employing the circumscribing yield locus analytical expressions. The chart is plotted for sandwich plates with reference materials properties, $\bar{\sigma}=0.05$ and $\bar{\rho}=0.1$.

Contours of the applied impulse $\bar{I}$ and non-dimensional mass $\bar{M}$ are displayed in Figure 10.25. Here

$$\bar{M} = \frac{M}{\pi R^3 \rho_f} = 2\bar{h}\bar{c} + \bar{c}\bar{\rho} \tag{10.57}$$

with M being the mass of the sandwich plate. The underlined values denote the non-dimensional impulse values while the arrows trace the path of the optimal designs with increasing mass.

10.4 Collision and Rebound of Circular Rings and Thin-Walled Spheres on Rigid Target

Collisions of moving deformable bodies on solid targets frequently occur in engineering, daily life, and sports. For example, a car crashes into a wall or a roadside guide rail, a helicopter accidentally drops to the ground, a spacecraft lands on the moon, a mobile phone falls to the floor, a tennis ball hits the playground and so forth. Here two typical cases will be studied.

10.4.1 Collision and Rebound of Circular Rings

When a deformable body/structure collides freely with a rigid/solid target or another deformable body/structure, the kinetic energy carried by the flying body is transformed into the internal energy of the deformable body during the compression phase, and then the stored elastic strain energy is progressively recovered into kinetic energy during the subsequent restitution phase through the contact force, which makes the free body/structure rebounding from the target have a lower velocity after the collision.

The coefficient of restitution (COR) plays a key role in measuring the global energy lost during the collision. COR is usually defined in one of the following three ways:

$$e_1 = \frac{V_r}{V_i} \tag{10.58}$$

$$e_2 = \frac{I_R}{I_C} = \frac{\int_{t_R} F dt}{\int_{t_C} F dt} \tag{10.59}$$

$$e_3 = \sqrt{\frac{E_r}{E_i}} \tag{10.60}$$

where e_1, e_2, and e_3 denote the Newtonian (kinematic), Poisson (kinetic), and energetic COR, respectively, V_r and V_i are the relative velocities of the colliding bodies after and before the collision, I_C and I_R are the impulses over the compression duration t_C and the restitution duration t_R, and E_r and E_i are the kinetic energies of the colliding bodies after and before the collision, respectively. Among these expressions, definition Eq. (10.58) is the most straightforward and widely applied one, and is therefore employed in the following analysis.

Model

A thin-walled free-flying circular ring made of elastic, perfectly plastic material impinging onto a stationary rigid target with initial velocity V_0 has been investigated experimentally and numerically (Bao & Yu, 2015a; Xu *et al.*, 2014).

From a dimensional analysis, it is evident that the dynamic performance of the ring is mainly governed by three non-dimensional parameters:

1) non-dimensional wall thickness of the ring $\eta \equiv h/R$, where h and R are the wall thickness and radius of the ring, respectively;
2) non-dimensional impact velocity $v \equiv V_0/V_Y$, where $V_Y \equiv Y/\sqrt{E\rho}$ denotes the yield velocity of the material, and E and Y are material's Young's modulus and yield stress, respectively;
3) yield strain of the material $\varepsilon_Y = Y/E$.

Figure 10.26 shows the schematic drawing of a finite element model. The radius is R = 0.1 m, the width is $b = R/5 = 0.02$ m, and $\eta = 1/20$ or $\eta = 1/30$. The material of the ring is assumed to be an aluminum alloy which is simplified as an elastic-perfectly plastic solid with Young's modulus E = 70.0G Pa, Poisson's ratio $\nu = 0.3$, density $\rho = 2700$ kg/m^3, and yield stress Y = 150.0 MPa. Due to the symmetry, only half of the ring along the x- and z-directions is modeled in finite element simulation; more details of the finite element simulation are given in Bao and Yu (2015a).

Simulation Results

Figure 10.27 depicts the relations of impact force with the displacement of the mass center and with time for the rings of $\eta = 1/30$ under the non-dimensional impact velocities

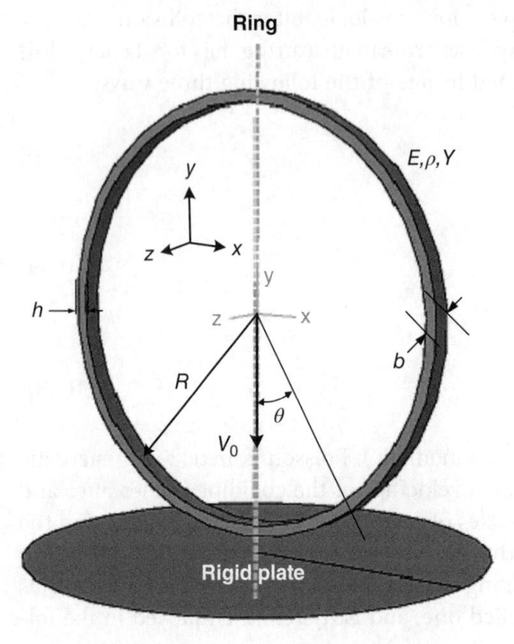

Figure 10.26 Schematic drawing of the collision of a circular ring on a rigid target.

(a)

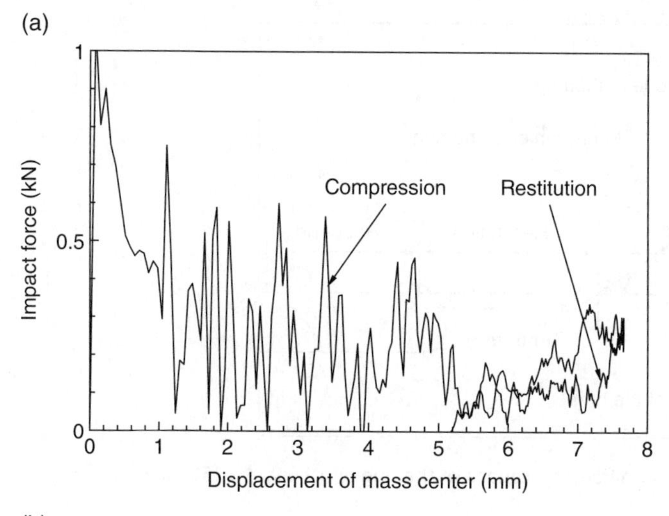

(b)

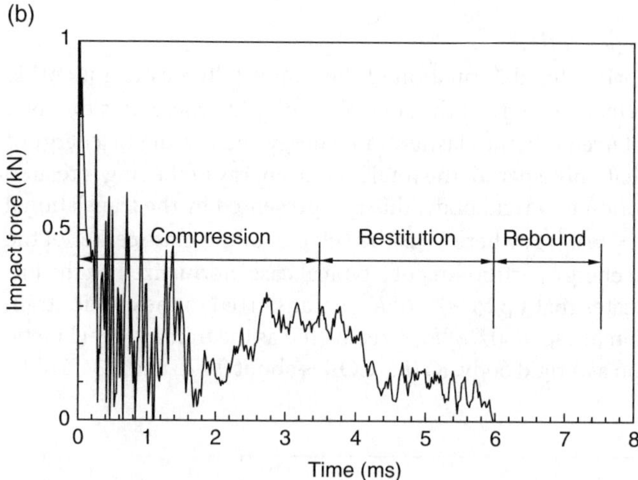

Figure 10.27 Impact force versus the displacement of mass center of the ring (a) and versus time (b). $\eta = 1/30$, $Y = 150$ MPa, and $v = 1.5$.

$v = 1.5$. Obviously, the ring collision is more complicated than the solid sphere collision, owing to the elastic vibrations and large deformation. The velocity histories at different locations on the ring are shown in Figure 10.28. According to the theorem of momentum, the velocity of mass center of the ring V_c should vary monotonically due to the compressive contact force. The instant of $V_c = 0$ is defined as the transition point from the compression phase to the restitution phase. At that instant, V_c becomes a constant $V_c = V_r$, i.e., the rebounding velocity, which marks the end of the restitution phase. Accordingly, the COR can be calculated by Eq. (10.58).

Based on the finite element simulations, the behavior of the rings is described in the following four regimes in terms of impact velocity.

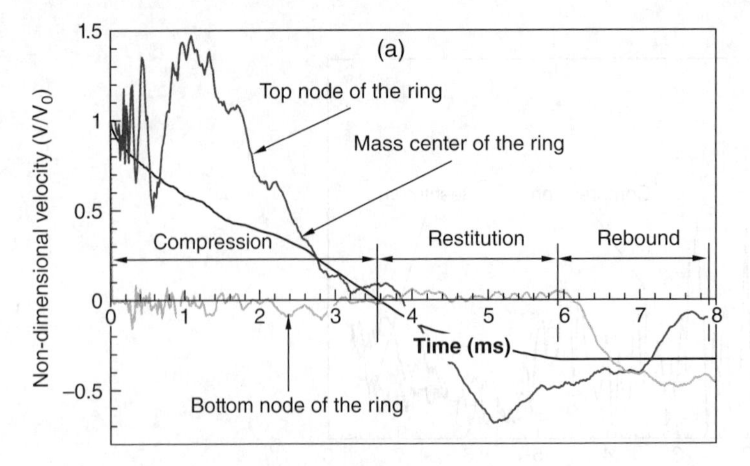

Figure 10.28 Velocity histories at different locations of the ring. η = 1/30, Y = 150 MPa, v = 1.5.

Regime 1: Fully Elastic Collision (v = 0.1)

When $\eta = 1/30$ and $Y = 150$ MPa, the deformation of the ring is fully elastic if $v < 0.15$. After the impact, the initial kinetic energy of the ring, $K_0 = mV_0^2/2$, where m is the total mass of the ring, will be transferred into the elastic strain energy and the kinetic energy of the ring. It should be noted that only a part of the total kinetic energy of the ring is related to the ring's translational motion as a rigid body; this is represented by the translational kinetic energy of the ring, $K_c = mV_c^2/2$, where V_c is the velocity of the mass center of the ring. Figure 10.29 depicts the energy partitioning of a typical case, normalized by the initial kinetic energy K_0. It indicates that up to 90% of K_0 is transferred to the elastic strain energy during the compression phase, ~60% of K_0 is recovered as the translational kinetic energy of the ring in its motion as a rigid body, K_c (i.e. COR is about 0.77), and > 34% of K_0

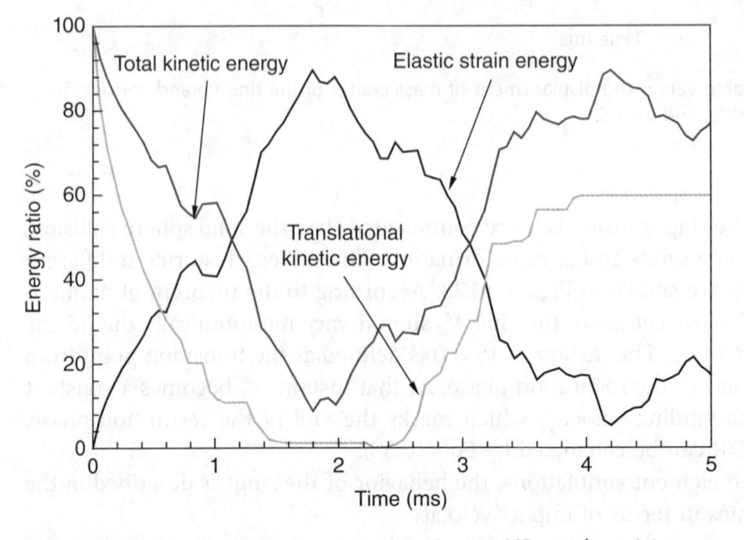

Figure 10.29 Energy partitioning: η = 1/30, Y = 150 MPa, and v = 0.1.

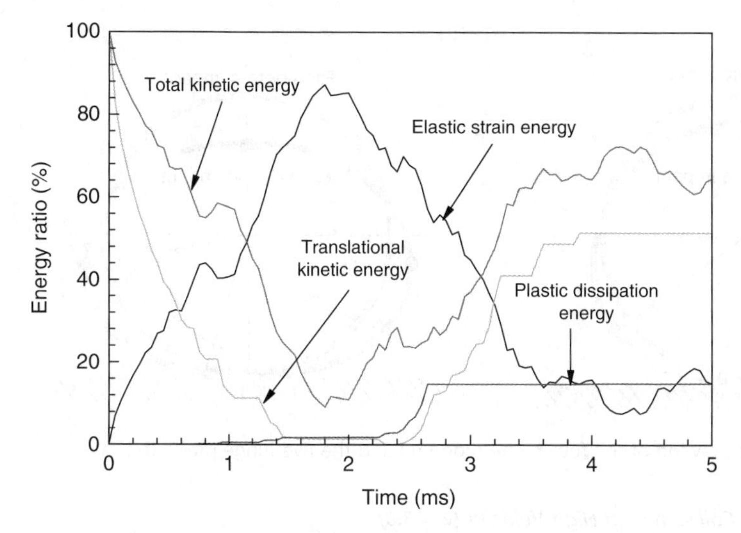

Figure 10.30 Energy dissipation: η = 1/30, Y = 150 MPa, and v = 0.3.

is retained in the ring in the form of elastic wave propagation and elastic vibration after the ring completely rebounds from the rigid target.

Regime 2: Elastic-Plastic Collision (v = 0.3 and 0.8)

When $v > 0.15$, the material near the impact point, i.e., near $\theta = 0°$, begins to yield. The distribution of non-dimensional bending moment M/M_e, where $M_e = bh^2Y/6$, along the ring's circumferential direction shows that $M/M_e > 1$ near the impact end, but $M/M_e < 1$ at other positions, which implies the plastic deformation appeared around the impact point. Figure 10.30 depicts the energy dissipation when $\eta = 1/30$, $Y = 150$ MPa and $v = 0.3$. In this case, ~15% of K_0 is dissipated by plastic deformation, and ~51% of K_0 remains in the form of the kinetic energy of rigid body motion after rebounding. With the increase of v, more plastic deformation is observed at various locations. When $\eta = 1/30$, $Y = 150$ MPa and $v = 0.8$, the bending moment near the impact point reaches the plastic yield moment $M_p = bh^2Y/4$, indicating the formation of a plastic hinge.

Regime 3: Fully Plastic Collision with Moderate Velocity (v =1.5)

When $v = 1.5$, two deformation modes, i.e., a four-hinge mode and a five-hinge mode, as shown schematically in Figure 10.31, are identified; they are significantly different from the corresponding static collapse modes of a ring compressed by parallel plates. As the entire ring is involved, this kind of dynamic response of the ring is regarded as a "fully plastic collision". It is found that a four-hinge collapse mode is triggered at the early stage of collision, and then all the four hinges travel along the ring wall, i.e., both angles α and β as depicted in Figure 10.31(a) increase, with the progressive crushing of the ring. Later, when β approaches $\pi/2$, the bending moment at the top of the ring also reaches the plastic yield moment M_p. Consequently, a five-hinge collapse mode is triggered (see Figure 10.31b). After a period of time, however, the top plastic hinge disappears, resuming a four-hinge mode until the restitution of the ring.

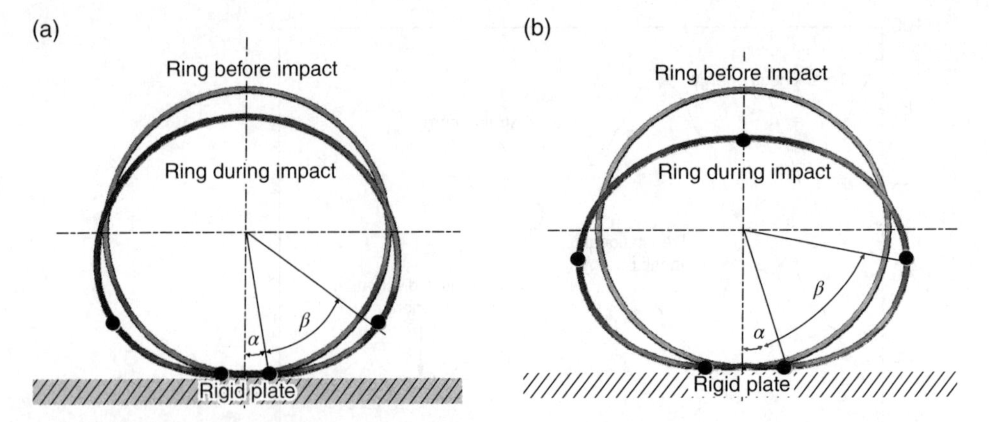

Figure 10.31 Schematic drawing of the four-hinge mode (a) and the five-hinge mode (b).

Regime 4: Fully Plastic Collision with High Velocity (v = 3.0)

For v = 3.0, Figure 10.32 depicts the ring's deformed configurations together with the bending moment per unit width at t = 0.5, 1.0, 1.5, and 4.0 ms, respectively. A four-hinge mode can be clearly identified at t = 0.5 and 1.0 ms, while a five-hinge mode can be

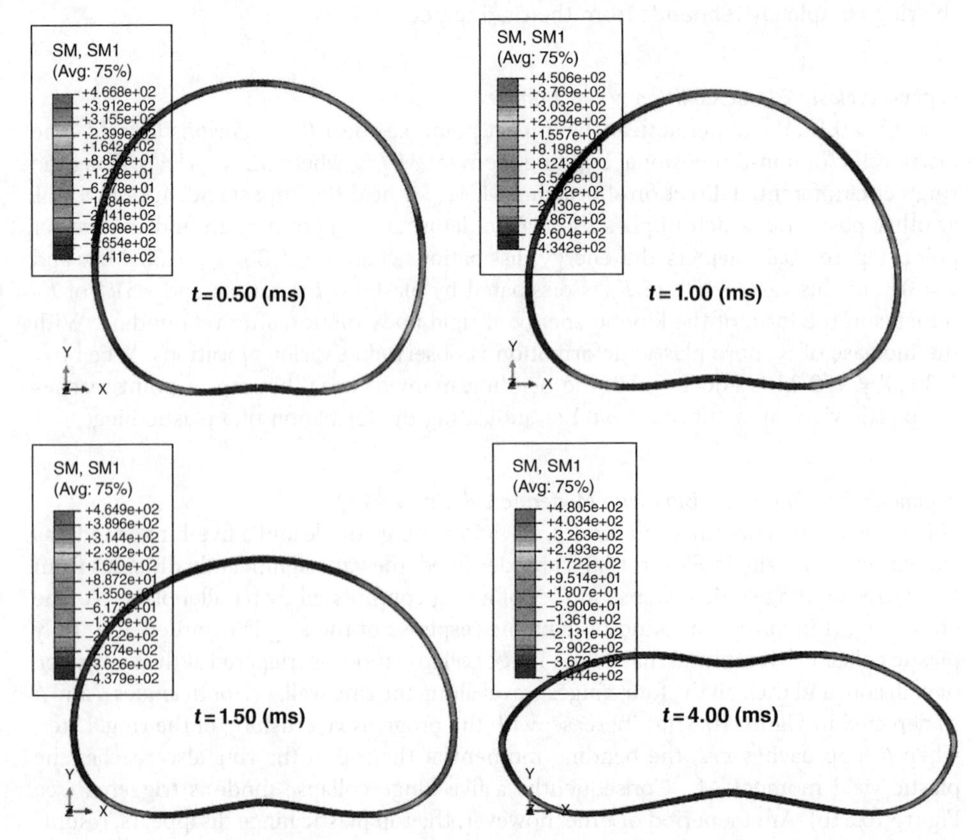

Figure 10.32 Deformed configurations with bending moment per unit width of the ring wall. η = 1/30, Y = 150 MPa, and v = 3.0.

Figure 10.33 Rebounding velocity (a) and the coefficient of restitution (COR) (b) as functions of the initial velocity.

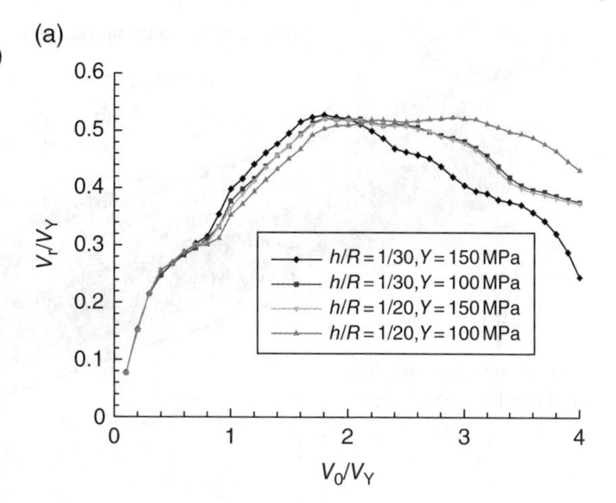

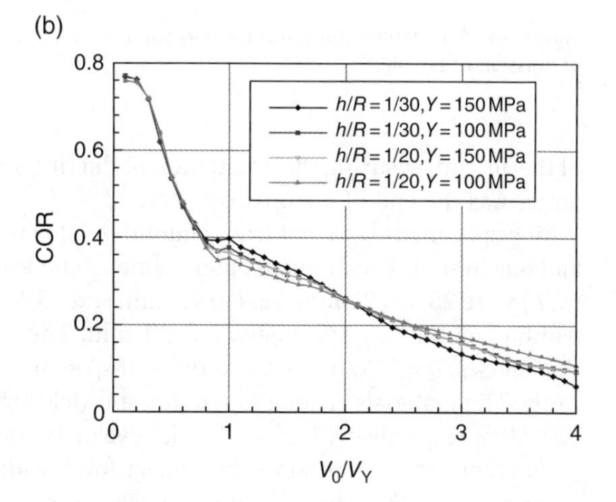

identified at t = 1.5 and 4.0 ms, respectively. It is found that > 95% of K_0 is dissipated by plastic deformation, while only about 1.8% of K_0 is retained in the form of translational kinetic energy of the ring, implying a very small COR in this case.

The non-dimensional rebound velocity V_r/V_Y and the CORs of the ring are shown in Figure 10.33 as functions of v. It is seen that V_r/V_Y is almost independent of the geometry h/R and yield stress when $v < 0.8$, and it increases with the increase of v when $v < 2.0$, and then decreases with the increase of v when $v > 2.0$. The maximum of V_r/V_Y is about 0.55. In summary, the COR remains at about 0.77 for a fully elastic collision, but it decreases quickly with the increase of v when the response is in the elastic-plastic regime (typically $0.2 < v < 0.8$), and then the decrease of the COR slows down when the plastic deformation dominates the ring's response (typically $v > 1.0$).

Experimental Verification

To verify the collision and rebound behavior of rings, an air gun system was employed to fire a ring specimen and make it impinge on a solid target (Xu *et al.*, 2015b). As shown in Figure 10.34, in the experiment setup is a specially designed fixture attached to the barrel

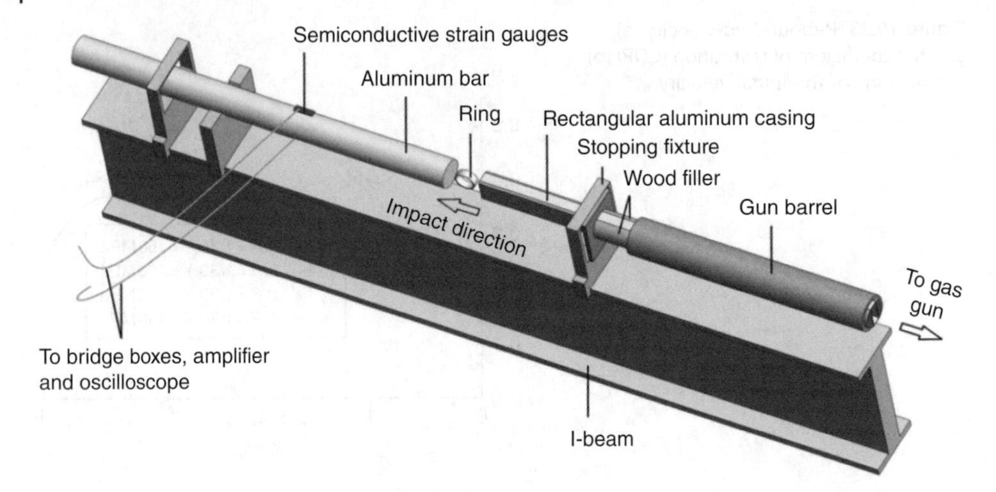

Figure 10.34 Experimental setup for ring collision. *Source*: Xu *et al.* (2015b). Reproduced with permission of Elsevier.

of the air gun, ensuring the orientation of the ring specimen before the collision, and the target was the end of a Hopkinson bar.

Ring specimens were cut from aluminum alloy 6061-T6 tubes of three different sizes and then tested. The nominal outer diameter and wall thickness of the three types of ring (R, h) were 28.5, 0.91 mm; 25.4, 0.91 mm; and 25.4, 1.28 mm, respectively. The nominal width of all the ring specimens was 9.1 mm. The corresponding non-dimensional wall thicknesses, $\eta = h/R$, of the three types of specimens were 0.064, 0.072, and 0.1, respectively. The materials' Young's modulus and yield stress were determined as 70 GPa and 302 MPa, respectively. Hence, the yield velocity was about 22 m/s.

To record the variation of the impact force with time, two sets of semi-conductive strain gauges with a gauge factor of 110 were mounted on the Hopkinson bar's cylindrical surface at a location 250 mm away from the impact end. The force history thus obtained was generally in good agreement with the numerical simulation described above, but the peak force recorded in the experiment appeared slightly later than that predicted by the numerical simulation, because in the experiment the target was an aluminum bar instead of a rigid one. It was found that the peak impact force was approximately proportional to the impact velocity.

A high-speed camera was used to capture the dynamic deformation history of the ring during the collision and rebounding. Detailed deformation histories obtained from the experiments were compared with the numerical predictions and excellent agreements were achieved – see Xu *et al.* (2015b) for more details. The final deformed shapes of a group of tested specimens shown in Figure 10.35 clearly demonstrate the dependence of the deformed shape of the rings on the impact velocity, which ranged from 17.2 to 114.8 m/s, i.e., $v = 0.78–5.2$. At low velocities ($V_0 < 40$ m/s), no concave portion was observed at either the collision side or the remote side. As V_0 increased, the collision side became concave. At even higher velocities ($V_0 > 60$ m/s), both sides became concave. These observations are consistent with the deformation modes predicted in the numerical simulation.

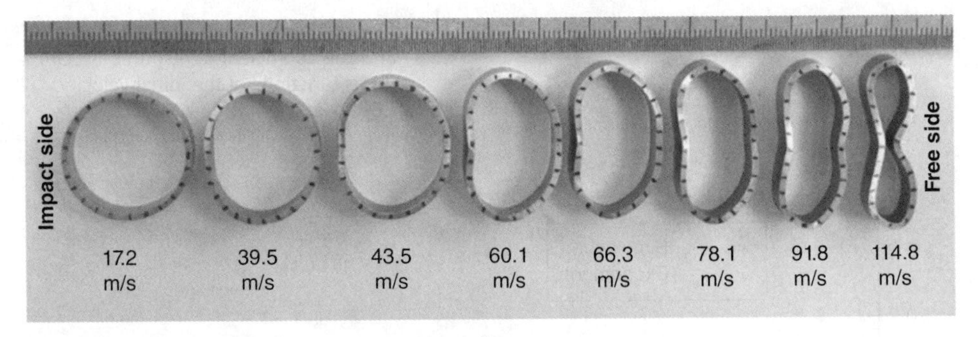

Figure 10.35 Final shapes of the ring specimens after collision with various velocities. *Source*: Xu *et al.* (2015b). Reproduced with permission of Elsevier.

Theoretically speaking, the rebounding velocity of a ring should be the translational velocity at the mass center of the ring. However, there was no physical point at the mass center whose motion could be captured by the high-speed camera. Therefore, the velocity of each point marked on the ring was first calculated based on the measurement of the displacements of the points in the images obtained by the high-speed camera, and then the average velocity of these velocities was obtained to represent the velocity at the mass center, i.e., the rebounding velocity of the ring when the ring had just departed from the target.

As shown in Figure 10.36, with the increase of V_0, the rebounding velocity V_r first increased to a maximum value of about $(0.5–0.6)V_Y$ at $v = V_0/V_Y \approx 2–3$ and then dropped; as it increased to $v \approx 1–5$, the CORs decreased almost linearly from about 0.5 to 0.05 and then remained at this level for higher velocities. As the rebounding of the rings reflects the elastic energy stored in them, the extremely low level of COR at high impact velocity implies that a major share of the kinetic energy during the collision was dissipated by plastic deformation. On the other hand, the wall thickness of the ring only had a minor influence on both the rebounding velocity and the COR in the studied range. Figure 10.36 also confirms the good agreement between the experiments and the simulation results regarding the rings' rebounding behavior.

10.4.2 Collision and Rebound of Thin-Walled Spheres

Dynamic Experiments

Quasi-static compression of a thin-walled spherical shell has been extensively studied since the 1970s, but the published studies of dynamic deformation of thin-walled spheres were very limited. Zhang *et al.* (2009) conducted experiments on both static compression of ping pong balls by a rigid plate and dynamic collision between ping pong balls and a solid target. The dynamic test setup is shown in Figure 10.37, allowing the evolution of the contact region to be observed with a high-speed camera.

With the increase of collision distance, the evolution of the contact area between the ping pong ball and the target plate can be divided into four stages:

1) a flattened area is formed and expanded;
2) the flattened contact area (as a cap of the ball) buckles inwards;

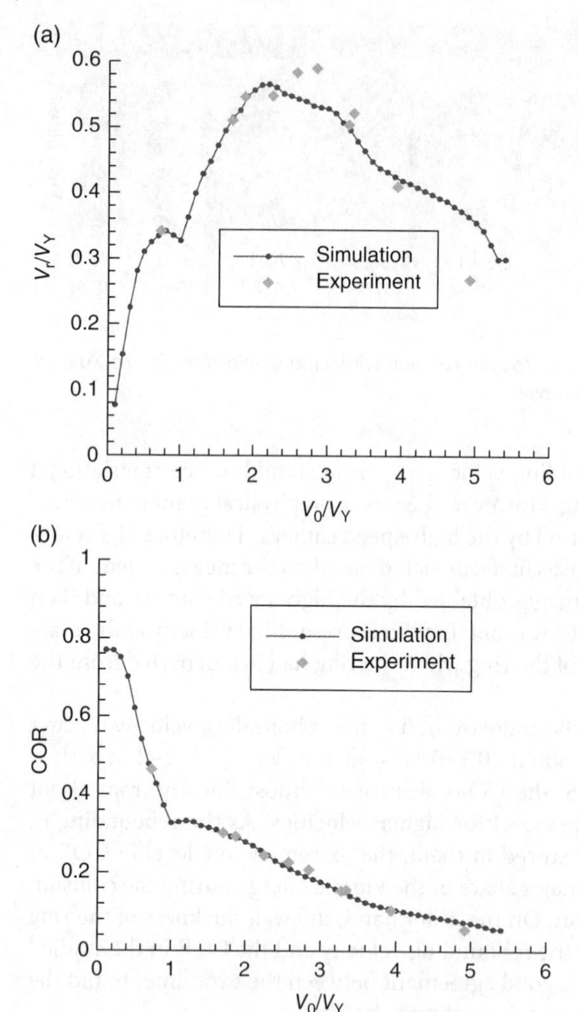

Figure 10.36 Rebounding velocity (a) and the coefficient of restitution (COR) (b) as functions of the initial velocity. COR, coefficient of restitution.

3) the buckled region continues to expand and the buckling may change from an axisymmetric shape to a non-axisymmetric one with three to eight lobes;
4) during the restitution phase the above steps would take place in reverse order but leaving a permanent dent in the ball owing to plastic deformation.

Figure 10.38 depicts the history of the contact diameter varying with time. The above experimental observation has led to a theoretical modeling (Karagiozova *et al.*, 2012), which successfully explained the differences between the dynamic deformed profiles of the balls and those under static compression.

Numerical Simulation

At the same time, Bao and Yu (2015b) conducted a systematic finite element simulation of this problem using ABAQUS/Explicit simulation. Both the collision and rebound processes of ping pong balls are simulated. The interaction between the ball and the rigid

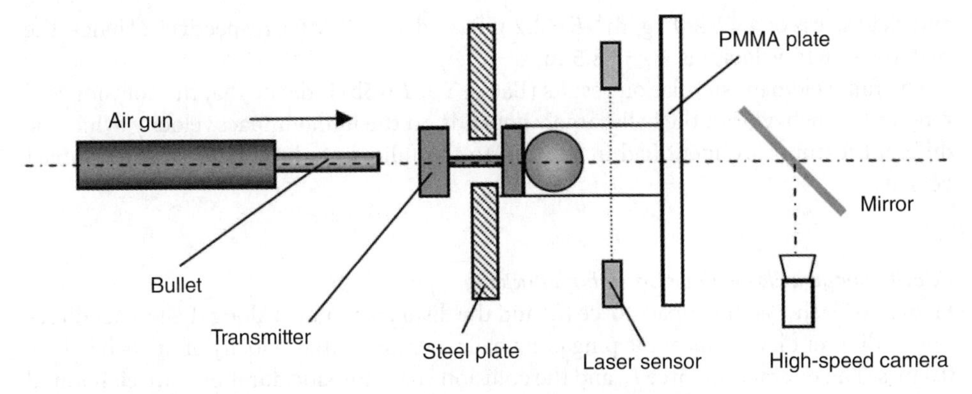

Figure 10.37 Dynamic collision of a ping pong ball onto a solid target. *Source*: Zhang *et al.* (2009). Reproduced with permission of SPIE.

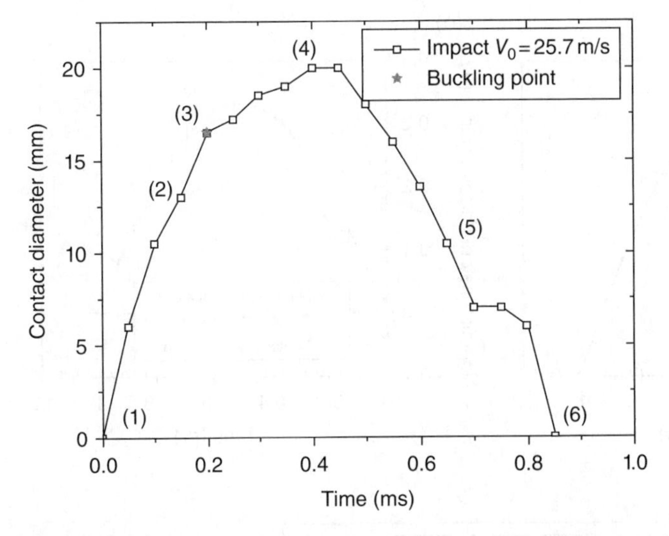

Figure 10.38 Variation of the contact diameter. *Source*: Zhang *et al.* (2009). Reproduced with permission of SPIE.

target is modeled as surface-to-surface contact pair with normal "hard" contact, and the tangential mechanical property is assumed to be frictionless.

Based on a dimensional analysis, similar to Section 10.4.1, three non-dimensional parameters can be identified as governing parameters of the collision and rebound process: the non-dimensional wall thickness of the ring $\eta \equiv h/R$; the non-dimensional impact velocity $v \equiv V_0/V_Y$, with $V_Y \equiv Y/\sqrt{E\rho}$ denoting the yield velocity of the material; and the yield strain of the material $\varepsilon_Y = Y/E$. Therefore, only the effects of these three parameters will be comprehensively simulated and discussed.

Most ping pong balls used in games are made of celluloid, which approximates to a linear elastic-perfectly plastic material (Ruan *et al.*, 2006) with a density, elastic modulus, and yield stress of $\rho = 1400$ kg/m^3, $E = 2.2$ GPa and $Y = 50$ MPa, respectively. Hence, the material's yield velocity is $V_Y = 28.5$ m/s.

The finite element simulation results (Bao & Yu, 2015b) indicate that the collision and rebounding behavior of the balls largely depends on the initial impact velocity, while the different features are identified in relation to the following three ranges of the impact velocity.

Velocity Range I: Elastic Collision with no Buckling

Figure 10.39 shows the impact force (F) and the displacement (u) along the impact direction at different locations of the ping pong ball when the initial velocity of the ball is $V_0 = 0.5$ m/s. The interaction force F_0 and the collision characteristic duration τ are elaborated in the following. Based on preliminary simulation results, the following assumptions can be reasonably introduced during the compression of the ball:

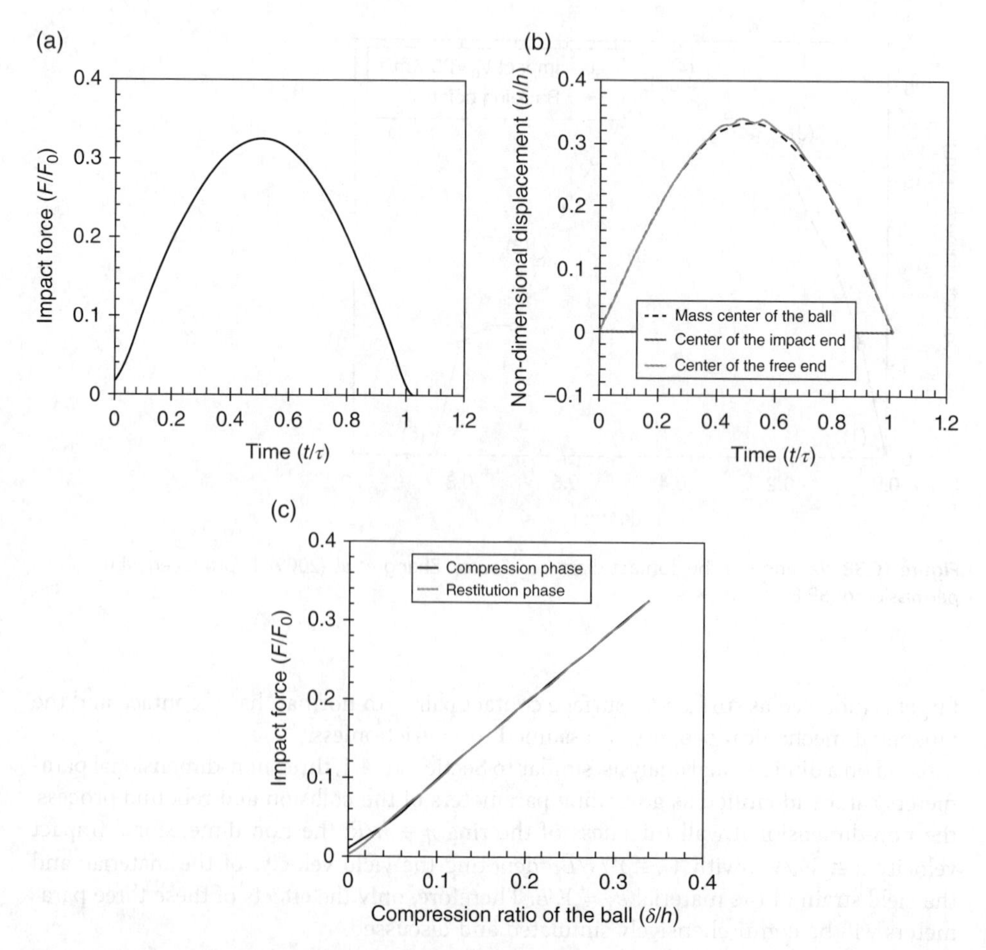

Figure 10.39 (a) Impact force versus time; (b) displacements versus time; (c) impact force versus compression ratio of the ball, when $V_0 = 0.5$ m/s.

Figure 10.40 Scheme of the ball flattened by a rigid target.

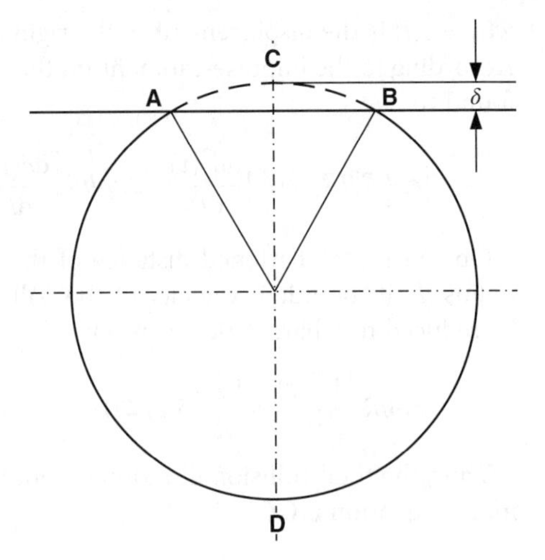

- A cap portion of the ball, ACB, as schematically shown in Figure 10.40, is flattened by the collision onto the rigid target and then rests on it.
- The deformation of the ball is locally constrained around the impact end (i.e., within the cap ACB), which means that the remaining part of the ball, ADB, as shown in Figure 10.40, remains undeformed and moves toward the rigid target at the initial velocity.
- The relationship between the interaction force and the displacement due to flattening remains the same as that under a static compression.

According to the solution given by Reissner (1949) for an elastic spherical shell loaded by a point force at the crown, the force F is

$$F = \frac{8D}{Rc}\delta \tag{10.61}$$

where $c = \beta h$ is the reduced thickness of the ball with $\beta = [12(1-\nu^2)]^{-1/2}$, which takes the Poisson's ratio of the material, ν, into account; $D = Ehc^2$ is the bending stiffness of the wall. In relation to the current collision model, a factor γ is introduced for the contact force to account for the effect of the loading by a rigid plate instead of a point loading; thus

$$F = K\delta \tag{10.62}$$

where $K = 8\gamma D/Rc$ and the modifying factor is about $\gamma = 0.9$ based on a numerical simulation of quasi-static compression.

In view of the assumptions (a) and (b) which describe the deformation mode of the ball, the momentum of the whole hollow ball at time t is

$$I(t) = 2\pi\rho hR[2R - \delta(t)]\frac{\mathrm{d}\delta(t)}{\mathrm{d}t} \tag{10.63}$$

where $\delta(t)$ is the displacement of the rigid plate (i.e., the flattening distance of the ball). According to the impulse-momentum theorem, $dI = -F(t)dt$, the governing equation is found to be

$$2\pi\rho hR[2R-\delta(t)]\frac{d\delta^2(t)}{dt^2} - 2\pi\rho hR\left[\frac{d\delta(t)}{dt}\right]^2 + K\delta(t) = 0 \quad (10.64)$$

Obviously, the flattened distance of the ball, $\delta(t)$, is much smaller than the average radius, R, if the initial velocity of the ball is not very high. Therefore, Eq. (10.64) can be reduced to a harmonic equation:

$$4\pi\rho hR^2\frac{d\delta^2(t)}{dt^2} + \frac{8\gamma D}{Rc}\delta(t) = 0 \quad (10.65)$$

Thus, the total collision duration is about a half of the fundamental period of the harmonic equation (10.65), i.e.

$$\tau = \sqrt{\frac{\pi^3}{2\beta\gamma}\frac{R}{V_e}\frac{1}{\eta}} \quad (10.66)$$

where $V_e = (E/\rho)^{1/2}$ is the elastic wave speed; and the compression duration τ_C and restitution duration τ_R are both equal to $\tau/2$. Hence, the total collision duration is about τ = 0.83 and 0.77 ms for 40 and 38 mm ping pong balls, respectively, as long as the material properties and geometrical parameters are as given earlier. These predictions agree well with the reported experimental measurements.

It can be clearly seen from Figure 10.39 (c) that F/F_0 is almost equal to δ/h during both the compression and restitution phases, which means that Eq. (10.62) obtained under static compression is valid for modeling the impact force when the initial velocity of the ball is low. It is quite different from the collision of a circular ring on a rigid target (see Section 10.4.1), where an oscillation of the impact force exists even though the initial velocity is quite low and the impact forces during the compression phase and the restitution phase follow very different paths, resulting in the COR being $<<$ 1.0 under ideal elastic collision.

Velocity Range II: Dynamic Deformation with Recoverable Buckling

With the increase in the initial velocity, the center of the impact end begins to move towards the interior of the ball at the end of the compression phase, and then oscillates during the restitution phase, which indicates the buckling or inward inversion of the cap. Numerical calculation indicates that the buckling condition $\delta = \alpha h$ obtained under static compression is still valid for the collision case when the velocity is low, while $\alpha = 2.2$ is a good approximation. Accordingly, the critical velocity, which makes the ball's cap buckle during the compression phase, should satisfy

$$V_c = \alpha\left(\frac{2\gamma\beta}{\pi}\right)^{1/2}\eta^{3/2}V_e \quad (10.67)$$

By using the material properties and geometrical parameters as given above and by taking $\alpha = 2.2$, the critical velocity given by Eq. (10.67) is about $V_c = 3.5$ m/s, which agrees with the numerical simulation results very well. As α, β and γ in Eq. (10.67) are almost all constant as given before, the critical velocity is almost proportional to the elastic wave speed of the material (V_e) and proportional to the $(3/2)^{th}$ power of the ball's thickness ratio $\eta = h/R$.

According to the numerical simulation in the case of $V_0 = 3.5$m/s, the maximum contact radius between the ball and the rigid plate is about $0.2R$ (i.e. 4 mm) at the end of the compression phase. This is very close to that measured by Zhang and Yu (2009), as well as in other experiments.

When the initial velocity becomes $V_0/V_Y = 0.30$ (i.e., $V_0 = 8.6$ m/s), the cap of the ball begins to buckle very quickly during the compression phase together with obvious oscillation of the impact force. The deformation of the ball is recoverable after the ball fully rebounds from the rigid target, but the rebounding velocity of the ball is far below the initial velocity, as a result of the energy transference to elastic vibration of the ball (especially that of the cap) and the possible plastic energy dissipation during the inward inversion of the cap. Owing to the oscillation of the cap, the ball and the rigid target may contact each other several times shortly after their first separation. This leads to the occurrence of the multiple impacts as well as the increase of the restitution duration.

Velocity Range III: Axisymmetric Buckling with Permanent Dent

When the initial velocity arrives at $V_0/V_Y = 0.46$ (i.e., $V_0 = 13.1$ m/s), the buckling of the cap will not be recoverable and will leave a permanent dent due to plastic deformation of the cap, whilst the multiple impacts during the restitution phase disappear. Moreover, the permanent dent after the ball completely rebounds from the target is found to be axisymmetric. As a good deal of the initial kinetic energy has been dissipated by the plastic deformation in and around the cap, the rebounding velocity of the ball is found to be lower than a quarter of the initial velocity of the ball, resulting in a COR of $< 1/4$. However, the initial velocity that makes the ball buckle permanently, as obtained by the numerical simulation, is much smaller than the experimental measurements. This may be attributed to ignoring the strain rate effect and internal gas in the simulation.

Velocity Range IV: Non-axisymmetric Buckling with Permanent Dent

As shown in Figure 10.41, the cap of the ping pong ball is flattened by the rigid target at the beginning of the collision, followed by axisymmetric inward inversion, and then the deformation of the ball is progressively transformed into non-axisymmetric mode, leaving a permanent dent which is no longer axisymmetric after the ball is completely separated from the rigid target. It is also evident that both the elastic/plastic energy distributions and the deformation of the ball are localized around the impact end, even though the initial velocity of the ball is high. This is notably different from the free collision of a circular ring as described in Section 10.4.1, in which the circular ring exhibits a global deformation mode even when the initial velocity is very low.

Restitution and Rebounding

Figure 10.42(a) plots the dimensionless rebounding velocity and the corresponding COR of the ping pong ball after it separates from the rigid target, and Figure 10.42 (b) depicts the dimensionless durations of the compression phase, restitution phase, and full collision, in which I, II, III and IV refer to the impact velocity ranges I, II, III and IV, respectively.

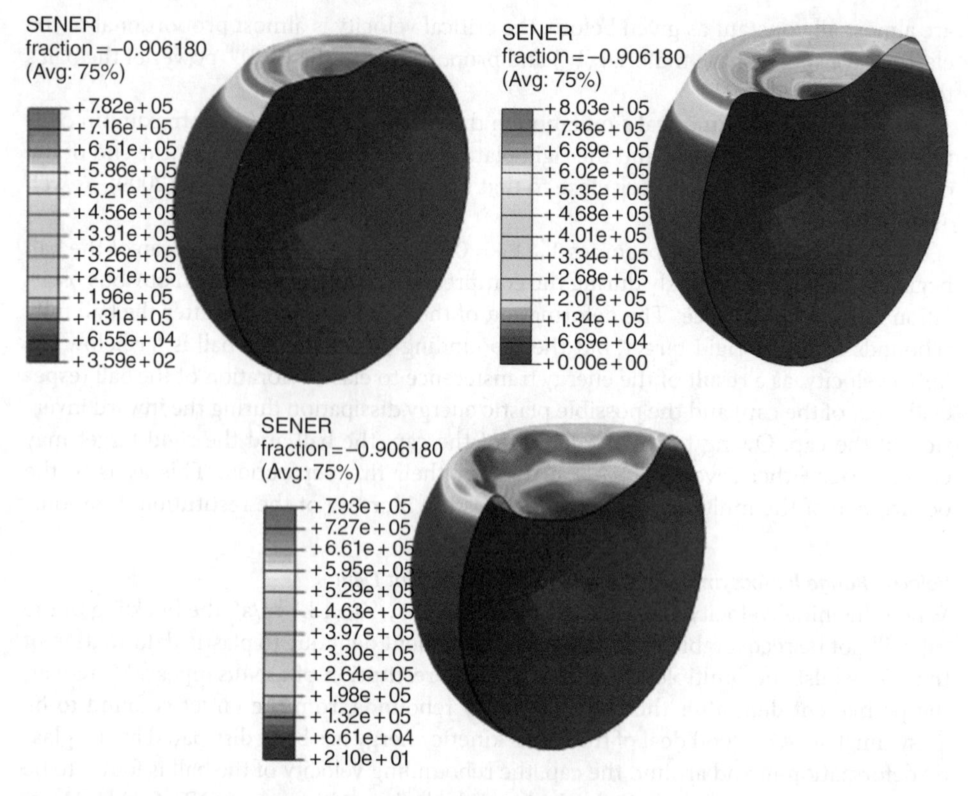

Figure 10.41 Distribution of the elastic strain energy density and configurations of a deformed ball at different time instants when v = 1.2, side view.

Within range I, i.e., $V_0 < 3.5$m/s and $V_0/Y_Y < 0.12$, the compression duration is approximately equal to the restitution duration as a result of harmonic vibration. However, with the increase of the initial velocity, the size of the cap resting on the rigid target is gradually enlarged, resulting in a decrease of equivalent mass as seen from Eq. (10.64); meanwhile the equivalent stiffness of the ball increases slightly. Consequently, both the compression duration and restitution duration decrease slightly. As the deformation of the ball is purely elastic and no buckling occurs, all the elastic strain energy stored during the compression phase can be progressively recovered during the restitution phase, and the corresponding COR is close to 1.0 within range I.

Within range II, i.e., $0.12 < V_0/Y_Y < 0.44$, the restitution duration is longer than the compression duration and it gradually increases with the increase of the initial velocity, as a result of the multiple impacts of the cap after it buckles. When the dimensionless initial velocity $V_0/\ V_Y$ increases from 0.12 to 0.44, a part of the energy is transferred into the form of oscillation of the cap and will remain in this form after the collision is completed. Furthermore, more and more energy would be dissipated by plastic deformation, and consequently, the corresponding COR decreases from 1.0 to about 0.55.

Within range III, i.e., $0.44 < V_0/Y_Y < 0.8$, the cap of the ball buckles inward, and the permanent dent makes the multiple impacts disappear during the restitution phase, so that there is a sudden drop in the restitution duration at about $V_0/Y_Y = 0.44$ after which it remains almost unchanged. However, the deformation of the ball remains

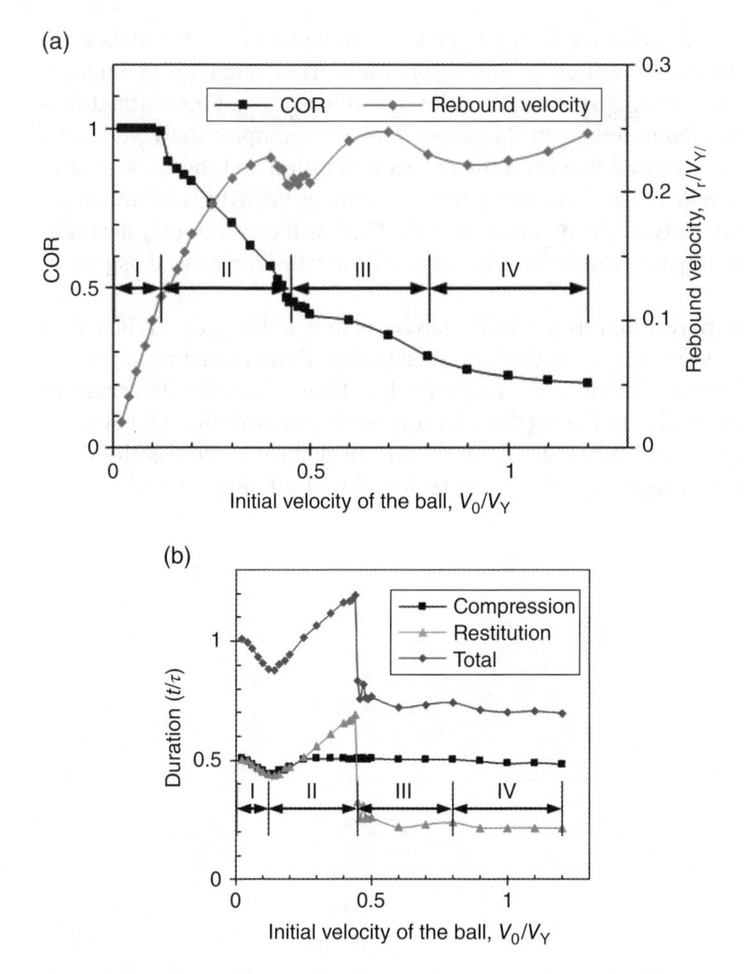

Figure 10.42 (a) Coefficient of restitution (COR) versus $v = V_0/V_Y$; (b) impact duration versus $v = V_0/V_Y$.

axisymmetric and the rebounding velocity of the ball is below one-quarter of the material's yield velocity as limited strain energy is stored during the compression phase, which indeed is the source of the kinetic energy being recovered during the restitution phase. The same phenomena can also be found in the free impact of a circular ring on a rigid wall, in which the limit of the rebounding velocity is about one-half of the material's yield velocity (see Figure 10.33a).

Within range IV, i.e., $V_0/Y_Y > 0.8$, a non-axisymmetric deformation mode occurs after the cap is buckled inward by the rigid plate, leaving a permanent non-axisymmetric dent after the whole collision event is completed.

10.4.3 Concluding Remarks

Different from conventional structures under dynamic loading, freely flying structures, such as the rings and thin-walled spheres analyzed in this chapter, do not have any support, so that apart from the impact contact region, all the surfaces are free of force, whilst the contact areas may vary during the collision as the result of an interaction between the impinging structure and the solid/rigid target. This fundamental character makes the dynamic behavior of moving structures impinging on a solid/rigid target quite different from the response of conventional structures under dynamic loading. One distinct feature of the former is the rebounding from the target. The two examples analyzed in this section have clearly demonstrated that both the collision duration and the COR greatly depend on the impact velocity. In particular, the rebounding velocity is bounded by about half of the yield velocity of the material, and the COR will continuously decrease with the increase of the impact velocity as a result of elastic vibration as well as plastic dissipation.

Another notable feature revealed from these analyses is that, although a circular ring and a thin-walled sphere seem only to be distinguished in their dimensional character (as two-dimensional and three-dimensional, respectively), their dynamic deformation modes are entirely different; that is, during the collision the ring deforms in a global manner, i.e., the entire ring is involved in the deformation mechanism, whereas the thin-walled ball's deformation is restricted in a small region close to its impact end.

References

Ashby, M.F., Evans, A., Fleck, N.A., Gibson, L.J., Hutchinson, J.W. and Wadley, H.N.G. (2000) *Metal Foam: a Design Guide*. Butterworth-Heinemann, Boston, MA.

Bao, R.H. and Yu, T.X. (2015a) Impact and rebound of an elastic-plastic ring on a rigid target. *Int. J. Mech. Sci.*, **91**, 55–63.

Bao, R.H. and Yu, T.X. (2015b) Collision and rebound of ping pong balls on a rigid target. *Mat. Design*, **87**, 278–286.

Børvik, T., Hopperstad, O.S., Berstad, T., Langseth, M. (2001) Numerical simulation of plugging failure in ballistic penetration. *Int. J. Solids Struct.*, **38**(34–35), 6241–6264.

Børvik, T., Dey, S., Hopperstad, O.S. Langseth, M., (2009) On the main mechanisms in ballistic perforation of steel plates at sub-ordnance impact velocities. Chapter 11, in *Predictive Modelling of Dynamic Processes*. Ed. Hiermaier, S. Springer.

Calladine, C.R. (1968) Simple ideas in the large-deflection plastic theory of plates and slabs, in *Engineering Plasticity*. Eds Heyman, J. and Leckie, F.A. Cambridge, 93–127.

Calladine, C.R. and English, R.W. (1984) Strain-rate and inertia effects in the collapse of two types of energy-absorbing structure. *Int. J. Mech.Sci.*, **26**, 689–701.

Campbell, J.D. and Ferguson, W.G. (1970) The temperature and strain-rate dependence of the shear strength of mild steel. *Phil. Mag.*, **21**, 63–82.

Campbell, J.D., Eleiche, A.M. and Tsao, M.C.C. (1977) Strength of metals and alloys at high strains and strain rates, in *Fundamental Aspects of Structural Alloy Design*, Plenum, New York, 545–563.

Chen, F.L. and Yu, T.X. (1993) Analysis of large deflection dynamic response of rigid-plastic beams. *ASCE J. Eng. Mech.*, **119**, EM6, 1293–1301.

Chen, W. and Song, B. (2011) *Split Hopkinson (Kolsky) Bar*. Springer.

Clifton, R.J. (1983) Dynamic plasticity. *J. Appl. Mech.*, **50**, 941–952.

Cox, A.D. and Morland, L.W.(1959)Dynamic plastic deformations of simply-supported square plates. *J. Mech. Phys. Solids*, **7**, 229–241.

Deya, S., Børvik, T., Hopperstad, O.S., Langsetha, M. (2007) On the influence of constitutive relation in projectile impact of steel plates, *Int. J. Impact Engng.*, **34**, 464–486.

Eshelby, J.D. (1949) Uniformly moving dislocations. *Proc. Phys. Soc. (Lond.)*, **A62**, 307–314.

Field, J.E., Walleyet, S.M., Proud, W.G., Goldrein, H.T. and Siviour, C.R. (2004) Review of experimental techniques for high rate deformation and shock studies. *Int. J. Impact Engng.*, **30**, 725–775.

Fleck, N.A., and Deshpande, V.S. (2004) The resistance of clamped sandwich beams to shock loading. *ASME J. Appl. Mech.*, **71**, 386–401.

Frost, H.J. and Ashby, M.F. (1982) *Deformation Mechanism Maps*. Pergamon, Oxord.

Introduction to Impact Dynamics, First Edition. T.X. Yu and XinMing Qiu.

Published 2018 by John Wiley & Sons Singapore Pte. Ltd.

Gao, C.Y., Zhang, L.C. (2012) Constitutive modelling of plasticity of fcc metals under extremely high strain rates. *Int. J. Plasticity*, **32–33**, 121–133.

Gao, Z.Y., Yu, T.X. and Lu, G. (2005) A study on type II structures. Part I: a modified one-dimensional mass-spring model. *Int. J. Impact. Engng*, **31**, 895–910.

Gibson, L.J., and Ashby, M.F. (1997) *Cellular Solids: Structure and Properties*. Cambridge University Press, Cambridge.

Gilman, J.J. and Johnston, W.G. (1957) *Dislocations and Mechanical Properties of Crystals*. Wiley, New York.

Greenman, W.F., Vreeland, T. Jr., and Wood, D.S. (1967) Dislocation mobility in copper, *J. Appl. Phys.*, **38**, 3595.

Hashmi, S.J., Al-Hassani, S.T.S. and Johnson, W. (1972) Dynamic plastic deformation of rings under impulsive load. *Int. J. Mech. Sci.* **14**, 823–841.

Hawkyard, J.B. (1969) A theory for the mushrooming of flat-ended projectiles impinging on a flat rigid anvil, using energy consideration. *Int. J. Mech. Sci.*, **11**, 313–333.

Hopkinson, B. (1905) The effects of momentary stresses in metals. *Proc. Roy. Soc. A*, **74**, 498–506.

Johnson, P.C., Stem B.A. and Davis, R.S. (1963) Symposium on the synamic behavior of materials. *Special Technical Publication No. 336*. American Society for Testing and Materials, Philadelphia, PA, 195.

Johnson, W. (1972) *Impact strength of materials*. Edward Arnold.

Johnson,W. and Mellor, P.B. (1973) *Engineering Plasticity*. Van Nostrand Reinhold, London.

Johnson, G.R. and Cook, W.H.(1983) A constitutive model and data for metals subjected to large strains, high strain rates, and high temperatures. *Proc. 7th Intern. Symp. Ballistics, Am. Def. Prep. Org(ADPA)*, Netherlands, 541–547.

Johnston, W.G. and Gilman, J.J. (1959) Dislocation velocities, dislocation densities, and plastic flow in lithium fluoride crystals. *J. Appl. Phys.*, **30**, 129–144.

Jones, N. (1976) Plastic failure of ductile beams loaded dynamically. *Trans. ASME. J. Eng. Ind.*, **98**(B1), 131–136.

Kaliszky, S. (1970) Approximate solution for impulsively loaded inelastic structures and continua. *Int. J. Non-linear Mechanics*, **5**, 143–158.

Karagiozova, D, Zhang, X.W. and Yu, T.X. (2012) Static and dynamic snap-through behaviour of an elastic spherical shell. *Acta Mechanica Sinica*, **28**(3), 695–710.

Klopp, R.W., Clifton, R.J. and Shawki, T.G. (1985) Pressure-shear impact and the dynamic viscoplastic response of metals. *Mech. Mater.*, **4**, 375–385.

Kolsky, H. (1949) An investigation of the mechanical properties of materials at very high rates of loading. *Proc. Phys. Soc.*, **B62**, 676–700.

Kondo, K. and Pian, T.H.H. (1981) Large deformation of rigid-plastic circular plates. *Int. J. Solids Structures*, **17**, 1043–1055.

Kumar, P. and Clifton, R.J. (1979) Dislocation motion and generation in LiF single crystals subjected to plate impact. *J. Appl. Phys.*, **50**, 4747–4762.

Lee, E.H. and Symonds, P.S. (1952) Large plastic deformations of beams under transverse impact. *J. Appl. Mech.*, **19**, 308–314.

Lee, L.S.S. (1972) Mode responses of dynamically loaded structures. *J. Appl. Mech.*, **39**, 904–910.

Lee, L.S.S. and Martin, J.B.(1970) Approximate solutions of impulsively loaded structures of a rate sensitive material, *J. Appl. Math. Physics*, **21**, 1011–1032.

Lensky V.S.(1949)On the elastic-plastic shock of a bar on a rigid wall. *Prikl. Mat. Meh.*, **12**, 165–170.

Lu, G.X. and Yu, T.X. (2003) *Energy Absorption of Structures and Materials*. Woodhead Publishing Ltd.

Martin, J.B. (1966) A note on the uniqueness of solutions for dynamically loaded rigid-plastic and rigid-viscoplastic continua. *J. Appl. Mech.*, **33**, 207–209.

Martin, J. B and Symonds, P.S. (1966) Mode approximations for impulsively loaded rigid-plastic structures. *J. Eng. Mech. Div., Proc. ASCE*, **92**, EM5, 43–66.

Menkes, S.B. and Opat, H.J. (1973) Broken beams. *Exp. Mech.*, **13**, 480–486.

Meyer, L.W. (1992) Constitutive Models at High Rates of Strain, in *Shock-wave and High Strain Rate Phenomena in Material*. Eds Meyers, M.A., Murr, L.E. and Staudhammer, K.P. Dekker, NewYork, 49–68.

Meyers, M.A. (1994) *Dynamic Behavior of Materials*. John Wiley & Sons, New York.

Meyers, M.A. and Chawla, K.K. (1984) *Mechanical Metallurgy: Principles and Applications*. Prentice-Hall, Englewood Cliffs, NJ.

Nonaka, T. (1967) Some interaction effects in a problem of plastic beam dynamics, Part 2: Analysis of a structure as a system of one degree of freedom. *J. Appl. Mech.* **34**, 631–637.

Owens, R.H. and Symonds, P.S. (1955) Plastic deformation of free ring under concentrated dynamic loading, *J. Appl. Mech.*, **22**, 523–529.

Parkes, E.W. (1955) The permanent deformation of a cantilever struck transversely at its tip. *Proc. Roy. Soc.*, **A228**, 462–476.

Qiu, X., Deshpande, V.S. and Fleck, N.A. (2004) Dynamic response of a clamped circular sandwich plate subject to shock loading. *ASME J.Appl. Mech.*, **71**(5), 637–645.

Qiu, X., Zhang, J. and Yu, T.X. (2009a) Collapse of periodic planar lattices under uniaxial compression, part I: quasi-static strength predicted by limit analysis. *Int. J. Impct. Engng.*, **36**, 1223–1230.

Qiu, X., Zhang, J. and Yu, T.X. (2009b). Collapse of periodic planar lattices under uniaxial compression, part II: dynamic crushing based on finite element simulation. *Int. J. Impct. Engng.*, **36**, 1231–1241.

Regazzoni, G., Kocks, U.F. and Follansbee, P.S. (1987) Dislocation kinetics at high strain rates. *Acta Met.*, **35**, 2865–2875.

Reid, S.R. and Peng, C. (1997) Dyanmic uniaxial crushing of wood. *Int. J. Impact Engng*, **19**, 531–570.

Reissner, E. (1949) *On the Theory of Thin Elastic Shells*. JW Edwards, Ann Arbor, Michigan.

Ren, Y., Qiu, X. and Yu, T.X. (2014a) Theoretical analysis of the static and dynamic response of tensor skin. *Int. J. Impact. Engng*, **64**, 75–79.

Ren, Y., Qiu, X. and YU, T.X. (2014b) The sensitivity analysis of a geometrical unstable structure under various pulse loading. *Int. J. Impact. Engng*, **70**, 62–72.

Ruan, D., Lu, G., Wang, B., and Yu, T.X. (2003) In-plane dynamic crushing of honeycombs-a finite element study. *Int. J. Impact Engng.* **28**(2), 161–182.

Ruan, H.H., Gao, Z.Y. and Yu, T.X. (2006) Crushing of thin-walled spheres and sphere arrays. *Int. J. Mech. Sci.*, **48**(2), 117–133.

Shen, W.Q. and Jones, N. (1992) A failure criterion for beams under impulsive loading. *Int. J. Impact Engng*, **12**, 101–121.

Singh, N.K., Cadoni, E., Singha, M.K. and Gupta, N.K. (2013) Dynamic tensile and compressive behaviors of mild steel at wide range of strain rates. *J. Engng Mech.*, **139**(9), 1197–1206.

Stein, D.F. and Low, J.R. (1960) Mobility of edge dislocations in silicon – iron crystals. *J. Appl. Phys.*, **31**, 362–369.

Stronge W.J. and Yu, T.X. (1993) *Dynamic Models for Structural Plasticity*. Springer-Verlag, London.

Symonds, P.S. (1967) *Survey of Methods of Analysis of Plastic Deformation of Structures under Dynamic Loading*. Brown University, Division of Engineering Report BU/NSRDC/1–67, June 1967.

Symonds, P.S. and Jones, N. (1972) Impulsive loading of fully clamped beams with finite plastic deflections and strain-rate sensitivity. *Int. J. Mech.Sci.*, **14**, 49–69.

Symonds, P.S. and Mentel, T.S. (1958) Impulsive loading of plastic beams with axial constraints. *J. Mech. Phys. Solids*, **6**, 186–202.

Taylor, G. I (1948) The use of flat-ended projectiles for determining dynamic yield stress I: theoretical considerations. *Proc. Roy. Soc. London*, **A194**, 289–299.

Vinh, T., Afzali, M. and Rocke, A. (1979) Fast fracture of some usual metals at combined high strain and high strain rates, in *Mechanical Behavior of Materials*. Proc. ICM. Eds. Miller, A.K. and Smith, R.F. Pergamon, New York, 633–642.

Whiffin, A.C. (1948) The use of flat ended projectiles for determining yield stress. II: tests on various metallic materials. *Proc. Roy. Soc. Lond.*, **194**, 300–322.

Wu, H., Ma, G., Xia, Y. (2005) Experimental study on mechanical properties of PMMA under unidirectional tensile at low and intermediate strain rates. *J. Exp. Mech.*, **20**(2), 193–199.

Xu, S., Ruan, D., Lu, G. and Yu, T.X. (2015b) Collision and rebounding of circular rings on rigid target. *Int. J. Impact Engng*, **79**, 14–21.

Xu, T.C., Peng, X.D., Qin, J., Chen, Y.F., Yang, Y. adn Wei,G.B. (2015a) Dynamic recrystallization behavior of Mg-Li-Al-Nd duplex alloy during hot compression. *J. Alloys Comp.*, **639**, 79–88.

Youngdahl, C.K. (1970) Correction parameter for eliminating the effect of pulse shape on dynamic plastic deformation, *J. Appl. Mechanics*, **37**, 744–752.

Yu, T.X. (1993) Elastic effect in the dynamic plastic response of structures, Chapter 9 in *Structural crashworthiness and failure*, ed. Jones, N. and Wierzbicki, T. pp.341–384, Elsevier.

Yu, T.X. and Chen, F.L. (1992) The large deflection dynamic response of rectangular plates. *Int. J. Impact Engng*, **12**, 603–616.

Yu, T.X. and Chen, F.L. (1998) Failure modes and criteria of plastic structures under intense dynamic loading: A review. *Metals Materials*, **4**, 219–226.

Yu, T.X. and Chen, F.L. (2000) A further study of plastic shear failure of impulsively loaded clamped beams. *Int. J. Impact Engng*, **24**, 613–629.

Yu, T.X. and Stronge, W.J. (1990) Large deflection of a rigid-plastic beam-on-foundation from impact. *Int. J. Impact Engng*, **9**, 115–126.

Zerilli, F.J. and Armstrong, R.W. (1987) Dislocation-mechanics-based constitutive relations for material dynamics calculations. *J. Appl. Phys.*, **61**, 1816–1825.

Zerilli, F.J. and Armstrong, R.W. (1990a) Dislocation mechanics based constitutive relations for dynamic straining to tensile instability, in *Shock Compression of Condensed Matter*. Eds. Schmidt, S.C., Johnson, J.N. and Davison, L.W. Elsevier, Amsterdam, 357–361.

Zerilli, F.J. and Armstrong, R.W. (1990b) Description of tantalum deformation behavior by dislocation mechanics based constitutive relations. *J. Appl. Phys.*, **68**, 1580–1591.

Zerilli, F.J. and Armstrong, R.W. (1992) The effect of dislocation drag on the stress-strain behavior of F.C.C. metals. *Acta Met. Mat.*, **40**, 1803–1808.
Zhang, X.W., Fu, R. and Yu, T.X. (2009) Experimental study of static/dynamic local buckling of ping pong balls compressed onto rigid plate. *4th Int. Conf. Exp. Mech.* 2009 (ICEM 2009), Singapore.
Zhang, X.W. and Yu, T.X. (2009) Energy absorption of pressurized thin-walled circular tubes under axial crushing. *Int. J. Mech. Sci.*, **51**(5), 335–349.

Index

Introduction to Impact Dynamics, First Edition. T.X. Yu and XinMing Qiu.

Published 2018 by John Wiley & Sons Singapore Pte. Ltd.